DRYSTONE – A GATHERING OF TERMINOLOGY AND TECHNIQUES

NICK AITKEN

ISBN 978-1-0686046-9-0

Published by Rymour Books
and Tippermuir Books
Perth
Scotland

A CIP record for this book is
available from the British Library
Thema Classification RGCU, NKP

Library of Congress CIP Data
Aitken, Nick
Drystone – a gathering of terminology and techniques
Library of Congress Classification S419

cover and book design by Ian Spring
Printed by Replika Press Pvt. Ltd, India

We are committed to the sustainable use of natural resources. The paper used in this book is approved by the Forest Stewardship Council

MIX
Paper | Supporting responsible forestry
FSC www.fsc.org FSC™ C016779

DRYSTONE – A GATHERING OF TERMINOLOGY AND TECHNIQUES

NICK AITKEN

RYMOUR BOOKS

TIPPERMUIR
· BOOKS LIMITED ·

A good drystone wall knows where it stands
and, what's more, it knows what it is: it belongs.
It has character and strength and it wears
its years well: like a castle or a cathedral.
A well-built drystone wall fits itself
as snugly to the landscape as good homespun

does to a man; it rides the townlands like
a boat rides the swell; you can follow it
for miles and miles as it takes the rough
and the smooth in its stride. Certain drystone walls
have been there so long they are bone once more
of the landscape's bone rooted as much

in myth and history as a standing stone.
Sometimes they look like the land giving birth.
One drystone wall I know in Friary
is so fleeced with grass and cushioned with plush
tussocks of moss that you'd be hard put
to catch a glint of stone in it at all…

from Francis Harvey, *Collected Poems* (2007)

FOREWORD

I am honoured to contribute this foreword to Nick Aitken's seminal encyclopaedic tome *Drystone – a gathering of terminology and techniques*: a work comprising twenty years of assiduous academic research by Nick, and far more as a skilled practitioner of drystone construction.

I have known Nick for twenty-five years and met when we both worked for the Highland Council, Scotland. At that time Nick was working in the Countryside Rangers section, specializing in drystone walling or 'dyking'. I became the Curator of the Highland Folk Museum and drawing upon Nick's skills soon realised that he, a true craftsman in drystone construction, was much more than a builder of walls intended to keep livestock penned. Subsequently, in addition to running drystone courses, Nick contributed to several Museum historic drystone projects including the rebuilding of the Museum's Lewis 'Blackhouse' and the re-construction of the early 1800s 'Highland Cottage'. Both latter drystone projects required skilled input from Nick as not only the structural and architectural features were addressed but also the mentoring of Museum staff who were assisting.

A critical aspect that helps define the value of this volume is that working with drystone is far more than just an occupation for Nick, it was and is a vocation. Although now based in Seattle, while still in Scotland Nick's passion variously took him in his own time overseas to both lead and contribute to major international drystone projects including in Ireland, North America and Europe. In these roles, that not only helped contribute to the content of this volume, where Nick contributed to buildings, structures, artistic works and landscape features, he was regularly also a teacher and mentor. Eminently qualified, this latter experience resulted in Nick becoming an examiner in the craft, encouraging both the knowledge of and the ongoing maintenance of high standards in drystone construction.

I understand the depth of Nick's knowledge from the times we have shared sources and information where our joint interests have crossed. Nick has striven to accurately check and validate the information within this volume. Apart from being an important documentary history of this craft, this is an invaluable resource for the historian or academic but perhaps more importantly an eminently practical volume for the embryonic or experienced practitioner who wants to understand this craft that has existed internationally since prehistoric times.

Therefore, I am pleased to commend Nick's work as a singularly important volume on stone, a natural material that matched with traditional skills has played an often understated role in our landscapes, social and working history.

Bob Powell, Working Horse & Agricultural Historian, Kingussie, Scotland.

COLLECTIONS ARE SELDOM COMPLETE

The following must only be taken as a contribution — and a slight one — to a department of constructive literature which is singularly deficient in complete works. Many causes, however, exist to render the task of compiling a full "dictionary of terms" a peculiarly difficult one... not only in... England, Scotland, and Ireland, but [also] in many of their separate counties or districts, a set of nearly totally different terms are used. Indeed, so marked is the discrepancy between the technicalities of one district and those of another, even in the same country, that many of the terms inserted in a specification fitted for the practice of the one would be quite unknown in the other.

From Henry Stephens and Robert Scott Burn. *The Book of Farm-Buildings, their arrangement and construction.* (1861).

CONTENTS

INTRODUCTION

The original intent of this research was to bring together English language terminology associated with drystone walling, or dyking as it is called in Scotland.

This inevitably expanded to include other stone structures because stone trades overlap and interlink. Masonry terms, applying to cut and mortared stone, are included where relevant or where they add interest to the text.

Research through books, bibliographies and theses led to the eighteenth and nineteenth century writings of Dr Singer, Reverend Samuel Smith and Bryce Johnson. Their descriptions of drystone dyking, in their own immediate area, set the standard for the next two hundred years. We should, however, be careful when we read those early descriptions. They come from people in the professional classes who did not have the practical experience. They described what they saw, reflecting their own prejudices and assumptions about manual work.

Drystone walling has been described as a 'dying art' since the 1930s, maybe even before then. Several writers realised it was important to record some practical details of the craft before they were lost, forever. Over the years there has been a progression from academic writings to hands-on descriptions. Early modern writers about drystone walling include F Rainsford-Hannay and W A J Prevost (both retired army officers) and Arthur Raistrick (a geologist and archaeologist). They wrote detailed accounts of the craft as it was practised. Anne-Berit Borchgrevink and Margaret Brooks allowed the craftsmen to talk about their work, in their own words, and leave the interpretation to the reader. Robert Cairns wrote an excellent account of his life as a dyker in southern Scotland. Most recent publications describe the historic development of drystone walls and different styles of drystone work, and outstanding modern construction, all round the world. The craft is no longer dying, it is adapting and surviving.

Drystone walling, as we know it, developed, partly, from ancient techniques used to build the likes of Maeshowe and Skara Brae, no drystone thesis can avoid mentioning these ancient structures.

The peak of activity for British drystone construction, at least for field boundaries, is reasonably stated as 1700 to 1900. This covers the period from the development of reliable drystone walls to the eve of World War I when a new era of social and technological changes heralded the decline of many manual trades. The choice of 1700 is based on Rainsford-Hannay (1957: 23) when he writes: 'Dry stane dyking, in a comprehensive way, was first undertaken on the lands of Palgown (The modern spelling for this farm name is 'Palgowan') in the west of the Stewartry of Kirkcudbright [southwest Scotland]. Somewhere about 1710… the brothers McKie… leased pieces of land free to people who would work for them in the summer.'

In the 1700s and 1800s, British advocates for agricultural improvement wrote passionately for the benefits of drystone walling. Drystone construction also flourished in Europe, especially retaining walls. New ideas flew in all directions and were discussed in farming journals half a world away. British and European emigrants took their skills to the Americas, Africa and the Antipodes, where they often met ancient indigenous styles in the form of fish traps, stone shelters or animal enclosures. Resourceful men adapted tried and tested techniques to suit foreign climates and landscapes.

This spread made for a lively exchange of information and development of local terminology. It varied from country to country, region to region, sometimes parish to parish, depending on language, dialect, landscape and the wall itself. Inevitably travel and distance brought about misunderstanding, and innovation. At one North American training event a participant wrote on the feedback sheet, 'learning the skill of drystone walling will take time but is fairly straightforward. The thick accent of the Scottish instructor was not so easy to follow.'

Nowadays dykers are seldom seen. Theirs is a solitary trade. Much of the work is still agricultural – maintenance of old dykes built by teams of men two hundred years ago – but the people who enter the craft today are just as likely to build public art pieces, impressive gateways or garden terracing and stairways. The language used by the modern dyker is more akin to that of the stonemason. It is not so regionally diverse as the historic vernacular gathered in this volume.

Many modern wallers have learned the craft as second careers, from national organisations, or from literature intended for international audiences, drawn from old sources specific to one area. The preservation and promotion of these craft skills is admirable but appropriation of local terms for national or international use often results in Scottish dykers using words from Yorkshire, or wallers in the south of England using phrases never spoken by their grandfathers.

Does this loss of language matter if the practical skills are preserved? Undoubtedly!

A record of drystone terminology is essential. These words and phrases are an important part of cultural history and crucial for a full understanding of the building processes, and the builders.

Arkell and Tomkeieff (1953: xvi) tell readers their volume was also a labour of love, and, partly, a distraction from stressful times. They had been working independently on the compilation of a glossary of mining and quarry terminology. Arkell considered it a pastime while in hospital and during lunch breaks in the London Library during 1941-43. Tomkeieff saw it as a hobby and by-product of his work on a Petrological Dictionary. In 1943 they combined their efforts to produce a book which was well received, and reviewed by Dr V A Eyles in 1954 (Eyles: 1954). The author hesitates before making comparisons, but hopes this volume is also accepted as time well spent.

HOW TO USE THIS BOOK

An index would be long and complicated. For simplicity the sections have been organised so that finding terms specific to an aspect of drystone construction should be straightforward, after consulting the contents page. Some information is duplicated, because, in the author's opinion, this simplifies matters and puts information where expected.

The text makes frequent references to 'double' and 'single' walls. These terms are fully explained in section 4.1.

The words 'dry stone', 'drystone walling', 'drystone dyking' and 'dyking' are used interchangeably, depending on the context.

Words in **bold type** indicate cross-references, these are kept to a minimum, usually within the same section unless a more specific destination is given.

Spellings of names and places and punctuation are as they appear in original sources. There will be errors in transcription, for which the author apologises.

Abbreviations are kept to a minimum. Most are commonplace, these are introduced for ease of reading:

SG—Scottish Gaelic
W—Welsh
DSL—Dictionaries of the Scots Language
JSD—John Jamieson's Etymological Dictionary of the Scottish Language
Dwelly—Edward Dwelly's Illustrated Gaelic-English Dictionary.

Full details of the dictionaries are given in the extensive bibliography which also allows the reader opportunities for further study.

GVA (General View of the Agriculture) is an abbreviation for the eighteenth and nineteenth century audits of agriculture in the counties of the British Isles, including Ireland. These surveys detail everything from farm buildings to soil types. The abbreviation 'GVA Perth (1794)', for example, clearly states the county and date for a source. A list, attached to the bibliography, expands the abbreviation to name the original author and publisher.

All images used in this volume are by the author, unless specifically attributed.

This volume could have included hundreds more illustrations but, with widespread internet access and constant updating of various North American, British, European and Australian drystone websites and social media, we shall leave it to the reader to pursue their own searches, increase their awareness of the drystone information 'out there', and realise the fundamental similarity of all drystone work, wherever it is in the world.

Both imperial (eg, yards, feet, inches) and metric (eg, metres, centimeters, millimetres) measurements are employed variously in the text, usually depending on the original source. Values have not always been converted to a metric or modern equivalent. In some case this would be awkward; for example, when trying to update early eighteenth-century descriptions. The reader is referred to Chapter 6, Weights and Measures, for some enlightenment on this complex topic.

Figure 1: The British Isles with approximation of places mentioned in the text. (Mapping courtesy of Drew McClelland.)

1. MATERIAL

1.1 Basic geology and quarrying

Rock dominates the landscape. Volcanic activity creates rock. Ice grinds it down. Heat and pressure transform it. This geological activity has, over eons, resulted in a myriad of rock types and a wide diversity of landscapes.

There is a difference between 'rock' and 'stone'. In geology the term 'stone' is not normally used except in words such as 'limestone' or 'sandstone'. Rock is a piece of mineral matter which constitutes part of the earth's crust. There are three main types – igneous, sedimentary and metamorphic, described in more detail below.

Rock is a stable material. King (2000: 29) proposes a column of limestone could be built 12,000 feet (2.27 miles) high before starting to crush under its own weight. It has a very high compressive strength which can vary with each sample. Stone structures also need to cope with lateral thrusts and tensions which damage stones well before they crush. Many problems with stone buildings can be traced back to bad foundations or poor management of water, particularly roof drainage.

The term 'rokk' was recorded in the works of Geoffrey Chaucer in 1385. The Scots form 'rolk' is frequently found in manuscripts, sometimes spelled as 'rokk'. JSD says this was because scribes found 'kk' easier to write and the 'l' was redundant because it was silent. Scots forms – 'roch', 'roche', 'rotche' – were derived from the French word *roche*, and, when used with a guttural 'ch' become an adjective meaning rough, course or unpolished. Hence the phrase 'rocht waw stainis' or 'rough wall stones', per JSD's definition for 'roch'.

Structures have been cut from the 'living rock' – a description of a mass of rock in the landscape, sitting where it was formed, perhaps a cliff face or mountainside. Not all are as spectacular as Petra in Jordan. Worldwide, many smaller caves and steps have been dug out with basic tools.

Individual pieces of rock in the landscape are generally called 'stones' but, more accurately, a stone is a small piece of the earth's crust which has been shaped or formed by natural or manual intervention to produce a viable building component or shaped object, for example a millstone or whetstone. Hubbell (2006: 100-101), defines the difference between a rock and a stone:

> The difference seems simple and straightforward to me. A rock is that hard material that will always be rock – until it is put to use by human beings. We change rock into a stone by removing it from its natural environment so that it can fulfil a human need.

The word 'stone' has had many spelling forms, including this British selection: stan, staen, ston, sten, steen, stoon, stane, stayne, stein, stien, stiene, ston, stone, stoan, steane, stain. The words may be applied to individual pieces of stone or to the general building material.

Some rocks are heavier than others. These weights are for comparative purposes only; there is a wide range within each rock type:

Water, pure	62.4 pounds per cubic foot = 999.6 kg per cubic metre
Earth, wet, excavated	100 pounds per cubic foot = 1601.8 kg per cubic metre
Sandstone, solid	145 pounds per cubic foot = 2322.7 kg per cubic metre
Concrete (with gravel)	150 pounds per cubic foot = 2402.8 kg per cubic metre
Limestone, solid	163 pounds per cubic foot = 2611 kg per cubic metre
Granite, solid	168 pounds per cubic foot = 2691.1 kg per cubic metre
Basalt, solid	188 pounds per cubic foot = 3011.5 kg per cubic metre

(Source: Thomas J Glover, Pocket Ref, second Edition, 2001, p. 427.)

Calculation of weight 'in the field' gets down to basics. An excavator driver can estimate the weight of a rock pile by pacing the margins in yards, estimating the height in yards, multiplying length by width by height (to get an approximate cubic yardage) and multiplying the number by 2 or 2.5 to get an approximate tonnage. Use a multiplier of 1.5 for wet soil.

The terminology surrounding stone is full of historic conjecture about its origin, for example the idea that 'rubble' was left behind by the biblical flood, or stones grew organically in the earth as a crop, like inconvenient potatoes.

The science of geology developed slowly. It took centuries to understand the physical and chemical processes, and extreme timescales, necessary to create rock and minerals. It seemed reasonable to assume rock grew from smaller material, like a tree grows from a tiny seed. Bush (1769: 61), ventures this opinion in a description of the basalt columns in the north of Ireland:

> This, however, I speak with diffidence, and submit to the judgement of more curious naturalists. That stones of many (and perhaps of all) kinds do really grow from a lesser to a larger size, is, at this time a well known truth.

Carreg is the Welsh word, *clach* the Scots Gaelic word and *cloch* the Irish Gaelic word for stone. *Loch* is an Irish Gaelic term for a large stone.

The following terms are descriptions of rock and stone, as raw material for construction.

Aqueous rock also known as **sedimentary rock**; they are formed by accumulation of deposits in water.

Arbroath slab / Arbroath stone / Arbroath pavement / Arbroath slate the export name for **Carmyllie flag**, a kind of hard flagstone from Carmyllie in Tayside, east Scotland.

Argillaceous a description of rock type which contains clay. Some limestones are argillaceous. They are used to produce hydraulic lime mortars; these set under water. This property is vital when building bridges or canals.

Armour rock large rock pieces produced by blasting in a quarry. Used in breakwaters, rough retaining walls and other situations where bulky irregular rock is required. Also known as 'blast rubble', 'shot rock', 'rap rock' or 'riprap'. The exact specification for this material may differ from quarry to quarry.

Baker (1913: 282) defines 'riprap' as 'Stone thrown in promiscuously about the base of piers, abutments, etc, to prevent scour, or placed on banks of rivers and canals to prevent wash.' In this instance 'promiscuously' has an older meaning, 'mingled together, confused', as defined by Bailey (675).

Bairdal (Aberdeenshire) soapstone, also known as 'steatite'. A soft rock which is easily carved.

Ballast stone loaded into the hold of empty ships to keep the ship lower in the water and on an even keel. Ballast was cast overboard at the end of the voyage and became a handy, cheap, source of building material. British stone found its way to North America in the holds of empty timber ships. The sandstone used for the base of the Statue of Liberty in New York Harbour came from Locharbriggs Quarry, near Dumfries in southwest Scotland, as ballast.

While in conversation with a quarry manager in Spittal, Caithness, the author was told Caithness flagstone was shipped to Boston, Massachusetts, as ballast. It was salvaged and used for paving.

There were superstitions about ballast. Gregor (1881:198) describes how white stones were not used in the fishing ports of northeast Scotland, nor would stones burrowed by Lithophaga (literally 'stone eaters'), a genus of small marine molluscs in the family *Mytilidae.*

Bargain in a Welsh slate quarry a 'bargain' was a section of the face allocated to a small group of men who blasted the rock, split the blocks and dress the slates. They paid for the privilege of working the bargain and hoped to earn a profit by producing enough roofing slates.

Basalt a hard, dense, dark, fine-grained igneous rock formed by the solidification of lava, often into vertical polygonal columns (Figure 2). The Giant's Causeway in Ulster, Ireland, is probably the best-known example with around 40,000 interlocking columns, mostly hexagonal. If one end of a basalt column is tapped with a stone, the ringing sound is clearly heard at the other end, hence an alternative name of **clinkstone**. A dull thud indicates a flaw in the column. Basalt is used extensively for paving and building. Also known as **trap** or **greenstone**.

Bathstone an easily worked pale oolitic limestone from Bath, England. A 'bathstone saw' is a handsaw used for cutting this stone. An old carpentry saw with the coarser teeth cut off serves the purpose.

Bedrock a mining term for the solid rock lying beneath the topsoil and upper layers of weathered rock.

Beenie blue (northeast Scotland) a local name for **hornblende**.

Beer stone a type of limestone quarried near the village of Beer in Devon, England. The stone consists of minute fragments of shells and is practically pure carbonate of lime (CaCO3). Beer stone weathers badly and is only suitable for internal work.

Binnie (Scots) a type of sandstone from quarries at Binnie near Linlithgow, West Lothian, Scotland.

Blinking stone (Shetland) white **quartz**, named for the sparkle of mica within it. DSL, under 'quartz', refers to a '...huge earth-fast 'blinking stone' of white quartz.'

Blue heathen an old Scots term for hornblende gneiss, described so because of its colour. Shades of grey are often called 'blue' because they have suggestions of blue or green in the colour.

Figure 2: Columnar Basalt. Oregon, USA.

Blue pennant a type of sandstone, typical of south Wales.

Bluestones the generic name for the stones used in Stonehenge. It is mainly dolerite from the Preseli Mountains of Dyfed, Wales, 240 miles away. Stonehenge is a stone circle in Wiltshire, England, estimated age 5,000 years.

Borrowpit a temporary quarry, a source of gravels or rock which will be refilled, perhaps with waste material, after works are completed.

Brathay flags (South Lakeland, English Lake District) a hard shale, originating in the Silurian period, 430 million years ago. It is ideal for paving. It weathers well and, like Caithness flag, is used for flagstone fences.

Breestane / brie-stone (central Scotland) sandstone.

Brownstone a reddish-brown sandstone with high iron oxide content, up to 250 million years old. Its use in north-eastern USA led to the description of many terraced houses as 'brownstone'.

The Burren in County Clare, Ireland (Figure 3), looks barren and bleak; within the rocky landscape there is fertility. An example of **karst** or **karstic** landscape. The name is derived from the Irish word *boirean* meaning 'a rocky place'. Ludlow (1751: 147) describes it thus:

> …we entered into the barony of Burren, of which it is said, that it is a country where there is not water enough to drown a man, wood enough to hang one, not earth enough to bury him; which last is so scarce, that the inhabitants steal it from one another, and yet their cattle are very fat; for the grass growing in turfs of earth, of two or three foot square, that lie between the rocks, which are of limestone, is very sweet and nourishing.

Figure 3: The Burren. Ireland.

Caen stone an oolitic, fine textured, whitish limestone, imported into England from Caen, Normandy, France, in the eleventh and twelfth centuries. It was used extensively in Gothic buildings. **Canestone** is the Scots variation.

Caithness flagstone / Caithness flag a hard, dense flagstone (Figure 4) formed 400 million years ago. The material is very heavy and comes in at around 2.7 tonnes (2,700 kg) per cubic metre. Geologically it is a sandstone. According to the Caithness Stone Industries Ltd leaflet for their quarry products '…its fine particulate structure makes it more comparable to a mudstone or siltstone with density and hardness akin to granite.' These flagstones have been fabricated into water cisterns and small lean-to house extensions, as illustrated in Figure 113. Large flags were used to make up a short wall, with a window opening cut into the stone. Water cisterns were formed by slotting sheets together with grooves, using white lead and iron bars to keep the seams watertight. Some Caithness quarries have sections rich in fossils.

Calcareous a description of a rock type containing calcium carbonate, one which contains lime or is chalky.

Calk / cauk / cawk a term used in the dialect of Midland and Northern counties of England for limestone or chalk.

Calp an argillaceous limestone (meaning it contains clay) particularly associated with Dublin, Ireland. This stone was extensively used for rubble walling, often behind the more

Figure 4: Stratification in a quarry. Caithness, Scotland.

expensive facades of ashlar stone and brick. It was also burnt in kilns to produce hydraulic lime mortars which can set under water.

Calley stone (Yorkshire) a hard argillaceous sandstone. Possibly named for similar material from Calais, France.

Canestone (Scots) a variation of 'Caen stone', an oolitic, fine textured, whitish limestone, imported into England from Caen, Normandy, France, in the eleventh and twelfth centuries. It was used extensively in Gothic buildings.

Carboniferous a geological period extending from 345 to 280 million years ago, a total of 65 million years. This period saw the laying down of most of our planet's oil reserves, coal and iron ore.

Carmyllie flag a hard, grey flagstone from Carmyllie in Tayside, Scotland, which was sold as **Arbroath stone** throughout the world.

Carrara a white crystalline marble from Tuscany, Italy. It is up to 98% pure calcium carbonate.

Carreg lwyd (W) literally 'grey stone', a generic term applied to any dark stone from the mountains.

Clinkstone / clink-basalt the basalt on which Edinburgh Castle sits, so called because it clinks, like iron, when struck. 'Clinkstone' is also applied to a hard form of feldspar (sometimes spelt 'felspar') which rings when struck.

Clint (south Scotland) a hard, rough or flinty rock; any reasonably large hard stone. Arkell and Tomkeieff (1953: 25) have this as a Yorkshire term for a '…variety of hard or flinty rock …adopted from the Danish and Swedish *klint*, rock.' DSL tells us a 'clint' is a cliff or a crag or a precipice. A 'clinted' sheep is one trapped on a ledge; it found its way down a rockface and cannot get back up. The term also applies to a crevice in limestone rocks. 'Clinty' ground is stony or rocky.

Cloch Irish Gaelic for 'stone'. The word sometimes appears as *lach* or *loch* and as *leac* in D'Arcy (2006: 29).

Clunch / clunsh a soft easily worked limestone, a dense mix of rock and soil which can be cut into blocks. Found in France and on the white cliffs of Dover, southeast England.

Clynestone a fine-grained, siliceous, sandstone, quarried in Brora, Sutherland, Scotland. This white sandstone was used to build local houses, monuments and drystone dykes (Figure 5).

Collyweston stone (named for a village in Northampton, England), a fissile limestone formed during the Jurassic period (140 – 190 million years ago). It is not a true slate though it can be split along natural cleavage lines. Until the nineteenth century it was used as a roof covering for most of the houses in the Collyweston area until the railway system imported lighter Welsh slate.

Conglomerate see Section 1.2 for a full description and local variations of this term.

Figure 5: Clynestone dyke. Brora, Sutherland, Scotland.

Crash / creach / creech (Rutland and Northamptonshire, England) a coarse surface-weathered limestone used for walling.

Culver (Somerset, southwest England) Arkell and Tomkeieff (1953: 34) note this as 'A blue stone used for steps …' but do not identify the rock type; it is probably a hard limestone. A 'culver' is a woodpigeon, a grey coloured bird. 'Blue' is frequently used to describe shades of grey.

Delf / delph (old English) a quarry or pit. A quarry for stone slates was a 'stone slate delph'. Etymologically related to 'delve,' meaning, in Middle English, 'to dig or plough.'

Devil's dough an English term for a hard grey-white siliceous rock containing silicon. According to Arkell and Tomkeieff (1953: 30) 'Attribution to the devil denotes extreme hardness.'

Dolomite a mineral, a calcium magnesium carbonate. The term 'dolomitic' applies to a sedimentary rock formed from this mineral. The Dolomite Mountains in northern Italy are dolomitic limestone.

Drip Stone an old name for stalactites (which hang down) and stalgmites (which grow up)

Dry wall stone (Minchinhampton, Gloucestershire, England) thin beds of limestone noted in Arkell and Tomkeieff (1953: 39) as being 'Suitable only for mortarless fencing walls.'

Dundry stone a yellow oolitic limestone from Dundry near Bristol, England, which was exported to Ireland from the twelfth century. Glendalough Cathedral, County Wicklow, southeast Ireland, was built from this stone. The churchyard in the village of Dundry has a large cube of limestone in the grounds, locally called the 'dole stone', measuring 5 feet (approximately 1.5 metres) on each edge. Steed (2011: 105) mentions the use of this stone for a new spire at St Mary Redcliffe church, Bristol, in 1491 and quotes the price for thirteen loads of Dundry stone. The stone cost £1.28 in modern decimal currency. The transaction included the cost of 20 pennies (approx. 8½p) for stacking the stone:

For xiii lode of frestonys	25s 7d
For dykng of stonys	20d

Dyke a geological term for a vertical vein of igneous rock in the earth's strata. This may be forced up through beds of sedimentary rock or coal.

Elvan / elvin (Cornwall) any dark coloured rock such as greenstone, basalt, chert or sandstone with a high content of Quartzose, which is a hard mineral composed of silica ($SiO2$). Also known as **whinstone**.

Erratic typically a large piece of rock which is quite different from the local rock types, usually moved there by glaciation. See Section 1.3 for more detail on erratics.

Faik / Fakes / fakey rock / fakey sandstone (Yorkshire and Scotland) mica-rich sandstone or siltstone which splits easily into thin sheets. Thin horizontal beds of shale-like sandstone or limestone are frequently found in quarries. DSL includes a 1789 reference:

> These thin grey strata are called by Scotch colliers grey fekes as well as grey bands. They are often found moderately hard and of a strong texture, so as frequently to make good flags and covers for sewers.

Fencepost limestone a type of **limestone** with beds 8-12 inches (20–30 cm) thick which has been quarried in Kansas since 1880 to make posts for barbed wire fencing.

Fissile is a description of a rock type which splits easily, for example **slate** or **flagstone**. In many cases this indicates rock which will be easily destroyed by weathering. It should therefore be used with caution, especially if the structure supports a load or is open to extreme weather.

Williams (2012: 31) gives the origin of three names for rock types which split easily into sheets:

> Slate comes from the French *esclater*, 'to split'; phyllite [a type of metamorphic slate] from the Greek *phyllon*, for 'leaf'; and schist from the Greek *schistos*, 'to split'.

Figure 6: Flagstone roof. Melness, Sutherland, Scotland.

Flagstone a geological term for 'a fissile micaceous sandstone', according to Whitten and Brooks (1979: 177). Generally speaking, flags or flagstones are any thin bedded sandstone or limestone which can be split into a reasonably flat surface. They are particularly useful for paving and flatwork. Flags are often mis-named 'slates' because they can be used for roofing, despite their weight and size. In Scots dialect 'flag' is closely associated with 'flake', whether it is a piece of turf, a snowflake or a slab of stone (Figure 6).

Freestone / freestane / freistane / free-stone early drystone walling commentators usually mean limestone or sandstone when they refer to freestone.

A concise definition is found in McMillan (1997: 20):

> A freestone is a massive medium to thick-bedded sandstone in which no internal lamination is apparent and which can be worked with equal ease in all directions. Historically this was

frequently referred to as liver rock. Often but not necessarily, the stone has a uniform appearance.

Liver rock is fully defined later in this section. Arkell and Tomkeieff (1953: 127) refer to 'wallend-stone' from Northamptonshire, England as 'An oolitic freestone used for cornices and chimney pieces, and also a good paving stone… Presumably 'willing stone', alias 'free stone', amenable to working.' 'Weather-freestone' is a Scots term for freestone which is hard and unaffected by the elements. 'Massive' is a geological description for rock with no grain or laminations, a 'freestone'. The sediments laid down to produce sandstone were sometimes irregular. This produced laminations and given names like 'ripple bedding' or 'cross-bedding', depending how they lay. Arkell and Tomkeieff (1953: 67) record 'knell stone' as a northern English term, defined as 'Freestone (sandstone).' In GVA Dorset (1812: 47) Dr Maton is quoted:

> …the name of free-stone is very expressive of the useful property of Portland stone, in enduring to be cut in any direction, whether horizontal, perpendicular, or parallel, to the site of the strata, while it bears the injuries of the weather equally well in every position.

Gabbro a coarse-grained igneous rock, similar to granite.

Geode a geological term for a plain looking rock which has a cavity lined with crystals projecting towards the centre. Known as 'potato stones' in Somerset, England. These nodules were formed in the Triassic period, 250 to 200 million years ago. The interior cavity of such a stone is called a 'vug'.

Gingerbread stone (Norfolk, England) a hard sandstone, named for its colour.

Gneiss a metamorphic form of granite or sedimentary rock with characteristic bandings of quartz, feldspar and mica.

Goshen stone a flat slab type of mica schist rock found in New England, USA.

Granite a common igneous rock composed of quartz, feldspar and mica, in shades of grey, pink, black, red and grey. The name is derived from the Latin *granum*, a grain. It is one of the hardest and most durable of rocks and tends to break into large pieces. Granite weathers well and its strength makes it ideal for building. It can also take a high polish which lasts far longer than on other rock types. Quarrymen historically used the term 'granite' for any hard granular rock such as diorite and dolerite.

Quarrying experience and granite products were traded between Scotland and New England for many years. Aberdeen, Scotland, is known as 'The Granite City'. Rubislaw is a granite quarry near Aberdeen which operated from 1740 to 1971. At one time it was the deepest hole in Europe at 466 feet (142m) deep and 395 feet (120m) diameter. Barre, Vermont, USA, styles itself the 'Granite Center of the World'; the quarries there have operated since 1812. Granite has the nickname 'ironstone' in Pennsylvania.

'Grenite' is a Scots form of 'granite'. 'Blue whin' is another term for granite from Fife in Scotland. The term **whin** is a generalisation, applied to any hard grey rock. 'Syenite' is a type of granite; it contains quartz, feldspar and hornblende, unlike other granites which have quartz, feldspar and mica. Like true granite syenite is very hard and takes a high polish.

Greek Arkell and Tomkeieff (1953: 51) include this 1886 Scottish term, defined as 'Any coarse-grained hard rock, such as coarse sandstone.'

Green stone a description applied to any freshly quarried stone containing a high percentage of natural moisture. Stone is easier to carve or cut in this relatively soft state, it becomes unworkable if it freezes. After drying out the stone becomes harder, more brittle and less easy to shape. It can also change colour. Fresh sandstones and limestones were often shaped with old carpentry tools. 'Quarry sap', per Hill and Birch (1994: 25), is a Cotswold term for the natural moisture and dissolved minerals in stone. 'Quarry-water' means the same and 'quarry bree' is the Scots equivalent.

Greenstone a hard igneous rock from Virginia, USA, difficult to work or shape. 'Catoctin' is a type of greenstone named for a mountain in the Blue Ridge Mountains, between Georgia and Pennsylvania. The term is also applied to **basalt**.

Greywacke / grauwacke / greywacke a bluish-grey gritty hard sandstone found in the Scottish Borders and southwest Scotland. Many of the dykes in southwest Scotland are built from hard greywacke sandstone; it is traditionally referred to as **whin** or **whinstone**.

Arkell and Tomkeieff (1953: 52) describe greywacke as '…Fine-grained, dark-coloured, Lower Palaeozoic grit. A loose lithological term now almost obsolete.' They also quote one 1837 writer who disliked the term 'greywacke' because it ' …had done much disservice to geology… [by merging] …deposits belonging to distinct periods of the history of the earth, we hope that the term may be henceforth discarded as a term of classification.'

Grikes fissures in the limestone in a **karst** landscape in northern England.

Grit a hard sandstone, found in north England, also known as 'gritstone' or 'millstone grit', the latter term because the coarse sandstone, composed of granulated quartz and quartz pebbles, was used to make mill stones. 'Greet-stone' is a Yorkshire term for coarse or gritty sandstone.

According to Arkell and Tomkeieff (1953: 53) 'In Yorkshire all sandstones, both course and fine, are often known as grits…'

Haggis rock (Scots) a **greywacke** from the Scottish Southern Uplands. According to the British Geological Society leaflet for Witch Craig Viewpoint, near Bathgate, West Lothian, 'This is a conglomerate deposited in a long lost ocean where gravel sized fragments of rock rapidly accumulated and became cemented together over 420 million years ago.' The name was given by quarrymen in the mid 1800s.

Heuch (Ayrshire, Scotland) a quarry or quarry face.

Holystone (Gloucestershire, England) limestone full of holes.

Hornblende a bluish green black mineral found in many sedimentary and metamorphic rocks. The chemical name is Calcium Magnesium Iron Aluminium Silicate Hydroxide. Beenie blue is a hornblende boulder from northeast Scotland, so called because of its bluish colour.

Howk (Scots) a verb, to dig or to quarry.

Hurlock a description of stony chalky soil from the limestone districts of England.

Igneous rock a rock formed from molten magma. Types include basalt (Figure 2), dolerite and granite. They are called 'primary rocks' because they are the primary source of the material on the Earth's surface.

Ironstone (Pennsylvania, USA) a nickname for **granite.**

Jacob's stones (Shropshire, England) calcareous ironstones burnt for lime.

Karr (Orkney, Scotland) DSL includes this, defined as 'A hard bed of rock, [especially] one which has to be cleared to get at flagstones below.'

Karst / karstic a description of limestone landscape, found in the English Mendips, the Burren in Ireland and the Yorkshire moors, where the surface has been weathered into 'limestone pavement'. This looks like large slabs with gaps between them.

The fissures on the surface (called 'grikes' in northern England) capture rainfall and lead it below the surface to where the water gradually wears the limestone, often creating underground caverns. The limestone pavement provides a good source of walling stone and, because it comes out in large pieces, leads to some unique walling styles.

Kingle / kennel (Fife, Scotland) a type of very hard sandstone.

Kleber / claber / clebber / kleber also known as 'steatite' or 'soapstone'. A soft easily carved stone. In Shetland it was used as a substitute for metal oxides in ointments.

The term 'clabber' is also applied to soft stone or mud or describes being covered in mud.

Knell stone Arkell and Tomkeieff (1953: 67) record this as a northern English term, defined as 'Freestone (sandstone).'

Limestone a sedimentary rock containing more than 50% calcium carbonate, sometimes fossiliferous (containing fossils). Varieties include 'oolitic', 'dolomitic' and 'carboniferous', each with its own characteristics. Limestone is often found as a 'freestone'. It has very little grain and can be worked freely in any direction.

'Sink stone', from Lincolnshire, eastern England, is a bed of limestone used for making sinks and troughs. Drystone wallers usually find limestone easy to work with, but it can weather badly. Stone from an old wall is often useless for a rebuild.

The burning or roasting of limestone is the first step toward making traditional lime masonry mortars and cements.

Lithic relating to stone, from the Greek *lithos,* meaning 'stone'.

Liver rock (Scots) a type of sandstone. Pride (1996: 49) defines this as 'a rock of homogeneous sandstone without any well-defined planes of stratification, therefore very suitable for masons to work.'

The Craigleith quarry in Edinburgh, Scotland, supplied 'liver rock', for ashlar blocks to build much of the Edinburgh New Town between 1756 and 1850.

Liver-stone (Derbyshire, England) bituminous limestone which emits a sulphurous smell when struck, also called 'stinkstone'. Many limestones release a fetid smell if crushed.

Marble a fine-grained metamorphic rock, a limestone which has been altered by heat and pressure. Historically the term 'marble' has been applied to any decorative stone which will take a polish. 'Marm-stone' is an obsolete old English term for marble. 'Maber', 'marbre', 'marbyr' and 'marvel' are all Scots variations of 'marble'.

Marl a geological term for a calcareous mudstone, a mudstone containing lime. 'Marlstone', by way of contrast, is defined as an argillaceous limestone, a limestone which contains clay.

Massive a geological term for rock with no grain or laminations, a **freestone**.

Metamorphic rock igneous or sedimentary materials which have been changed by heat or pressure. In simple terms, one material is changed to another by a massive cooking process. Limestone becomes marble; shale morphs to schist; mudstone changes to slate; sandstone transforms into quartzite; granite converts to gneiss.

Mica a clear flaky silicate material with perfect basal cleavage found in many igneous and metamorphic rocks. 'Cat-siller', 'sheep's siller' and 'craw-siller' are Scots terms for this shiny mineral. A 'craw' is a crow, a bird of the genus *Corvus*. 'Siller' is the Scots word for 'silver', used as a synonym for 'money'.

Mica schist a metamorphic rock with a high proportion of mica and quartz.

Mine (Cotswolds, England) an underground stone quarry.

Figure 7: Exposed rock, a good source of building stone. Scottish Highlands.

Sandstone was mined at Drumkeelan Mines near Donegal, Ireland from the twelfth century until the 1950s. Tunnelling avoids the effort of removing huge amounts of overburden; extra care is essential to ensure the roof does not collapse when rock is removed.

Mountain limestone (north of England) a term used by the old colliers to differentiate between the limestone of the Pennine hills and the limestone found in the coal mines.

Mulatto stone Arkell and Tomkeieff (1953: 78) include this term from Antrim, Ireland for a 'Green speckled sandstone with a calcareous cement …So called on account of its mottled appearance.… adopted from Spanish *mulato,* young mule, hence a man of mixed race.'

Nebraska marble a colloquial term for sod, as used to build **sod houses**, described in Section 4.4.

Obsidian a very hard volcanic glass, a valuable resource for tools and decoration.

'Pitchstone' is a geological term for a glassy igneous rock, similar to obsidian.

Oolitic limestone a soft rock with small round calcareous grains, common in the Cotswolds. The name comes from the word 'oolite' because the grains look like fish eggs.

Figure 8: Metamorphosed sedimentary rock. Inverness-shire, Scotland.

Before the introduction of the word 'oolite' the name for oolitic limestone was 'spherical grit', according to Arkell and Tomkeieff (1953: 80).

Peasie whin / peysie-whin / pacey whin / peasywhin (northeast Scotland) a rock, usually granite, with a marled granular surface. It is named for the resemblance of the spots in it to pease porridge, unlike other hard rock with more even colouring. Pease porridge is made from pea meal.

Pitchstone a geological term for a glassy igneous rock, similar to obsidian.

Porphyry a hard fine-grained igneous rock, reddish in colour, which contains feldspar. The name comes from the Greek word *porphyros* — the colour purple.

Potato stone a Somerset (southern England) term for a geode.

Plies / ply a mining term for a thin layer of hard rock separated from another hard layer by, say, clay or mica.

Post a British mining term for a thick layer or seam of sandstone or limestone, as per Arkell and Tomkeieff (1953: 88). The authors believe an old term for a horizontal beam had been transferred to describe a layer of stone.

In central Kansas, the local creamy-buff limestone breaks easily into long pieces with a square cross-section. These were used for fenceposts in the nineteenth century because timber was scarce and there was a demand for barbed wire fences. Muilenburg and Swineford cover this topic in their book 'Land of the Post Rock' (1975).

Primary rocks those rocks, usually igneous, created from the first material extruded from the middle of the planet. They have not been formed by sedimentation or altered by any geological process.

Pumice / pummy (Sherbourne Abbey, Dorset, England) a type of tufa, a porous limestone used in the town and for the vaulting in the abbey. 'Pummy' is a corruption of 'pumice'.

Purbeck marble a hard limestone from Dorset, southern England.

Quar (South Wales) a hard sandy shale, a type of sandstone.

Quarrel / quarrel / quarrel / quarrel-heugh (Scots) variations on an old term for a stone quarry or the stone taken from a quarry. The verb 'to quarrel' can also mean to raise or dress stones in a quarry. JSD introduces shades of meaning – 'to *win*, is to select and gather: to *quarrel*, is to dig and raise and shape however roughly.'

Querrell-hollis literally 'quarry holes', old quarries filled with water.

Quarry a source of stone. It can be as simple as a cutting into a hillside (Figure 8). Ruined buildings such as castles and abbeys have been used as quarries. The earliest quarries in New England, USA, were glacial boulders and outcrops of rock. In the granite areas of Ireland surface boulders were an accessible source of good stone. These sources avoided the expense of removing topsoil and substandard rock before finding good material.

Stone from demolished buildings can be recycled; it is not always cheap. The 1842 Statistical Account for Elginshire refers to stone from the Castle of Burgie in the parish of Rafford — 'The strength of the building was such that a detaching of the stone from the mortar or lime cost more than the quarrying of new materials would have done.'

There is also a possibility the stones from a ruin have been weakened by exposure to weather or fire.

King (2000: 22) describes how the city of Rome became a quarry. After the 1348 Black Death the population of Rome was reduced to 20,000; they retreated to a corner of the city. Many fine buildings deteriorated and were scavenged for the stone. Marble statues were burned in kilns to produce quicklime or fertilizer. Some of the stone ended up in Westminster Abbey, London, England.

In Britain and Ireland castle ruins, ancient structures such as Hadrian's Wall and brochs were regarded as an easily worked 'quarry'. Many tons of good stone were removed from these sites with little effort.

The uninitiated may regard old stonework as 'fair game', free to take away, but, however dilapidated, it is part of our heritage. Theft of stone in the English Lake District has been thwarted by spraying oil over the wall – as illustrated in Rollinson (1998: 41).

The double copestones (inserted in two halves, one from each side) in some Kentucky walls are cemented down the middle to make sure they are not easily removed. Kentucky introduced a $500 fine for every foot of wall stolen. New England states have also introduced legislation to protect their drystone heritage.

Quarrying is a careful and considered process, requiring skill. McMillan (1997: 32) writes:

> Good quarrying practice ensured that stone was carefully handled during and after quarrying. Dropped or knocked stones, for example, may develop unseen small fractures at the site of the shock which may open up when the stone is exposed to weathering.

The exposed working area of a quarry is called the 'face'. It can be a vertical or 'benched', meaning 'stepped'.

Clach-shlochd is Scots Gaelic for a stone quarry. The word 'Slochd' appears in some old place names.

Quarry sap / quarry water / quarry bree various terms for the natural moisture found in freshly quarried stone. Such stone is called 'green stone' and is softer: easier to work, before it hardens.

Quartz a crystalline form of silicon dioxide. Quartz is present as speckles in granite and as bands in rocks such as gneiss and hornblende. In the twentieth century, quartz was a valuable component of many emerging technologies. An article in 'The Scotsman' newspaper, for Tuesday 16th June 1936, records the finding of quartz in Uganda. This was a strategically important find for '...the growing demands of industry for apparatus required in wireless telegraphy, television, depth-sounding at sea, and in detecting the presence of icebergs, rocks and other dangers to shipping, ... '

Previous sources for this mineral were outside the British Empire.

White rocks, those rich in quartz, have been used in structures for thousands of years. Apart from their decorative value, they were thought to have supernatural powers.

Quartzite a hard, brittle sandstone consisting mainly of quartz grains cemented with silica. It contains up to 90% quartz.

Quhynstane / quhin / quhyn / quhin-stane older Scots spelling for **whin.**

Rag / rag-stone / ragg / ragstone English terms used for various coarse hard rocks which are easily broken up into walling stone. 'Rag' in this context suggests tattered, ragged and small pieces. One limestone from Kent and Oxford which contains fragments of fossil coral is called 'coral rag'.

Barnack ragstone is an oolitic limestone from near Stamford in Lincolnshire, England. It was used to build Ely Cathedral. Construction started in 1083; the main structure was completed by 1189. One local type is known as 'Kentish rag'.

Rhyolite a volcanic rock, similar to **granite**.

Roach a term, according to Arkell and Tomkeieff (1953: 97), applied to various kinds of rock in different parts of England, for example conglomerate and Old Red Sandstone in the Lake District. Arkell and Tomkeieff go on to detail how 'roach' is a varietal spelling of 'roche', a Middle English term derived from Old French *roke, roque, rocque* and applied from the thirteenth century to large masses of rock or a boulder. In Ireland the term 'roach' was commonly used for 'quicklime' and is occasionally still used today.

Road-stone a common name for inferior stone used to repair roads, from the days before tarmacadam when the surface was formed with hard packed stone.

Rubble rough untrimmed rock, stones with no obviously good faces, often relegated to backfill behind an ashlar face. According to JSD:

> In E. [England] *rubble-stones* are said to 'owe their name to their being *rubbed* and worn by the water, at the latter end of the deluge, departing in hurry and with great precipitations.' ...The term *rubble* itself is used as denoting rubbish... In S. [Scotland] the term is applied to rough stones, of any description, such as are commonly employed in building, without being polished, but merely as hewn by the hammer.

Figure 9: Limestone dyke. Note the lichen growth. Sutherland, Scotland.

A good waller who knows the material, and how to shape it, can still make a nice wall with rubble.

Sandstone a sedimentary rock, typically porous, formed from sand cemented together with silica or iron oxide. The reddish-brown colours come from the iron. It is easy to carve and split depending on the bedding planes. If it has no bedding planes it can be called 'freestone' because it can be cut freely in any direction. In old accounts of Scottish dyking many references to 'freestone' meant 'sandstone'.

New Red Sandstone is the geological term for sandstones laid down between the Carboniferous and Jurassic periods — between 195 and 280 million years ago. Old Red Sandstone originates from the Devonian period – between 345 and 395 million years ago.

Sandstones with larger, and mixed, particle sizes are better described as 'conglomerate'.

In GVA Angus or Forfarshire (1813: 19) the Rev. James Headrick notes a type of sandstone ' ...commonly called bastard freestone, or rubble-stone, because, owing to its hardness it cannot be cut with the chisel, and is only fit for what masons call rubble-work.'

Looking at the face of a sand or sandstone quarry is interesting. The quarry face can look like a marble-cake with all the swirls and layers of different coloured sand: lying as it was laid down in the original desert.

Schist a metamorphic rock made up of parallel layers of minerals including iron and mica, created by pressure. It readily splits into sheet or slabs. 'Schiste' is a French form of the word. Slate, shale and schist are distinct rock types though, to the non-geologist, they look the same.

Sclate-stane is a Scots term for a piece of slate, or stone which looks like slate, because it is flat and cleaves into thin sheets.

Scourstone from Durham, England is a bed of massive sandstone, a freestone with no beds. It was worked for hearthstone, hearths and sills. With no chance of laminating this stone is ideal for exposure to reasonable heat.

Scrap (Somerset, England) a three foot bed of stone in Ham Hill, Somerset, England. Arkell and Tomkeieff (1953: 103) declare it is '...only fit for rough-walling... adapted from Old Norse *skrap,* [meaning] remnants, fragments.'

Scurdie / scurdy / skurdy a type of **whinstone** or **basalt** used for street paving. From Scurdie Ness, a headland near Montrose, Tayside, Scotland. Basalt type rocks are called 'skurdy' in the north of Scotland, probably from 'scur', an old Scots mason's term for the rough projecting surface of a stone. Scurdy Ness was probably named for the rock rather than the rock named for the place. JSD notes 'skurdy' as a kind of moorstone, stone found on the moors, 'a dark blue colour and so closely textured that water cannot penetrate it.'

Sedimentary rocks those rocks which start as layers of sediment built up at the bottom of a lake or sea. The process of 'lithification' creates sedimentary rocks by compaction and cementation of the sediments. 'Lithification' stems from the Greek word *lithos*, a rock. Limestone, sandstone, mudstone and shale are typical examples.

Shale a soft sedimentary rock made from compressed mud or clay. It is like a sandstone though the grains are much smaller, usually dark grey to black in colour. 'Paper shale' is a geological term for shale which splits into very thin friable sheets. 'Penny shale', from Derbyshire, England, produces durable flakes, like penny coins, when it weathers.

Shell rock (Kansas, USA) a bed of chalky limestone, rich in fossils, in north central Kansas, USA.

It can be broken out in large sheets which Muilenburg and Swineford (1975: 33) describe as stood on end to build a corral '...six to eight feet high enclosing a sloping hillside area of about 200 square feet... '

Shillet shales from Devon and Somerset, southwest England. Derived from the Old English words 'sheel', 'shell', 'shale' or 'shill' meaning 'to shell off.' From this comes the word 'shilly', an angular scree.

Siliceous containing silica, the most abundant mineral in the Earth's crust. Silica is found in a wide range of precious and semi-precious stones and in more humble materials such as sand.

Skarn a geological term for an impure limestone.

Sklaffirt / sclaffer (Banff, Scotland) a rock lying horizontally in thin beds.

Slate a fine clay siltstone metamorphosed by heat and pressure. The stone is dense and splits easily into thin pieces with a flat, parallel top and bottom bedding which makes it ideal for its main use as a roof covering. Slate is used for roofing, billiard tables and tombstones. It is especially suited for the latter purpose because fine detailed carving stays sharp. In Scotland, the slate quarries at Ballachulish (in Lochaber, near Glencoe) and Easdale in Argyll are the best known. There were also quarries in Aberdeenshire and Banffshire. Slates were mined in the English Lake District and Wales. Few remain working, these quarries are now a shade of their former glory. Britain now imports roofing slates from Europe, Brazil and Newfoundland.

This word is an excellent example of one with a multitude of old, almost phonetic, spellings which were eventually standardised. DSL includes these variations:

sclate; sclaytis; sclait; slaite; sklate; sclatt; sklatt; sclayt; sklaitt; sklaytt; sclett; schlett; skleatt; skleit; sklethe; slat; slayt; sleat; sleatt; slett.

Snatch quarrying a modern term for the reopening of old quarries for short periods, for repairs and new build. Many old quarries have closed or been incorporated into the cities they helped build. Sooner or later maintenance is required on the structures and, if local planning conditions inhibit the opening of the original quarry, other sources must be found. As an example, stone from a quarry in Hopeman, Moray, Scotland, is sent to Barcelona to continue construction of *The Basílica I Temple Expiatori de la Sagrada Família.*

Soapstone a soft metamorphic rock with a greasy, soapy, feel. It is easy to carve because it is mostly talc, a mineral composed of hydrated magnesium silicate. It is also known as 'steatite' and 'kleber' in Shetland.

Trap Arkell and Tomkeieff (1953: 120) include the definition:

> An old Swedish term adopted in Great Britain in the eighteenth century. Originally in Sweden this name was applied to dolerite but later it was extended to cover different varieties of dark compact igneous rocks such as basalt.

JSD defines 'trap' as 'whunn' and gives no more definition. 'Whunn' is a variation of **whin**, a name frequently applied to any hard, dark coloured rock. Thorson (2005: 33) gives 'traprock' as the local New England term for basalt.

Tuff / tufa / tuph / tuft a geological term for compacted volcanic ash. It often contains larger particles up to small pebble size. Variations of the word, for example 'tufa', 'tuph' and 'tuft', have been used in the past for any porous or soft stones.

Urbanite (North America) a modern term, not geological, for recycled broken concrete or paving. It is arguably very similar to a **conglomerate** because urbanite also contains aggregate and a cement. The walls near Kilkeel, County Down, Northern Ireland are a good example. When the runways of RAF Greencastle were lifted in the 1960s, farmers used the broken concrete to build field walls. (See Section 1.2 for more detail on conglomerate).

Vug Arkell and Tomkeieff (1953: 123) identify this as the internal cavity of a geode.

Walling bed / walling-stone Arkell and Tomkeieff (1953: 124), have this as beds of limestone from Lincolnshire and Northampton which were '…fit only for walls.'

Whin / whunn / whinrock / whinstone / whunstane / whyn / quhynstane / quhin / quhyn / quhin-stane generic terms, mainly from Scotland and northern England, commonly applied to any hard grey rock such as basalt, or diorite, or to describe hard **greywacke** sandstone. DSL includes an 1835 definition — 'The term whin is applied in Berwickshire to all rocks except freestone and quartz.' In this context 'freestone' means limestone or sandstone. JSD defines 'whunn' as 'The stone called Trap.' Other references to whin include: 'Bastard whin' – whinstone mixed with limestone; 'Black whin' – a variety of dolerite; 'Whin boul' – a hard nodule of whin embedded in sandstone; 'Whin brush' – whinstone dust or small rubble; 'Whin craigman' – a person who quarries whinstone; 'Whin-flag' – a flag or flat slab of whinstone; 'Blue whin' – a Fife (east Scotland) term for granite; 'Peasie-whin' – a term applied to granite because of the colouring, and spots likened to peas.

The British Engineering Standards Association recommends the term 'whin' should be restricted to types of basalt. GVA Roxburgh (1794: 22) confirms whin's suitability for dyking:

> Whinstone, from its angular fractures, is capable of being built very compactly and firm, and is probably the best kind of stone for dry dykes. Care, however, should be taken, that it is not of the kind that crumbles down by exposure to the air.

'Whin' is also a term for the Common Gorse (*Ulex europaeus*), a thorny bush historically used in conjunction with stone or earth banks to give a livestock proof fence (Figures 81 and 91). Whin was once ground down for animal feed.

Willen stone Arkell and Tomkeieff (1953: 126) define this as 'An oolitic freestone used for cornices and chimney pieces, and also a good paving stone… Presumably 'willing stone', alias 'free stone', amenable to working.'

1.2 Conglomerate

Conglomerates are sedimentary rocks with inclusions of rounded pebbles, and other fragments, bound together with natural cements. The word is derived from the Latin *conglomeratus* meaning 'heaped or pressed together' and has been used as a rock term since the early nineteenth century. Conglomerates are not always easy to cut or shape. They are good for drystone walling because the stones are rough and friction 'sticks' them together. (Figure 10).

Bibbly rock (Staffordshire and Leicestershire, England) conglomerate.
From the local word 'bibble', a pebble.

Breccia a conglomerate with angular rather than rounded fragments. Arkell and Tomkeieff (1953: 12) connect the term with the Italian *breccia* meaning 'rubbish of broken walls' or the French *brèche* meaning 'broken'. Most conglomerates are made up of fragments which have been rounded by water or glacial flow then cemented together with minerals.

Breedingstone (Hertfordshire, England) so called because the smaller stones were supposedly 'breeding' within the larger mass.

Bull Rock (Shropshire, England) conglomerate.

Bowels / bowel-stones (Buckinghamshire, England) a quarryman's term for a type of course conglomerate found in sandstone which looks like a cluster of grapes. The term is derived from an old French word for 'guts' or 'intestines'.

Box stones a geological term for concretions of sediments around a nucleus.

Cornstones a geological term for concretions of limestone.

Dogger Arkell and Tomkeieff (1953: 38) define these as 'Ironstone nodules, usually of iron carbonate, found in the rocks of Jurassic age in Yorkshire… Any large rounded or lumpy mass of sandstone in any part of the country.'

The authors also quote a 1670 source which refers to '…Doggers, or Cats-heads…'

Figure 10: Conglomerate and sandstone boulders set as protection for the beach. Dunrobin Castle, Sutherland, Scotland.

Gravelstone a stone consisting mainly of larger sized sand, pebbles and river gravels fused together with a natural cement. Note the similarity of 'gravelstone' to the word 'sandstone'. The different terms indicate the size of the particles.

Greet-stone (Yorkshire, England) coarse or gritty sandstone, a form of the word 'gritstone'.

Growan / grouan / growan-stone (Cornwall, England) granite or any course gritstone or sandstone.

Growingstone (Hertfordshire / Buckinghamshire) a local name for the Hertfordshire Puddingstone, with the alternative name of 'breedingstone', as noted by Arkell and Tomkeieff (1953: 54) who quote from a 1942 collection of Wiltshire folklore telling us 'The older generation firmly believed, and do still, that the loose stones in the ground grew there, rather like potatoes.'

Grozzle (Leicestershire and south Derbyshire coalfields) an alternative name for conglomerate, possibly from the Welsh word *gro* – gravel.

Motherstone / Modrestone Arkell and Tomkeieff (1953: 77) define the term as the 'Local name of the Hertfordshire Puddingstone where it was quarried *in situ* at Radlett, Herts.'

Radlett lies just north of London. 'Modrestone' is an old form of 'motherstone'.

Pimp rock / pimple-stone a Shropshire dialect term for conglomerate. The latter is a variation of 'pimble-stone' which is, in turn, a variation of 'pebble-stone'.

Plumcake (Nottinghamshire, England) a breccia conglomerate, with angular rather than rounded inclusions. The stone will have been named for its similarity to the plumcake which has clusters of fruit and an inconsistent texture.

Plum-pudding stone / Plumb-pudding stone / Pudding stone beds of glacially deposited pebbles and sands, bound together with siliceous cement. A term found throughout England for conglomerate.

Popple-rock (Devon and Somerset, southern England) a conglomerate of 'popples' or pebbles.

Rombly limestone (Northumberland and Durham, England) carboniferous limestone or sandstone containing pebbles.

Yittle (Shetland) an induration or nodule in a stone, especially of quartz in sandstone. A small pebble will have landed in the sediment, become enveloped, and part of the solid mass.

Yolk / yowk (Scots/English) a hard nodule in softer rock, like a yolk in an egg. 'Yolky stone' is therefore a conglomerate stone. According to GVA Angus or Forfarshire (1813: 19) course pudding-stone, gravel stone or breccia is also known as yolky stone '...because it is composed of a vast number of rounded pebbles, resembling yolks of eggs, which are bound together by a ferruginous sandy cement...'

1.3 Erratics, travellers and strangers

A large rock standing on its own becomes a focus of attention. Many were dragged there by mile-high ice and left stranded when the glacier retreated. As far as humans are concerned, the rocks have been there 'forever'. Myths and legends grew in their shadows (Figure 11). Dykers were not always so romantic. A large boulder lying on the proposed line of a dyke would be incorporated into the structure or broken down to build a section of 'single'.

Cowstones (Devon and Dorset, England) large concretions of hard sandstone, so called because they resemble cattle lying on the slopes along the cliffs.

Devil's putting stone a common name for large boulders or glacial erratics whose placement is attributed to the Devil by local superstition or legend.

Dowie stones DSL defines 'dowie' as '...[a] term commonly applied in Eskdale [southwest Scotland] to large granite stones deposited during the glacial age.'

Dowie means 'sad and melancholy'. A dowie stone is therefore lonely, separated from the herd.

JSD defines to 'lie sad' as lying close to the ground. DSL has various definitions for 'sad' suggesting something compacted or low lying.

Druid stones an alternative name for **sarsen stones,** so called because these large erratic boulders were used in Stonehenge and other supposedly Druidic temples and circles.

Eart-fast (Shetland, Scotland) literally 'earthfast', a description of a large rock firmly imbedded in the earth. Heavy stones will sink with time. Also known as a 'sitfast' or 'yird-fast'.

Erratic / glacial erratic a glacially deposited rock, often of considerable size, and usually out of context with the local geology. Blocks of stone which originated in Norway have been found in Caithness, northern Scotland. Granites from the north of Ireland are found on the limestone of the Burren. The phrase 'Connemara Stranger' or 'Connemara Visitor' has been used for glacial erratics around the Burren and Aran Islands in the west of Ireland. This picturesquely describes granite boulders captured by a glacier, transported south and left stranded on warmer Galway limestone.

Bloc erratique is the French equivalent. *Erratique* literally translates as 'wanderer' or 'vagrant'.

Foundling-stones (Scots) erratics, from the word 'foundling' – a deserted infant.

Glacial erratic a rock left behind after the retreat of a glacier which originally pushed it.

Grostane (Scots) a large surface boulder of grey stone. 'Gro' here means 'grey'.

Greywethers (England) large boulders, so called because from a distance they look like a flock of sheep. A 'wether' is a young sheep or a young castrated male sheep.

Heathens / haithens / (Aberdeenshire, Scotland) gneiss boulders, stone rubble from fields, often of glacial origin. They are not easy to break.

The DSL includes a reference to 'heathens' from Aberdeenshire in 1845:

> Immense boulders of blue sienite were, at one period, scattered over the soil. The farmers term them heathens, probably from the incessant contest they and their forefathers have waged with them from time immemorial.

'Sienite' or 'syenite' is an igneous rock like granite, with less or no quartz content. Another

Figure 11: The Split Stane, beside the A836 road, marking the old boundary between Caithness and Sutherland, northern Scotland. Reputedly cut by the Devil's tail.

reference comes from Banffshire in 1928 — 'That's a haithen; it'll naither mak nor brak; try't wi' a shot o' poodher.' This is translated to: 'That's a heathen, it won't cut or break, try gunpowder.'

Hen-stones (West Yorkshire) groups of stones on the ridge of high moors. They look like a line of hens on a perch.

Laird (Scots) a large earth-fast boulder in a field which must be ploughed round. It dictates the use of the ground, like the laird, the owner.

Logan stone / logging stone (Cornwall, England) a rocking stone, according to Arkell and Tomkeieff (1953: 72) it is a '...block of stone so finely balanced on its base that it will easily 'log' or 'rock' from side to side... [from the] Cornish *log*, to rock.'

Moore-stone (Dartmoor, southwest England) according to Arkell and Tomkeieff (1953: 77) these are large granite or gritstone boulders lying out on the moors. They are an easily accessible source of good stone for splitting into lintols, quoins etc.

Outliers (Scots) large rocks, often lying some distance from any obvious source of that rock type... They have been broken away from a larger mass, perhaps moved by glacier or flood, and can be exploited as a source of stone without the effort of quarrying.

DSL has an extensive definition. Outliers are defined as 'Stones found loose above ground, not quarries... Outliers as they are called, abounding in quantity, are more used than quarries, because they are both more easily got at, and taking them away clears the ground... Stone dykes, marches, enclosures for cattle, etc. in rural districts are generally built of outlers [sic], gathered from fields, burns, and streams.'

Piedras cansadas (South American Spanish) literally 'weary stones', ones intended for Inka building projects which did not reach their destination. In Quechua, the language of the South American Andes, these stones are called *sayk'uska*. Dean (2010: 50) writes:

> ...the sympathetic Inka, recognising the stones' anguish at being relocated and their refusal to cooperate with the Inka builders, left the weary rocks where they can be found today. We might say that sayk'uska, renitent and tearful, were stones that the Inka first moved (literally), and then were moved by (emotionally)

There are British equivalents. If a millstone fell off the cart, after it left the quarry, it was abandoned as too much trouble, if there were no convenient way of lifting it back on. These stones slowly sank into the ground and become curiosities.

Ploughing boulders according to Wrightham and Kempe (2006: 81) these boulders have moved downhill, faster than the soil in which they are partly imbedded, leaving a furrow in their wake and pushing up a ridge of soil in front, like a bow wave. The typical ploughing boulder is 0.5 – 2 metres (20 inches – 6 feet 6 inches) across, easily recognised by the furrow behind them, up to two metres in length. They are common on many Scottish mountains, especially on fine silty soils derived from weathered schists. The movement is caused by ice forming under the boulder. When it melts the boulder drifts on the saturated soil. The boulders move from a few millimetres to a few centimetres in a year.

Roches Moutonnées a French term for a glacial erosion feature, a rock outcrop smoothed by moving ice. Named by Horace-Bénédict de Saussure in 1786. He saw a resemblance between these rocks and the wigs which were fashionable amongst French gentry. The wigs were slicked over with mutton fat (hence *Moutonnées*). This term is often incorrectly translated as 'sheep rock'.

Rocking-stane a large boulder balanced on smaller rocks, undoubtedly an example of a glacial erratic. JSD informs us it is:

> a stone so poised by art, as to move at the slightest touch. The popular opinion, which supposes them to be inhabited by a spirit, coincides with of the ancient Icelanders, who worshipped the demons, which they believed to inhabit great stones. It is told in the *Kristnisaga*, chapter 2, that the first Icelandic bishop, by chanting a hymn over one of these sacred stones, immediately after his arrival in the island, split it, expelled the spirit, and converted its worshippers to Christianity.

The belief that demons lived inside big stones ties in with a comment, from Ireland, by Evans (1979: 107) — '…until recently there was a reluctance, for superstitious reasons, to break up large stones.'

Rugs defined by Arkell and Tomkeieff (1953: 99) as a '…local name for blocks of weathered Chipping Norton Limestone resembling sarsens, found on the surface in the valley of the Sars Brook, Sarsden, Oxfordshire…'

The authors think the term may derive from a local description of the rock as 'rugged' or 'rough'.

Sarsen / sarsen-stone / sardens / sarsdon stones large boulders or blocks of sandstone found scattered on the surface of the chalk Downs, especially in Wiltshire, England. They are sometimes called 'greywethers' because they look like sheep on the landscape. A 'wether' is a young sheep. Arkell and Tomkeieff (1953: 101) give a full description and offer options for the origin of the term. The first opinion is the name comes from 'Saracens', the Arabian nomads who once harassed their neighbours. From this developed the meaning of something outlandish and therefore applicable to glacial erratics — rock outside its normal context.

The other option bases the word 'sarsen' on the Old English adjective *sār* — '…meaning 'sore grievous, troublesome', from the trouble caused by the stones to agriculture on the downs: but there seems to be no written evidence for this.'

Travel / traivel / traivelled a Scots term describing stones, deposited by nature or humans, which end up at some distance from their origin.

1.4 Natural rock and building stone

This section describes the broken rock and smaller sized stone used to build drystone walls.

Birlie steens an old Scots word for pebbles. 'Birl' means 'to turn' and 'steens' are stones. The term therefore applies to small round stones looking like marbles.

Bitlag / booklag (Caithness, Scotland) a hard, round stone, particularly one with no laminations.

Blunkart (northeast Scotland) a small block of stone.

Boulder / bool / boolder / boul / bowlder / bowlder stane a large piece of rock, typically rounded in shape and too big for one or two men to lift. In geological terms a boulder has a particle size of more than 256 mm (10 inches) following the Wentworth-Udden scale, as used by Whitten and Brooks (1979: 340).

Thorson (2005: 17-18) writes:

> The word 'boulder' derives from the Old Norse *bullersten*, literally 'noise stone'...Rounded stones on bedrock ledges can often be tipped back and forth, resulting in a low, rumbling noise. These are called 'thunderstones' in New England and are alleged to have been used as communication devices by the Indians, not unlike the use drums and smoke signals further west.

At times of spate stones are heard rolling down the river bed. This action produces rounded boulders.

Arkell and Tomkeieff (1953: 11) define 'boulder' or 'boulder-stone' as 'A large water-worn or weather-worn rounded stone or a cobble; frequently used for paving and building.'

'Bool', 'boolder' and 'boul' are Scots variations.

The term 'bowlder', in England, describes a round pebble. The common root of 'boulder', 'bowl', 'boule' and 'ball', is clear.

In Selkirk (Scottish Borders) naturally rounded boulders were used as gravestones and known as 'bulls'.

'Boul-faced' is a Scots term for a building stone with a slightly convex face, literally 'bowl faced'.

'Bull nose' is a term for the rounded front edge of, for example, a step tread.

JSD records 'bowlder-stane' as a term from Perthshire (Tayside, Scotland), 'The name given to the large single stones found in the earth by those who made early roads'.

'Clay and bool' is an old building method using boulders and clay, detailed in Section 4.1.

Some boulders are too large to handle. The 'Scientific American Supplement', number 110, page 1753 of February 9th, 1878, records the finding of a large 'bowlder' in the west edge of Nottingham, New Hampshire at the base of Pawtuccaway Mountain. It measured 62 feet long, 40 feet wide, and at least 50 feet high with a capacity estimated at 70,000 cubic feet, giving a weight of nearly 6,000 tons.

Other large 'bowlders' were nearby. The same article tells us 'The largest bowlder in Vermont is called the Green Mountain giant, lying on a hill in Whittingham; it contains 40,000 cubic feet.'

Boulder field an area of ground covered with boulders, often glacially deposited. A 'single' dyke, built one stone wide, from base to cope is often the best use of this material.

Brash loose broken rock lying beneath the soil and above the solid bedrock, a corruption of the French word *breche* meaning 'break' or 'broken'.

Brashy (Gloucester, England) stony ground.

Bullet / billet (Scots) a large stone, any rounded boulder found in clay or accumulated glacial sediment. A 'bulletstane' was a round stone, most likely the size of lead shot used in a muzzle-loading rifle.

Bumlie / bumlak / bumlock (Aberdeenshire, Scotland) a small shapeless stone.

Figure 12: Slate quarry, Ballachulish, Scotland. Note the terracing (also known as 'benching') which allowed easier access to the quarry face. Slate was quarried in this area from the late 1600s to 1955.

Calyou / caildo / kaylowe / kaildo (England) pebbles used for rubble walling, a corruption of the French word *cailloux*, a small stone or pebble. Small stones are useable for mortared walls if the corners and openings of the structure are formed out of larger blocks, like sandstone, because it is easily cut and shaped.

Catbrain / catsbrain (England) soil consisting of rough clay mixed with stones. Fossil corals in limestone districts of England were called 'brain stones'.

Chuckie / chuckie stane / chuckie stone (Scots) a small stone, a small pebble, a quartz crystal rounded by water action and friction on the beach. Prevost (1957: 97) defines chuckie stanes as 'stones composed mainly of quartz crystals.'

JSD links 'chuckie-stanes' to poultry from their use as a nest-egg. Egg-shaped stones put under broody hens will encourage their instinct to lay and sit on fertile eggs until they hatch.

Cleff (Scottish Borders) an adjective meaning 'thin or flat', as in 'cleff stane'. A 'cleffie' is a flat thin stone thrown so it skims across water. There are many old dialect words for flat stones used in such games.

Clod DSL has this as a term for a lump of earth or similar material and includes an interesting phrase from Argyll in 1937 — '…*to clod the clover*, to gather stones from a field from which a hay crop is to be taken…'

This reduced the damage to scythes.

Clumper (Shetland and Orkney) a large fragment of rock, for example a boulder on the beach.

Clonker (southwest Scotland) a large stone, used for the upper part of a drystone dyke.

DSL includes an entry from northeast Scotland, defining 'clunkart' as 'A very large piece of anything ... [for example] ...a 'clunkart o' a stane'...'

DSL quotes Rainsford-Hannay (1957: 34) using the word 'clonks', defined as 'A large stone in the upper single-thickness part of a dry-stone wall below the cope-stones.'

It is cited as a variation of the Norwegian word *klunk*, a lump.

The word 'clonk' is dealt with in more detail in section 3.2 (iv), discussing covers and copes.

Cobble / coggle / cobblestone / cogglestone / copple-stone a rounded water-worn stone. In geological terms 64 mm to 256 mm (2 ½ to 10 inches) following the Wentworth-Udden scale used by Whitten and Brooks (1979: 340). These egg-shaped stones were packed, length into the ground, to form paths.

Craig / crag / craige / kraig / kraeg / craigstane (Scots) a projecting spur of rock or a large detached rock. These were easily accessible sources of building stone, especially for dykers working through remote moorland.

Creek rock / creek stone / creek limestone (Kentucky, USA) smooth water-worn stones from a stream bed. Known elsewhere as 'river rock'.

Crust slabs (France/Germany) defined in Mader and Zimmermann (2011: 66) as rough faced stone from the quarry, a by-product of extraction and cutting. These are thin slabs with one side cut, and one side left rough with the original weathered surface. Often used for low retaining walls or replication of 'natural' rockscapes, without the bulk of the whole rock.

Cungle / kungle (Orkney) a large water-worn stone, rounded by the action of the sea. 'Cungly' means gravely, covered with shingles or roundish water-worn stones, as in 'a cungly shore'.

Day-stones Arkell and Tomkeieff (1953: 35) have this as an English dialect term for '... loose stones on the surface...'

Deaf (Gloucestershire, England) a light infertile soil. The term also appears as 'douff' in the Scottish Lothians, around Edinburgh.

Dockie (Tayside and Fife, Scotland) a rounded stone from the seashore used in road making and for bordering the pavement water-channels.

Fast country / fast ground an old miner's term for solid or undisturbed ground.

Feidíns defined by D'Arcy (2006: 18) as cobble-sized stones 'cleared from the ground to protect the plough...'

Field rock (Kentucky) rocks gathered from the field after ploughing.

Field stone / fieldstone (New England, USA) angular rocks collected from the fields, not quarried. They are flat or slabby, having been detached from the substrata. These are ideal for dry stone walls. Clearing stone takes time and effort (Figure 13). Stones which have been in water or moved by glaciation are rounded and not, strictly speaking, fieldstone.

A New England proverb declares 'Maine's number one crop is stones.'

Field stones are driven to the surface by the freeze / thaw cycle and erosion of the topsoil. Stones which appear in the fields in the spring, after winter's frosts are called 'frost pickings' in North America.

Stilgoe (1982: 175) explains how New England farmers '...piled the rocks not in heaps but in rows equidistant from the center of their rectangular fields, along each edge.'

Stilgoe goes on to explain why New England fields were small. It took effort to clear stone to the edge of a small space, less than to the edge of a large field. Europeans thought the walls were untidy. They did not understand the 'fences' were primarily field clearings rather than formal boundaries.

Wallers, naturally, try to get stone from the closest and most convenient source.

GVA Derbyshire (1813: 85) tells us '...sometimes the farmer [provides] a horse and cart... to prevent the Wallers digging holes in improper places in the Fields.'

Corangamite (1999: 91) advises there was so much stone on the volcanic plain west of Melbourne, Australia and '...it was said that a new wall could be built 50 metres inside the old walls still using surface stone'.

Figure 13: Stone cleared by machinery while preparing the field for a crop of carrots. The shape of some stones suggest they might have come from an ancient ruin. Dalcross, Scotland.

Sometimes stone was replaced in the fields. Forsyth (1805: 70) describes how farmers in Roxburghshire and in Galloway (southern Scotland) removed stone from their shallow, porous soil. This had a detrimental effect on the crops. Farmers realised the stones kept moisture in the ground and warmed the soil. Crops improved when the stones were replaced. This was in the early days of agricultural improvement. The land may have been better left in permanent pasture instead of being ploughed for a crop of grain. A spread of crushed limestone would also have helped the soil.

Frost polygons a phenomenon from Arctic regions. Frost heave produces a landscape full of rounded shapes, several feet in diameter, with a wide margin of small stones on the outer edge. These patterns can look like a series of collapsed walls, if viewed from high above. On the ground the landscape looks like a **scree**.

Galti / golti (Orkney and Shetland) a bulky round shaped rock, for example in a ploughed field.

Gerrock (Scots) an outcrop of rock, especially on a steep hillside. The term is similar in pronunciation to *garbhach*, Scots Gaelic for 'a rocky place'.

Girdles (Durham and Northumberland, England) thin hard stones, possibly named for their similarity to oatcakes or the griddle they are cooked on.

Girt muckle (Shetland) a large stone, a stone with a big girth. These are most effectively used sitting upright in the foundation of a wall, taking up a sizeable proportion of the wall's height or width.

Glacial till an accumulation of material, small pebbles to sizeable boulders, created by a glacier as it grinds over the landscape and left behind when the ice melts. These deposits are a good source of ornamental stone and aggregates.

Glidder (Roxburgh, Scottish Borders) a small loose stone, typically found in large numbers below a rocky hillside after frost and weathering had broken them off. This accumulation is also known as 'talus' or 'scree'.

Gravel geologically defined as having a 2 mm to 4 mm particle size following the Wentworth-Udden scale as used by Whitten and Brooks (1979: 340). According to the scale sand has a smaller particle size, up to 2 mm; pebbles have a larger particle size, 4 mm to 64 mm (⅕ to 2 ½ inches). Engineers and builders use the term more loosely than geologists, applying it to river and glacial gravels which may vary in size from sand to small boulders. In practice, gravel is considered to have a particle size from 5 mm to 60 mm. Small 3/8 inch (1 cm) crushed quarry stone is called 'pea gravel'.

Gravels are used by the construction industries for aggregates in concrete. Arkell and Tomkeieff (1953: 1) inform us the word 'aggregate' derives from the Latin *aggregatus,* united as a flock, joined together. Gravels are seldom used in drystone work.

Pleasant (2002: 49) calculates one ton of ¾ inch diameter gravel will cover 120 square feet to a depth of 2 inches. In metric terms this means 1016 kg of 19.05 mm diameter gravel will cover 11.15 sq metres to a depth of 50.8 mm.

Hagstone naturally occurring beach pebbles with small holes through them. Throughout Europe they were used to keep witches away, often strung in garlands. A similar Scottish superstition was 'adder-beads' or 'adder stanes'. These were crystal or agate gemstones, near spherical, supposedly made by adders when they sloughed off their skins and plastered the skin with saliva to form a bead. Shepherds regarded these as a talisman of good luck and dipped them in water to cure the bite of the adder.

A 'whorly stone' is a naturally perforated stone thought to have magic properties. The word will have the same root as the modern word 'whirl'.

Hardscabble / hardscrabble (Ireland) hard stony ground with little agricultural value.

Hirst (Scots) a piece of rough stony ground, too stony to cultivate. A bank or mound of stones or earth, perhaps created by a river.

Hooloch / hurloch (Scots) a heavy fall of rock, a hurl of stones, an avalanche.

Hulter / holter / holder / hilder (Shetland) a large boulder or lump of rock, a heap of boulders resulting from a fall of rock.

Hulster (Orkney and Shetland) a shapeless block of stone.

Hummer stones (Yorkshire lead mines) quartz pebbles or stones from the millstone grit, a rough sandstone. Possibly a corruption of 'hammer stones', an old use for these stones.

Hurd (Shetland) a big boulder, a large piece of rock, especially by the seashore.

Junrel (southwest Scotland) a heavy fall of rock, for example in a quarry.

Klak (Shetland) a boulder, a rugged piece of stony ground.

Figure 14: This dyke marches along the skyline, through glacial deposits. Glenshee, Scotland.

Knibloch / kniblack / kniblock / knibblach / knublach / knublock (Scots) a small rounded stone, a boulder. 'Knibblockie' describes a rough road with small stones on the surface.

Klak (Shetland) a boulder, a rugged piece of stony ground.

Lan-bool (Caithness, Scotland) a large rounded stone taken off the fields, similar to another Scots term — 'land-stane' — a stone turned up by digging or ploughing. This rounded material is not ideal for dyking, as noted by J C Loudon (1825: 443): 'The wall or dike made with round or land-stones, by labourers and covered with a coping of turf, is a very indifferent fence… their benefit is not permanent, as the wall is perpetually tumbling down'.

Leacag (SG) a little flat stone.

Ledge rock a North American term for material which can be broken away from the exposed face of a quarry or hillside with simple tools, and less effort than full scale quarrying.

Ledge rock is typically sedimentary in origin. It has been called 'strip rubble' when collected and cut into narrow pieces for veneering.

Maiden steen a fresh building stone. DSL includes a 1951 definition from Aberdeen – 'A building stone newly hewn, that has not been used before. It is unlucky to use in a new building stones that have been salvaged from an old one.' Perhaps quarries wanted to restrict the sources of building stone. It is also a caution against recycling stone which has been stressed by impact, fire or exposure to the weather.

Man-sized rocks the Marenakos Rock Center in Issaquah, Washington, USA, include definitions on their website (https://marenakos.com, consulted 5 October 2021) for this term 'which can get thrown around loosely in our industry'. Boulders are sized by 'measuring how many people it would take to *manuever* (sic) it into position by rolling/pushing/using a rock bar, not in any means *lifting the boulder*.'

Marenakos quantifies the different sizes. A ½ man boulder weighs less than 150 pounds; a 1 man 150-300 pounds; a 2 man 300-1000 pounds; a 3 man 1000-2000 pounds and a 4 man 2000-4000 pounds.

Clach-ultaich, literally 'lift stone' in Scots Gaelic is a stone one man can lift comfortably. *Ulbhag* is a Scots Gaelic term applied to a stone which is bigger than a man can handle.

A 'hand-stone' or 'hand-stane' is a small stone easily lifted by hand. Suitably shaped stones were among the earliest tools.

JSD, defines 'hand-stane' as a '…term formerly used in S. [Scotland] for a small stone, or one that could be easily lifted and thrown by the *hand*, in contradistinction from which required much greater exertion.' JSD goes on to include a quotation from a description of Galloway, southwest Scotland — 'There is a cairn, or great heap of small *handstones*, with five or six high stones erected.'

Thorson (2002: 124) refers to the early New England settlers moving 'two-handers' (stones which required two hands to lift them) and leaving 'one-handers' on the ground.

Molag (SG) small round stone, especially seashore stone.

Muckle chuckies (Scots) large stones.

Naked stone (Mendips, Southern England) a description of the freshly exposed surface of a stone, after the outer face is broken off.

Pebble geologically defined as a size from 4 mm to 64 mm following the Wentworth-Udden scale as used by Whitten and Brooks (1979: 340). The name comes from the Old English word 'popel' or 'popel-stan' with variations of 'pipple', 'popple', 'puble' and 'pebble'.

Pick-up rock (Kentucky, USA) field rock which can be collected off the ground.

The term 'pick-ups' is used by D'Arcy (2006: 27) when referring to the easily collected smaller stone used for walls in the Burren, western Ireland. Most of the rock comes in the form of large flat slabs which must be broken out.

Raab (Scots) a description of a fall of rock from a cliff face and the noise it makes.

Rachill / ratch / ratchel / ratchell Arkell and Tomkeieff (1953: 92, 94) record these terms — 'Used for loose surface stones in Yorkshire, Derbyshire, Nottinghamshire, Herefordshire and Radnorshire [east Wales].'

Ramblin-stone (Northumberland, north England) a loose stone on the ground. 'Rambler' also applies to an overlooked potato, left in the ground to grow the next year.

Ramgunshoch (Ayrshire, Scotland) rocky, stony soil.

Reuch / roughstane (Scots) a natural boulder, hence 'reuch stane dyker', one who used stone in its natural state.

Rhab (South Wales) shaly subsoil.

River rock / river jacks / river slicks / creek rock (USA) stones which have been smoothed flat or rounded by the action of water. These are not ideal building material.

Rockle (Ayrshire, Scotland) a pebble, hence 'rocklie' – abounding in pebbles.

Rocks (County Dublin, Ireland) rocky places with large granite boulders on the surface or where the bedrock came to the surface. These were accessible sources of good, easily worked stone.

Rublie (Moray, Scotland) a big stone, as in 'rubble'.

Rudestone Arkell and Tomkeieff (1953: 99) record a 1525 reference which defines this as '…stones (or slates) in the rough.' This describes a raw material which can be shaped or

Figure 15: Storm damage by the sea, showing how much stone goes into a wall. Donegal, Ireland. (Photo courtesy of Louise Price).

split to advantage, rather than rough stone with limited potential.

Sand an accumulation of ground down rock geologically defined under the Wentworth-Udden scale, per Whitten and Brooks (1979: 340), as having particle sizes from 1/16 mm (very fine) to 2 mm (very course) and distinguished from gravel by having a smaller particle size. In practice the particle size varies from dust to 5 mm diameter.

Building on sand can be tricky, the foundation is unstable and easily washed out. Rabbits can easily undermine a wall built on sand.

Sand and gravel are important components of mortar and concrete mixes. The colour and texture of a mortar depends on the sand – vital when trying to match a repair with existing work.

Coarse sand, sometimes called 'concrete sand' or 'sharp sand', produces mortar which binds better than mortar made from finer sand with fewer points on each grain.

'Rabbit sand' is a picturesque Irish description of sand thrown out of a rabbit hole. It is not good for mortar because it contains a lot of soil. This term came from a Tipperary man quoting his father.

Scabbert (northeast Scotland) a stony piece of ground.

Scarnach / skarnoch (Scots) heaps of loose stones on hillsides, scree, a bed of detritus, waste or broken rock. The word is very similar to the Scots Gaelic word *sgairneach* meaning heaps of stone, the noise of falling scree or a howling rumbling noise.

Scots pebble a semi-precious stone, a type of coloured quartz, called 'Cairngorms' because they were found in the Cairngorms mountain range. The stones were popular in Victorian times, often set in silver to make brooches.

Scree the mass of loose rock on a hillside or at the foot of a steep face.

Arkell and Tomkeieff (1953: 103) note the word derives from the Old Norse word *skritha*, a landslip. Scree is a valuable source of stone for building dykes and for splitting down to make building stone for mortared work.

'Colluviation' is the geological term for a build-up of rock and mineral deposits at the base of a slope or cliff. 'Hill creep' is the geological term for the movement of loose material down a slope.

The Scots word 'raab' means 'a fall of rock from a cliff face, an avalanche', and the noise it makes.

DSL connects the word with the Old Norse *hrapa*, to tumble, to fall down.

In Derbyshire 'slither' describes angular limestone rubble, also known as 'sliding gravel.'

Self-stones GVA Derbyshire (1811: 143) differentiates between rounded 'bolders' and 'self-stones' which are '…possessing still the original shape and angles of the block…'

The angular shape makes them easier to fit together and build with.

Skrüid (Shetland) stony ground in general, a land-slide, an avalanche, a steep slope where the soil has eroded, or where a landslip has occurred. A type of scree.

Shoddie (Scots) a natural stone which is roughly dressed and used for building.

'Shoddie' is used in English and Northern Irish quarrying and ore mining districts to mean loose or fragmented stones or lumps of ore. The form 'shoddoo' from Orkney, meaning a large chunk or lump of anything, is probably a variation of this word.

Stac / stack a tall column of rock rising out of the sea, detached by sea erosion, for example Stac an Armin in the St Kilda group of islands or The Old Man of Hoy in Orkney. Shetland has another term for them, 'drong', from Old Norse *drange*, a detached pillar of rock.

Stacs are described in the poem 'Under Milk Wood' by Dylan Thomas (1914-1953) as:

A bit of stone with seaweed spread,
Where gulls come to be lonely.

Figure 16: Scree from the cliff has provided for this dyke. Storr, Isle of Skye, Scotland.

Stany (Shetland) a description of ground as 'stoney', full of stone.

Stanners / stenners / stanirs (Ayrshire and Clydesdale, Scotland) shingle, rounded water-worn stones, pebbles and gravel in the bed or on the banks of a river or on a sea beach.

Steenie a Scots term for a little steen, a small stone. Generally speaking, the addition of

'-ie' or '-y' to the end of Scots words is an affectionate reference to something small and friendly.

Steiniegate (Aberdeenshire) the place where stones were collected after they were gathered off the fields.

Stony ground has its disadvantages, but it drains well and retains heat better than damp clay.

Surface ledge (New England, USA) bedrock, exposed above ground, and therefore easily won.

Talus an accumulation of fallen rock below a slope. A good source for building material.

Toman / tomhan (Scots) from the Scots Gaelic *toman*, a knoll or hillock. Often a glacial moraine in the upper reaches of a Highland glen. In folk-lore these sites were associated with fairies and tales of the supernatural. These moraines are good quarrying sites for sand, gravel and round boulders.

Waly-stane (Clydesdale, Scotland) DSL, under 'wallie', describes this a nodule of quartz, used as a plaything by children. A 'waly' is a toy or gewgaw. From a general meaning of 'wallie' as fine, excellent, pleasant or beautiful.

Wild stone (Ireland) big granite boulders lying close to, or on, the surface.

Yattle (Scots) a description of ground covered in small stones.

1.5 The awkward squad

These stones require ingenuity, skill and discretion to realise their potential. Some are beyond redemption, best used in a lobster pot.

Many racist and derogatory terms have been used to describe material which is inferior or unsuitable for building. The most offensive are not included in this list.

Ballish stane / ballisten (Shetland) rounded water-worn stones.

Barr (Aberdeenshire) a second-rate material near the surface of the granite deposits. This is cleared to expose good quality rock.

Bastard stones (Durham / Northumberland, England) usually a whinstone or very hard igneous stone, with no grain, which is difficult to cut or trim. The term appears variously as 'bastard freestone', 'bastard rock' or 'bastard whin'.

Bitch Arkell and Tomkeieff (1953: 7) have this as a 'solid irregular sandstone' bed at Collyweston, Northamptonshire.

Blaffen (Fife, Scotland) the loose flakes or laminations of stone.

Book the description of a type of shale stone, thinly laminated, split into its strata from weathering. Like a closed book in which the individual pages are visible. Such shales can be easy to work into a wall but often disintegrate after a winter or two.

Bowlders an old English form of 'boulder', once commonly applied to the description of a round pebble.

Brackly stone (Northamptonshire, England) brittle stone. Broken or full of cracks.

Cam (Tayside, Scotland) soft disintegrating rock lying below the surface soil and on top of the workable rock in a sandstone quarry. After this waste is cleared the quarrier has access to useable stone.

Figure 17: A piece of sedimentary stone has weathered, and failed the pressure test.

Cank / cankstone (England) close-grained sandstone from the Yorkshire and Midland coalfields. 'Cank' is also a mix of clay, stone and iron in west Yorkshire quarries. In Northamptonshire the term is found as 'canky' — a description of rotten, decayed stone.

Cats-heads / catheads (Durham and Northumberland, northern England) nodular ironstone. An awkwardly shaped stone, difficult to work with.

Cleckin-stane an old Scots term for stone which breaks up on exposure to the air. It looks solid when first quarried and deteriorates after weathering (Figure 17).

Walker (1812: 105) advises caution — 'The snap dyke should never be built of any stone that either splits or moulders in the air.'

Cleek (Scottish Borders) to weather into thin sheets. Stone which deteriorates like this is unreliable for building. This happens with soft sedimentary rock, sandstone or shales.

Cloddings / field cloddings (Scots) poor quality fieldstone.

Clue (Tayside, Scotland) an unworkable piece of quarried sandstone. A variation of an old English word meaning 'a ball of wool'.

Codlin head / coddlin head / codling head (Orkney and Shetland) an awkwardly shaped building stone. In the Scottish context 'codlin' refers to the cod, a fish. Further south the 'codlin' is a type of apple. These awkward stones have many dialect terms, including 'grice heads', 'cat heads', 'dog heads', 'horse's heads', 'bastard stone', 'hardheads', 'gubble stones'.

'Grice' is an old word for 'pig'.

Crapply / crapel / crappely / craply a northern English dialect term for brittle, crumbly, friable rocks.

Creek stone (North America) stones from a river bed, water-worn stones.

Cull reject material which falls below a required standard because of flaws, for example poor colour, size or strength. 'Culling' is the sorting of material with preset standards in mind. Shards, flakes and misshapen pieces are of limited use to the formal stonemason but may be all that the drystone waller has to work with.

Dog's heads another term for a stone with no definite shape, not ideal for building.

Goulder (2013) in his 'Notes on Building a Cairn' confirms 'Any builder or mason can create a fine structure using good, flat or block stone, but to do it with a heap of dog's heads and highwayman's hats takes REAL SKILL.'

Goulder also advises his reader that '...cannonballs, dog's bottoms, and dinosaur heads...' are only useable as infill for the cairn.

Dogger (England) a large lumpy mass of sandstone, a type of conglomerate.

Dry (Caithness and Aberdeenshire, Scotland) a crack in a flagstone which reduces its strength and usefulness, a flaw. The 'dry' might not be obvious until a lot of effort has gone into preparing the stone.

Dry seam a pressure crack in rock which has not subsequently filled with other igneous material.

Erosion stone deterioration occurs through weathering or chemical attack. The effects of heat and frost are understandable; other causes of deterioration are more subtle. Rainwater falling on limestone creates a very weak acid. If limestone is built above sandstone, it can cause erosion in the sandstone as the water drips down. Sandstone is also vulnerable to frost damage and discolouring if roof drainage is poor and water runs down the face of the wall. Using harder stone in the lower courses of a building protects the walls from rainwater and impact.

Fissile the description of rock which splits easily, for example slate or sandstone.

In many cases this indicates material which will weather badly and produce a short-lived structure.

Foliation an effect of metamorphism whereby pressure and heat produce parallel planes of similar material, found in schist and gneiss. The experienced dyker will take advantage of this type of rock and know instinctively know how to split it out into pieces more suitable for cheekends, lintols or throughbands.

Exfoliation is the breaking down of rock into thin flakes. This is caused by weathering in the natural situation, and aggravated by use of inappropriate mortars in building facades. Exfoliation is also known as 'onion-skin weathering', 'flaking', 'lamination' or 'delamination'. Deterioration is especially bad where the bedding planes of sedimentary rocks are laid on edge instead of horizontally in, for example, some sandstone memorials. 'Foliation' comes from the Latin *folia* – leaf or page (of a book). Slates and shales are particularly prone to this type of deterioration.

Free / frie an old Scots term for 'brittle', used when describing stones, wood etc.

Ganil an old English term for a brittle limestone.

Grice heads awkwardly shaped stones. 'Grice' is an old word for 'pig'.

Gubble-stone / gub Arkell and Tomkeieff (1953: 55) have this as an Oxfordshire dialect term for a 'rough round stone which will not lie evenly in a wall or building' and notes it as a variant of 'cobblestone'.

Hardheads (Michigan and Vermont, USA) hard rounded stones, probably basalts or granites. Vivian (1976: 103) describes '... those boulders so common in Vermont, worn smooth and round by centuries of water action.'

Heather strain (Scots) a fault in building stone, supposedly caused by roots of the heather plant (*Calluna vulgaris*). The fault appears as white streaks on the face of the stone and allows damp to penetrate.

Horse's head (Scottish Borders) a stone which is round and will not sit steady.

Robert Frost in his poem 'Mending Wall' mentions this type of stone when describing repairs to his walls in Derry, New Hampshire:

And some are loaves and some so nearly balls
We have to use a spell to make them balance:
'Stay where you are until our backs are turned!'

Wiggins (1840: 284) quoting from a New York source notes that round stones '...will not stay long in their place...' and thinks they '...should only be used in what are denominated half walls, which are to be crowned with wood.'

Hulbie an old Scots term for a large unwieldy object, as in the phrase 'a hulbie o' a stane.'

Knab / knablick (northeast Scotland) a boulder, a large awkward stone, one which is knobbly, rough or lumpy.

Kugl (Shetland) an awkwardly shaped and unsteady rounded building stone.

Lamination / delamination a characteristic of sedimentary rock with thin layers (lamina) which separate easily. This is not always a good thing in building stone. It is the exact opposite of 'freestone' which has no laminations or internal bedding, and cuts in any direction without fear of splitting.

Lump DSL includes this as a Shetland term for 'a large shapeless stone used in dry-walling.'

Mealie (Scots) describing rock as friable.

Mongrel (Lyme Regis, Dorset, England) irregular grey limestone.

Mumbly (west Somerset, England) describing awkward stones with no flat surface, ones which are shapeless and crumbly.

Offcasting (old Scots) the weathering or crumbling of soft building stone.

Onions spheroids of weathered basalt from Rowley Rag Quarries, English Midlands.

Paper shale a geological term for shale which splits into very thin flexible sheets.

Pendle (England) a term applied to the top beds in a quarry. They are exposed to weathering and are shattered, rotten, soft and fissile.

Rachill / ratchel (Derbyshire, England) loose weathered rock on the quarry surface.

Raffey (Spittal, Caithness) the poorer quality rock above the good stone in a quarry. The spelling is an educated guess, the term was used in conversation with a local quarryman who had no idea how the word was spelt.

Rag / ragstone within the limestone areas of England this term describes broken stone or stone which contains a lot of fossil shell, not easily worked. This contrasts with 'freestone' which is easily worked because it has no grain or flaws.

Ramme / rumel (northern and midland England) loose stone, rubbish stone. A stone type unfit for building because it cleaves or shales into many small irregular pieces when quarried. The term also applies to waste or broken stone left after construction work.

River rock a description of water-worn stones. Their rounded smooth shape means they are difficult to build with.

Rottenstone (Oxfordshire, England) any decomposed rock.

Rough skinned (Annandale, Scottish Borders) Prevost (1957: 100) defines this as 'a rough skinned stone is one which cannot be broken.'

Roundheads boulders of fossilised coral found in limestone rubble in Yorkshire, England.

Scale thin lamina or paper-thin sheets of loose rock. Examples are obvious on some old sandstone gravestones. This is unfortunate because the original lettering and stone edges were crisp. In masonry, scaling or delamination is often a result of poor construction choices or maintenance, especially use of stone with thin bedding planes or inappropriate use of Portland cement.

Scrap (Somerset, England) a bed of stone fit only for rough walling.

Figure 18: Rough round boulders form a 'single' dyke. Wire now assists the stonework. Glentruim, Scotland.

Derived from the Old Norse word *skrap* – remnants or fragments.

Sea-stane water-worn stones from the seashore.

Shellat (Devon, southwest England) weathered slate at the top of a quarry.

Sheep's head an awkwardly shaped stone.

Skelder (Orkney and Shetland) thin flaky stones, unsuitable for building purposes.

Skelf / skelb / skelp / skelve (Northern Scotland) a very thin flagstone slab, a thin flat fragment or slice, a flake. As a verb it describes stone weathering into flat slices or fragments. A stone is said to 'skelve' if thin layers flake off, either intentionally with a tool, or on exposure to the air.

Skelfer (Orkney) a thin lamina or flake of stone. A stone is 'skelfery' when it flakes easily.

Spale / spall to break or split, especially with effects of weather or frost. Spalled stone, or stone which is inclined to spall, is not good for building. It has no long-term stability or strength.

Stammerl (Fife, Scotland) friable stone, a building stone in which the grain is irregular and runs in various angles through it. This material is difficult to cut into any regular size.

'Stammerl' also describes the broken fragments accumulated at the bottom of a quarry.

Stinkstone / swinestone many limestones give off a bad smell when struck or broken. Arkell and Tomkeieff (1953: 115) define stinkstone as 'Fetid limestone, found in the Carboniferous Limestone, Lias, Portland Sand, &c.'

King (2000: 71) notes the fifteenth century Italian masons tested newly quarried stone by smell. The stronger the freshly quarried sandstone or limestone smelt of rotten eggs the better the quality of the stone.

'Liver-stone', from Derbyshire, England, is bituminous limestone which emits a sulphurous smell when struck.

Swimming stones a problem associated with individual stones, or courses of stones, set in wet mortar. They tend to move and 'float' because of the reduced friction.

Swirlie (Shetland and Caithness and Aberdeenshire) describing rock with uneven grain. DSL defines 'swirl' as '[a] twist, twirl, coil, a twisted or tangled state.'

Throw off when a stone is soft or brittle, and tends to delaminate or disintegrate, it literally 'throws off' its face. If a stone is laid on edge this exposes the bedding planes to erosion and delamination.

Tumblers large boulders. Arkell and Tomkeieff (1953: 121) quote this term from Derbyshire in 1747, describing them as 'All sorts of stones that will not go into a corse [sic] without they be broken less.'

Twang (Ireland) semi dressed rock used for rubble walling.

Water-worn stones those smoothed and rounded by glacial action or tumbling in water. Building with this material is difficult because of the reduced friction between them. They are called 'river rock' or 'creek stone' in North America.

Stones from a river or glacial moraine are not ideal for building, even with mortar, because they are too smooth and tend to be of regular size and shape. They are easier to build with after splitting.

Margaret Brooks (1977: 101) advises us – 'If the available stones are too regular – water worn stones for example – they should be split with the hammer.'

Rainsford-Hannay (1957: 35) includes an 1856 specification for a 4 foot 6 inch dyke, including a definite condition – 'Water-worn stones are not to be used.'

Mitchell (1992: 52) relates what a Yorkshire waller thought of stone from a beck (a river):

> Beck-bottom stone, washed smooth by the water, sometimes looks like 'a big egg'. A hapless waller, who was provided with a cartload of beck-bottom stones by the farmer was complimented, when the farmer returned, for the speed with which he had got rid of them. He replied: 'I drowned t'lot', meaning that he had returned them to the beck.

'Sea-stanes' are defined in Pride (1996: 68) as building stones from the shore or cliffs which, probably because of their salt content, show dampness in wet weather. They are very hard and difficult to cut and do not make good building stones. This suggests they were blocks of freestone rounded by the action of time and tide.

'Ballisten' or 'ballish' stane is a Shetland term for rounded sea-worn stone. Probably named for the likeness of the stones to 'ballis', thick round oatcakes, or balls. 'Ballisten' is also the term for the stone on a hearth in front of the fire where the 'ballis' were baked.

Weeping stones according to McAfee (2009: 27) these can occur on the internal face of an external wall. They are hard stones, shining with condensation. This may be because the stone is a throughstone which reaches from side to side of the wall, acting as a bridge to conduct the colder outside temperature to the warmer interior wall face.

White horses (Ireland) white veins of quartz or pegmatite in granite. A flaw which, according to Ryan (1992: 23), renders the stone 'fit only for rubble and twang' because the stone easily shatters along these veins.

'Hard bar' is another term for a vein of quartz or hard material running through a stone. This is not welcome when trying to carve detail into otherwise soft stone.

A 'glass seam' is a crack in rock which has filled with calcite or similar material and therefore strengthened in the geological process.

'Knots' is an English term for nodules of pyrites or concretions or any harder inclusions in a rock.

Yagg (Orkney) a layer of useless stone lying above good stone in a quarry.

1.6 Mud

Mud is geologically defined by Whitten and Brooks (1979: 309) as '[a] wet mass of clayey material in a liquid or semi liquid state…' It is an inevitable and loyal attendant to stonework; it clogs the site and slows up progress, during any season of the year.

According to the U. S Geological Survey (https://www.usgs.gov/special-topic/water-science-school/science/rain-and-precipitation, consulted 5 July 2021), 'One inch of rain falling on one acre of ground is equal to about 27,154 gallons and weighs about 113 tons.' That could create a lot of mud.

The wide variety of muds, and their different properties, makes it a very versatile building material. Mud brick, mud mortar and mud plasters feature in various architectural styles.

There are numerous descriptive dialect words for mud. The very sound of them evokes memories of smell, texture or taste.

Gloves are basic safety equipment though some dykers prefer not to use them. Margaret Brooks (1977: 102) tells of a waller wearing a glove only on his left hand because he felt it 'would hamper his hammering'. Presumably he was right-handed. Dykers from the Scottish Borders used to wrap gauze strips around fingers to give some protection. Cairns (1975: 7,8) describes this system and how: 'Old Tom Watt tried rubber rings from old lemonade bottles but they were much too tight and no 'give' in them.' These rubber rings were part of the stopper, used to ensure a leakproof seal.

Cairns wore a piece of leather or motor inner tube on the thumbs as protection. A modern equivalent is electrical insulation tape.

Corangamite (1995: 138) includes an extract from 'The Age' (a Melbourne newspaper) dated 2nd September 1992:

> For many builders, stone-walling was part-time work because their hands could not cope with the wear and tear for long periods. They would do general farm work while their torn hands recovered, often soothed from the lanolin in the sheep's wool they handled. Some tried hardening the skin with brine and vinegar, but found the best remedy was to urinate on their hands last thing at night.

This remedy was recommended by masons who used mortar. The urine reputedly neutralised the effect of caustic lime. Nowadays vinegar is equally efficient.

In Corangamite (1995: 82) Bill Harlow says wallers were unable to use gloves because 'You can't feel your stone and plus the fact you'd get bad nerves.'

It was also thought to be safer without gloves, the wallers could feel a stone fall and move their hand safely away. Harlow also said walling was not done in cold weather because '... you'd just bust your hands.'

Mitchell (1992: 80) speaks from a colder climate:

> Such men [Mitchell quotes Teddy Holmes from Rathmell Moor, Yorkshire] were inclined to work in bad weather to ensure they had a reasonable income. 'The old people used to think they could keep warm when working in any sort o' weather, but they suffered because they didn't have proper waterproofs. A waller would be seen wearing an ex-Army greatcoat or with an old sack round his shoulders. Any waller who donned gloves or mittens was considered a cissy'.

A 'jib' is a Scots term for a protective leather covering for the stonemason's left thumb, the hand which normally held the chisel. Cotswold wallers used leather palm pads for protection.

In the Cotswold stone quarries the limestone slurry produced during the cutting process was valued as a poultice on cuts and grazes, per Hill and Birch (1994: 34).

Loss of a hand was not always a handicap. In Corangamite (1995: 121) we are told Roy Gardner '...learned stone walling watching 'old blokes' Jack Hogan and the one-armed Mick Hickey while he was a schoolboy in the 1920's.' No more detail is given.

Prevost (1957: 93) tells of Archie Proudfoot, a Moffat dyker with no right hand; it was amputated at the wrist. 'This did not seem to handicap him in any way ...two steel hooks had been fitted. He was able to work with facility and at an amazing speed.'

Eyes have always been at risk, especially from cumulative damage caused by dust and chips. Stonemasons and dykers learned to 'hit and blink' to protect the eyes.

Anyone who works with stone, especially craftsmen who cut or grind, are in danger of silicosis. This is less so with drystone wallers. They don't work in a constantly dusty atmosphere; and it rains.

2.1 (ii) Lines, frames and centering

A-frame / frame / batter frame a wooden or metal frame, normally used in pairs, with lines stretched between them, to establish the outer and upper edges of the stonework (Figure 19).

Also known as a 'template', 'gibby' or 'profile'.

A **story pole** is a North American variation of an A-frame. These poles are set to ensure work is vertical or following the batter. In effect they are one leg of an A-frame. Poles are ideal when building a straight retaining wall, only the front face is built strictly to line. Lines between the poles declare the face of the work.

Figure 19: Placing the throughstones on a new dyke. Note the cheekend with frames, plumbob, spade, hammer, the aprons and one person each side. The scene could apply to modern times, if the clothes were changed. Extracted from Stephens and Burn (1861: 244).

If a wall must go around a slow curve, frames or poles will be set at intervals to establish the basic curve and a line set to confirm the level. This gives a series of straight lines making up the general sweep of the curve. Stone is placed, mostly by eye, between poles, to ensure the curve is 'sweet'.

Batter frame a common name for the 'A-frame', named for its shape.

Centering / centres / centreis (Scots) a temporary framework to support lintols or arches during construction. Also known as 'falsework'. A pile of earth or turf can also act as centering; it is dug away after the stonework is completed.

Devaling / develing (Scots) according to Pride (1996: 29), these are terms for the 'narrow planks or boards covering the framework used in the construction of arches.'

Gibby an alternative name for the 'A-frame'. Searches through DSL looking for the origin of this term gave a variety of spellings. One is a name applied to a castrated male cat. Another is a name given to Old Father Time. One likely origin is 'gibble' meaning tools or working paraphernalia. The word may also have connection with 'gibbet' the frame of a loom or a chimney bracket for hanging a pot over an open fire.

Lines thin cotton or synthetic strings stretched between frames or poles as a guide to the proper placement of stones (Figure 19). A properly set line dictates the profile of the wall and keeps courses level. In general masonry it is important to be flat, level, parallel, plumb and square in various combinations, sometimes all at once. Using a string line and spirit level makes this possible.

Wallers and dykers usually work with two lines, one each side of a freestanding wall. The lines mark the foundation and are lifted on completion of each course. The distance the line is lifted depends on the thickness of the stone.

One variation is to set a line for the foundation and another at the top of the 'double'. A further variation is to set individual lines for the heights of the foundation, the throughstones and the top of the 'double'. These are set before work starts. In both these variations the builder creates the profile of the wall by laying the stones on a sightline between the strings.

Wood (1977: 373) quotes a New England waller who '... didn't use a string or anything. I just laid it up by eye, and sometimes by curses.'

Charles Fothergill, in Romney (1984: 115), describes the use of lines attached to a frame '...as a guide to the workmen' during his 1805 tour of Yorkshire.

Profile another name for the 'A-frame'.

Skainie / skenye are old Scots terms for a mason's line.

Story pole normally a piece of timber, set up for attaching lines to, as a guide for the line of a structure. Often used when the normal 'A-frame' is impractical.

Trig a small stone sitting on top of a long line to act as a guide, eliminate sag or reduce the effects of wind. In Dartmoor the term applies to a small stone put under a bigger stone to steady it.

2.1 (iii) Hammers

There are many types of hammers (Figure 20).

In some regions local wallers have created an ideal hammer, with help from the local blacksmith.

Wallers in areas with hard stone, for example the Pennines or Highland Scotland, tend to use the hammer less than wallers using softer limestone in the Cotswolds.

The pointed hammer used by Mallorcan wallers, for their local limestone, is of little use on the granite boulders of Maine.

Beetle (Aberdeenshire, Scotland) a heavy mallet used for beating or shaping.

Bullset a hammer whose head has a broad chisel-like face and a flat face. The chisel end is laid on a stone and another hammer strikes the flat end. It is used to split or roughly shape rock.

Catchie / catchie-hammer / catch-hammer (Scots) a small mason's hammer, weighing two to four pounds (0.9 – 1.8 kg). DSL defines them as 'One of the smallest hammers used by stone masons for pinning walls, etc.'

Forehaimmer / forehammer (Scots) a sledgehammer, used with both hands, held in front of the bearer and seldom lifted above head height.

Hibbert (1891: 275) records the islanders of Foula, off Shetland, finding their entertainment with workaday items:

> It was formerly the custom of the inhabitants, after divine service, to repair to the church-yard, for the sake of a competition in strength, — who should throw to the greatest distance a large stone, named a *putting-stone*, — the old men standing by to tell what feats they and their ancestors had done, and to lament the feeble powers of a newer and degenerate race.

Throwing the hammer and putting the stone are now formalised Olympic sports.

Fuge (old Scots) a pickaxe.

Gelloch Rainsford-Hannay (1957-97) mistakenly defines 'gelloch' as 'a 14 lb hammer'. It is actually a crowbar or a pinch bar, also described as 'gellock', gelloch', or gelleck'.

Goat's head a large round granite boulder weighing about 56 pounds or 25.4 kilograms, used as a hammer by Padraic Póil on Inis Oírr, Aran islands, Ireland, to break up the limestone pavement. Field walls are built from the broken rock.

After a few years of adding sand and seaweed, to create soil, the field is fit for cultivation. This aspect of life on the Aran Islands is featured in the 1934 Irish film 'Man of Aran', directed by Robert J. Flaherty (1884-1951).

Kavel-mell / keval mell / keval hammer / keveller (Scots) a sledgehammer, a large hammer for breaking stones. DSL quotes weights of 18½ pounds (8.4 kg) for these heavy hammers.

'Kevel' is also listed as 'work which is clumsily or badly executed …'. This suggests a keval hammer is used for its effect, not for precision.

Knapper (Yorkshire, England) a walling hammer. The verb 'knap' means 'to strike sharply' and is best known in 'flint knapping', where lumps of flint are struck together to produce sharp flakes.

In Scots 'knap' also meant to break down to smaller pieces, specifically for road mending, using a 'knap`hammer' or 'knapping hammer' – a light hammer with a long shaft for maximum impact.

A 'knapper' was a person who earned by breaking down stone. (See the entry in Section 5.1 (stoneworkers) for more detail.)

Lamb's foot the slightly thickened end of a hammer shaft, especially in small hammers. It keeps the shaft comfortably, and securely, in the hand.

Lump hammer a basic metal-headed hammer. From an original meaning of 'lump' – to beat or thrash.

Mash hammer / masher hammer / mashie (Scots) a heavy two-faced hammer used for breaking stone. DSL links the word 'mash' to the French '…*masse*, sledge-hammer, … with influence from *mash*, to smash to pulp.'

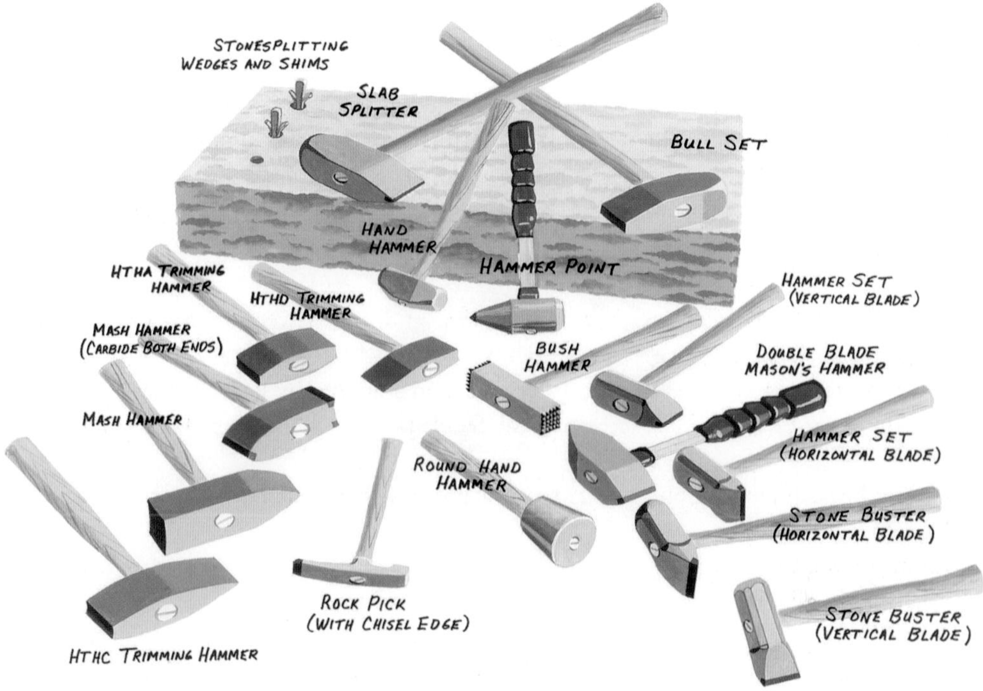

Figure 20: A selection of stone hammers, some with specialised purposes.
Image courtesy of Trow & Holden Company.

Prevost (1957: 99) records the mash hammer as a 4½ lb to 5 lb hammer used for breaking stone, compared to the everyday hammer weighing only 2½ to 3 lb. Callander (1986: 43, 85) refers to the mash hammer weighing around 12 lb and defines it as a heavy hammer used for crushing stones for hearting. Prevost describes dyking in the Scottish Borders, Callander writes about dyking with Aberdeenshire granite. A heavier hammer is required to break granite.

Mell (Scots) a hammer, defined in JSD as a maul, mallet, or beetle. A 'querrell-mell' was a large sledgehammer, up to 14 pounds (6.4 kg) in weight, used in quarries.

Muckle hammer (Scots) a description of a large heavy hammer, used on the hardest stone. From 'muckle' meaning 'big' or 'heavy'.

Reel DSL records this term from Scotland, and USA in the early 1900's for a medium weight quarry hammer with two oblong faces used to roughly dress stone, for example granite. The name may have come from another meaning of 'reel' – '… noisy, crashing, full of din or uproar…', a description of the tool in use.

Revel a southern English tool used on Portland limestone, as described in GVA Dorset (1812: 52):

> …the quarryman with a tool called a *revel*, which is at the one end a hammer, and the other an axe, whose edge is so short and narrow, that it approaches to the shape of a pick, by a repetition of sturdy blows, soon reduces a piece of stone, by his eye, to the largest

square figure it will admit; and blocks are thus formed, from half a ton to six or eight tons weight…

Rifter (USA) a heavy hammer used to break stone along the grain; it is laid against the rock and hit with a sledge hammer.

Scabbel hammer a hammer used particularly to dress the large boulders used in some styles of drystone walls.

Scutch hammer a hammer with a detachable comb-like blade used to remove surplus material with more finesse than a larger hammer. The tool is sometimes referred to as a 'scotch', derived from 'scutch', meaning 'to hew or dress', rather than from 'Scots'.

Scutch chisels have a similar arrangement of teeth.

Sledge hammer / sledge mell / sledge mill weighing, typically, 7 lb to 14 lb, used for breaking or splitting stone. Sledge mell or sledge mill are old Scots terms for this heavy hammer. Muilenburg and Swineford (1975: 35) describe 'sledging' as:

> …a common method of quarrying building block before the feather-and-wedge technique …involved raising an uncovered slab with a crowbar or lever, propping it up with small pieces of rock, and breaking off various lengths with a sledge.

Spalling hammer a light stonemason's hammer, typically around three to four pounds weight, used for rougher work, the first preparation of a stone.

Triffle (New England, USA) a four pound stone hammer, described in Wood (1977: 365) as '…used primarily in granite quarries for trimming granite blocks.'

2.1 (iv) Chisels

Bolster a wide chisel made in a variety of shapes, sizes and weights. Used in the initial stages of dressing, where finesse is not essential.

Broach / broch a mason's sharp pointed chisel, also known as a 'puncheon' in Scotland.

Chipen defined in Callander (1986: 84) as 'an uneven sided chisel for putting batter on a stone.' This puts an angle on the face of the stone, making sure it is consistent with the slope in the face of the wall.

Chisel / chessel / cissell pointed or broad-edged tools, made in many shapes and sizes, for splitting or shaping stone (Figure 21). Ryan (2005: 27) suggests the date of AD600 for the earliest iron chisel in Ireland. The Aztecs used chisels made from hard quartzite or basalt to 'peck' the shape into a stone. The Egyptians had chisels made from copper.

'Chessel' and 'cissell' are Scots variations of 'chisel'.

Devil (Ireland) a slang term for a punch or chisel used to split rock.

Jib (Scots) a leather cover to protect a stonemason's left thumb, the hand which normally holds the chisel.

Pincher / Pitcher / pitching chisel a blunt chisel with a broad bevelled face used for removing large pieces of waste stone. Hence 'to pitch' – to give a stone a preliminary rough dressing.

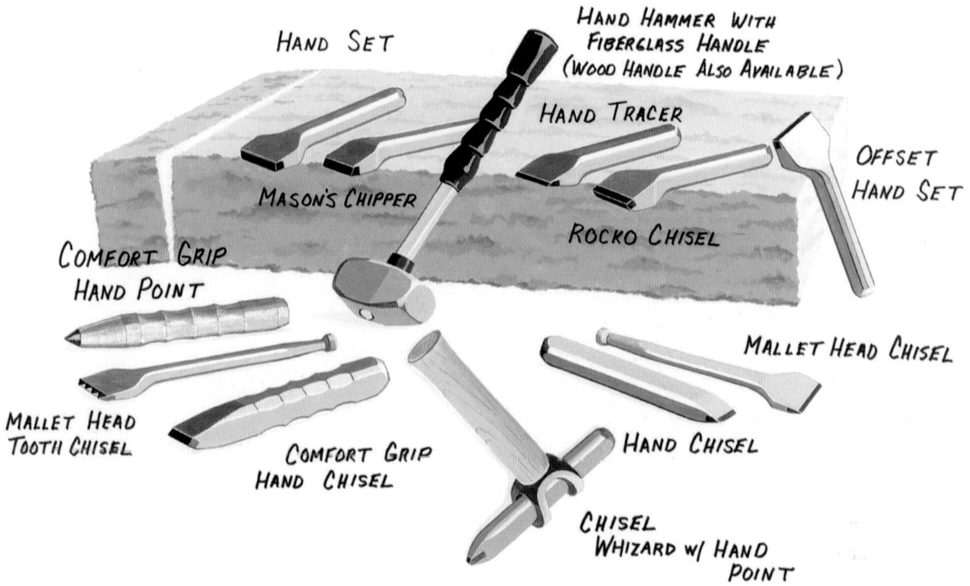

Figure 21: A selection of stone chisels.
Image courtesy of Trow & Holden Company.

Point a basic chisel with one point, used to take high points off the surface of a stone.

Puncheon (Scots) a pointed chisel used by masons.

Sned (Scots and northern England) to hew or polish stones with a chisel, derived from a more general meaning of the word – 'to trim or cut back'.

Stug / stog (Scottish Borders) to dress stone roughly with a pointed chisel.

Tracer a wide bladed chisel used to mark a line and break a stone, particularly where there is a good bedding plane

Weathering can add as much detail as a chisel. Murphy (2005: 91) quotes one of the Irish masons – 'The weather will give texture and effects that a fire-sharp chisel in the hands of a good man will never get.'

2.1 (v) Pinch bars, guddles and jumpers

For thousands of years levers, and long-handled tools for drilling or moving stone, have saved hands and backs from injury. The power and flexibility of levering principles is well known to anyone who has tried to move a large rock.

Beak a long-handled bar for moving large blocks of stone, a type of crowbar.

Craw iron / craw-steel (Scots) a crowbar.

Gaud / gad (Scots) a steel wedge, driven into cracks and crevices to break out rock in a quarry.

Gavelock / gewlock / gewlick / gavluk / geblick (southwest Scotland, northern England) a variety of spellings and pronunciations for a term describing a crowbar or lever, used to quarry rock or make a hole for a fencepost. The Scots word 'gavelock' also refers to the earwig (forficulo auricularia), species of beetle and centipedes. There is a similarity between the curved end of the lever and the shape of the earwig's pincers. The Scots Gaelic word geamhlac – an iron bar or pinch bar – sounds very similar to the Scots word.

Gellock / gelloch / gellek (Scots) an iron crowbar or pinch bar. A variation of 'gavelock'. Rainsford-Hannay (1957-97) mistakenly defines 'gelloch' as 'a 14lb hammer'.

Guddle (Scots) a crowbar. Derived from a Scots meaning of the word – 'to bore, prod or poke'.

Jumper / jamper (Scotland and Ireland) a long, pointed chisel used for boring holes into rock. It varied in length from 15 inches to several feet. The tool was held in the hands of one man and hit with a sledgehammer by another man. The man who held the jumper and the man with the sledgehammer had to have absolute trust in each other, any distraction or ill feeling between the men could have disastrous results.

Ryan (1992: 125) describes the drilling:

> It took a long time to drill a deep hole, normally they drilled 3 inches per hour, pouring in water to help clear the hole as they went. Each time a new jumper was used and each time it was sharpened, the point was made a little narrower to ensure it didn't jamb.'

Ryan (2005: 29) details granite extraction in the granite quarries of Dalkey in County Dublin in 1874:

> 3-inch jumpers, used in boring holes from 9 to 15 feet deep; 2 men striking, and 1 man holding and turning the jumper, bored on average 4 feet in a day or 5 feet with a 2 ½ – inch jumper, which was frequently used for boring the same depth.

Fairweather (1988: 7), detailing work in the Ballachulish slate quarries, explains the use of hazel branches, or lengths of hosepipe twisted around the wedges, to keep them on the line of cleavage while being struck with the sledgehammer. This was much safer than holding the wedges by hand.

Merrill (1903: 403) includes a different type of 'jumper' used in English quarries. It was a pointed metal bar with an iron ball welded to the middle. One hand held the bar above the ball, the other hand below the ball. The tool was dropped repeatedly on one spot to produce a small hole suitable for plug and feather splitting.

Kinsh / kinch (Scots) a lever used in the quarrying of stone, according to DSL it is 'an alteration of 'pinch'.'

Piddling tin a container used in a quarry to dribble water on to a drill bit or saw blade. This cleared the dust and kept the edge cool. Hill and Birch (1994: 34) say it was '... often made from an old treacle tin.'

Piel (Scots) an early nineteenth century term for an iron wedge used to bore holes in stones.

Pike (Scots) a pick, a pickaxe.

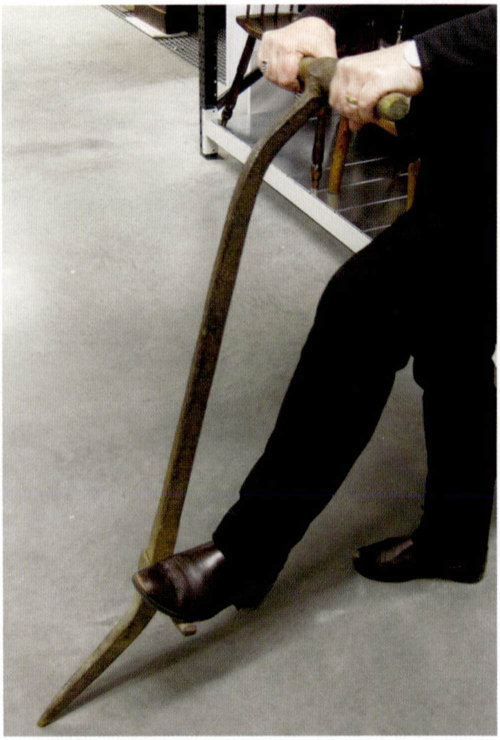

Figure 22: Tramp pick.
Photo courtesy of Bob Powell.

Pinch / Pinch bar a metal bar for moving large stones. They can be several feet long, and quite heavy, depending on the size of the job. 'To pinch' is defined in DSL as 'To move (a heavy object) by a succession of small heaves, to lever.'

Prise a plank typically seven feet long and ten inches wide originally used to ramp the boulders up above the foundation course. Oram and Stelfox (2004: 86) describe the use of this tool in the Mourne Mountains, County Down, Ireland, when building boulder walls.

In Scotland a 'prise' is a lever.

Pry bar a hand tool, a 'pinch bar', used for lifting and separating bedded rock or moving large blocks. Known as a 'shale' bar in the USA.

Star drills a four-ribbed punch, sharpened at the end, used with a hammer for drilling holes, one-quarter turn after each hit.

Tramp pick / trampick (Scots) a levering tool, like a long crowbar, with a footrest and wooden handle, something like a foot plough (Figure 22). In GVA Angus or Forfarshire (1813: 267) the Rev. James Headrick declares he has not seen the tramp-pick anywhere else in the kingdom. We are grateful to Norman Haddow, master craftsman dyker from Crieff, for confirming the tool is also known as a 'Lumphanan Auger', after a village in Aberdeenshire.

Torrachan (Aberdeenshire, Scotland) a tool used by granite cutters to bore holes in stone, an auger. The word derives from the Scots Gaelic word *torachan*, an auger or gimlet.

Tyne bar / tine bar (Scots) a pinch bar, a spike.

Wa-iron (Scots) a crowbar. Literally 'wall-iron'.

Wummle (Scots/north England) an auger, a drill for boring through soil and rock.

2.1 (vi) Tool maintenance, a brief note

Tool maintenance is important; it extends their life and efficiency, and keeps them safe. Operators should always be 'sharper' than the tool they are working with.

Bleezed is a Scots description of a hammer or chisel where the hit or hitting face has a ragged edge (Figure 23).

Curly-heidit, literally 'curly headed', is a Scots description of a chisel shaft or hammerhead 'mushroomed' by constant use. Note thatThis can be dangerous if one of the 'curls' breaks

Figure 23: A 'curly-heidit' or 'bleezed' hammer handle.

off when hit.

Hammer handles Dan Snow (2008: 103) includes Norman Haddow's guidance for wooden handles:

Norman follows his father's advice for treating a newly installed hammer handle. He soaks it in linseed oil once a day for the first week, one day a week for the first month, and one day a month for the remainder of its life.

This treatment ensures the wood swells and keeps tight in the 'ee' — the eye-shaped hole in the head of the hammer.

Sharpening Maxwell (1743: 279) gives advice for sharpening edges, quoted here with original spelling:

A receipt for hardening Edge-tools. Take the Juice of Horse-Raddish, and common Earth-worms, equal quantities. The instruments to be quenched in that Composition.

As an aside, and to prove there is nothing new under the sun, the copy of Maxwell consulted includes a note — 'This book is entered in Stationers Hall; and further to prevent pyrating, I have signed each genuine Copy on the Back of the Title-page.'

Often things don't work out.

DSL collected an apt phrase from Roxburgh in the Scottish Borders in 1951: 'I dout it's the tottlin pin that's no richt.' This is defined as 'A jocular expression used when a workman finds fault with his tools, to indicate that it is not the tool or implement but the operative that is inept.'

2.1 (vii) Carrying stone

Moving stone can be easier than anticipated. Difficulties arise at each end of the scale – how to move a huge rock and how to carry many small stones.

Throwing a handful of pebbles under a large boulder reduces friction, levers can then move the boulder. Baskets or carts move the smaller material.

Transport was a substantial proportion of the cost of a dyke, especially if stone was carted any distance. Historically the stone accumulated as a by-product of ploughing or pasture improvements, or it was brought in from nearby quarries and rocky outcrops.

Nowadays a dyker may bring in stone from a particular quarry to suit the colour and texture preferences of his client.

Cart to carry, to 'cart stone to a site' by truck, trailer or wheelbarrow.

Stone carts are described by Prevost (1957: 100):

> Carts used for hauling stone were specially made, with low sides. It is said that three of these cart-loads were required to build one yard of stone-dyke. Another estimate averages two farm-cart loads of stones for every yard, 'one load each side'. Two stone carts are included in an inventory of farm implements, made on Craigbeck in Moffat parish in 1827.

Winter was the best time to move stone. Carts and sledges could reach a boggy site if the ground was frozen hard.

Ball (2008: 256-60) explains 'The Cult of Carts' in twelfth-century France. Citizens in a state of pious hysteria dragged carts of stone to the cathedral building sites. This was orchestrated by church officials who created an Act of Penance, and reduced haulage costs.

Clood (Banff, Aberdeenshire, Scotland) a small load, to carry away in small loads. Perhaps because the ground is too soft for a heavy load.

Come-along a hand operated ratchet lever winch with a wire cable used for pulling boulders. This is often used in footpath construction when large boulders are dragged into place to create steps.

Handbalming (Canada) moving a pile of stone to another location by hand, like the Scots term 'handballing'.

Haut (Ettrick Forest, Scottish Borders) to gather stones together with the fingers, as if with a garden rake. In Scots a 'hot' or 'hott' is a small heap of stones, gathered together before carting away.

Paddy barrow (Scots) a wheelbarrow with no sides, used for carrying large stones.

Puddock barrow / paddock barrow / poddock barrow (northeast Scotland) a flat wooden triangular platform, shaped rather like a flattened frog — a 'puddock' in Scots. Used for transporting heavy loads such as stone or hay.

Rudge / rodj / ruddged (Orkney and Shetland) to gather loose stones on a piece of arable ground into small heaps before carting away.

Slipe / slype (Scots) a wooden platform in the form of a sledge used for moving heavy loads, pulled by a man or a horse. Callander (1986: 86) gives 'slipe draughts' as measurements of stone, each draught being equivalent to three hundredweight (336 lb, 152 kg).

Patterson (1880: 93) defines 'slipe' from Ulster, Ireland, as 'a triangular framework of wood on which large boulder stones are drawn out of the fields…'

Stone boat a flat-bottomed sled used to transport stone and other heavy materials. In Robert Frost's poem 'A Star in a Stone Boat', he is referring to the chance of a fallen meteor, carried on a sled and built into a wall:

> Never tell me that not one star of all
> That slip from heaven at night and softly fall

Has been picked up with stones to build a wall.

Wood (1977: 364) quotes a New England waller who tells of old-timers using a '…drag or a stone boat …Made up of six or eight planks…'

Truckamucks Toulson (1983: 15) quotes a late nineteenthcentury source, moving granite on truckamucks in Dartmoor, southwest England:

> These were a sort of cart with the ends of the shafts carved out and dragging on the ground. In fact the cart was nothing but two young trees, and the roots dragged, and the tops were fastened to the horses. When they wanted to move a heavy weight they used four trees and lashed the middle ones together.

These are similar to 'travois', a term, with French-Canadian origins, for a framework used by native American peoples. The travois was hauled by people, dogs or horses.

2.1 (viii) Walling machines

Captain Boomer of the whaling ship Samuel Enderby was delighted with his prosthetic arm — 'all spanking new and scientific' (Melville: 1851). There is joy in novelty; mechanical innovation promises hope of new, elegant, ways to perform heavy or repetitive tasks.

Machines have been proposed for building drystone walls. They mainly assist with the lifting of the stone. The initial selection, and final placement, still relies on the skilled eye and hands of a craftsman.

One Swiss experiment went further. Johns et al (2020) detail an experiment which used a '…customised autonomous hydraulic excavator' to build 'a 3-m-tall wall composed of 40 stones with an average weight of 760 kg.' The wall is functional rather than decorative. This mechanised process, with no human involvement, apart from programming and pressing the start button (after delivery of stone to site), may have applications in hazardous situations or for building rough retaining walls.

Hubbell (2006: 105-107) includes a 1904 image, from Maine, of a four-legged wooden frame straddling a wall and lifting large stones. This device appears to hold the stone with a grab, hinged like a scissors. A hand-cranked block and tackle system lifts the stone on to the wall, with some gearing to move the stone from side to side. In the background is a simpler block and tackle system, on a tripod, pulling stones from the ground. The work looks backbreaking. Hubbell quotes William Bunting – 'Certainly this farm possessed a great wealth of stone. What is not so certain is whether viewing this photo would more likely inspire a Maine farmer to innovate or emigrate.'

In a separate image Hubbell shows a gin pole derrick lifting stones onto a wall.

Oram and Stelfox (2004: 86) describe how granite boulder walls were built in the Mourne Mountains, County Down, Ireland. The 'prise', a plank typically seven feet long and ten inches wide, was used to ramp the boulders up above the foundation course. MacWeeney and Conniff (1998: 60-61) describe more modern times when drystone walls, ten or eleven feet high, are built with the assistance of JCB hydraulic shovels. Hydraulic power replaces muscle, moving large stones up on to the wall, where they are edged into their final position by men wielding pinch bars.

Machines have been more effective when used to destroy walls.

The 'fence eater' was a portable stone crusher used in Kentucky in the 1840s. It crushed old limestone walls into surfacing material for the new road system. This represented the landowners' contribution to the cost of the road. Lexington-Fayette (1990: 18-19) quantifies this destruction – 'Professor Karl Raitz estimates that over 90% of the stone fences which once were a part of the Bluegrass region are gone, and that 'nearly all were victims of 'progress."

Hubbell (2006: 69) describes how, in New England, after the patenting of barbed wire in 1874, stone walls became redundant:

> Suddenly an inexpensive form of fencing was available. Almost overnight the value of stone walls plummeted. Thousands of miles of wall soon disappeared – for many by undignified burial or by being thrown into the stone crusher.

2.2 Working the stone

2.2 (i) Basic shaping

A dyker applies minimum effort to shaping stone, especially when repairing an old dyke; any shaping was likely done with the original build, and most of the stone is on site. Shaping with a hammer is quick and 'good enough' for building a field boundary on a high moor. Dykers became skilled in recognising the attributes of each type of stone and how to use it to best advantage.

Wood (1977: 374) quotes a New England waller who prefers the natural look of stonework:

> When you do stonework, you try to avoid scarring. Some of these stone walls on these big estates are all trimmed and fitted. They did a lot of trimming. More than we would bother with. We want it natural.

Chap / chapping (Scots) to hammer lightly. The Scots phrase 'I have chapped my finger' means a finger has been accidentally hit, perhaps with a hammer.

Daise according to JSD is 'The powder, or that part of a stone which is bruised in consequence of the strokes of the pick-axe or chisel.'

Prill a Cornish hedging term for a small corner or edge which is chipped off a stone before it fits tightly to its neighbour.

Punish (Caithness) DSL defines this as 'to reduce [a stone] much in cutting and dressing; a term used by workmen.'

Scur (Angus, Tayside, Scotland) a mason's term for the rough projecting surface of a stone.

Shap (Shetland) to break off with a hammer.

Skirp (Argyll, West Scotland) to chip or knock a splinter off a stone while dressing it.

Souff (North Scotland) to strike. One stone is said to 'souff' on another when dashed against it. Hitting one stone with another stone causes less stress. There is less chance of internal fractures than when using a hammer.

The shock of a hammer hitting stone is immense. Striking a rock in the middle of the river with a sledgehammer was once a common way of poaching because it stunned the fish, as described in Cairns (1975: 14).

2.2 (ii) Splitting stone

If a stone is too big for the purpose intended it can be broken down with a hammer or, more precisely, with feathers and wedges.

Evans (1979: 107) makes an interesting comment regarding Ireland, without enlarging on it — '...until recently there was reluctance, for superstitious reasons, to break up large stones.'

This may stem from a belief that spirits, good or bad, lived inside.

Rock looks solid and impenetrable but there will be a weakness somewhere.

Any natural, or manmade, crack is the ideal place to insert water, iron or explosives and force the rock apart.

Frost can split stone because water, trapped within, expands inexorably by approximately 9% as it freezes. Over years of freeze and thaw cycles the largest rock will succumb to the relentless force of expanding water. Slates and sedimentary rocks were split by leaving them outside, on edge, over the winter. Moisture in the cracks, or drilled holes, froze and split the rock. This method was used in British slate quarries and Kansas limestone quarries. The process is natural, time consuming, and wasteful.

When Hannibal crossed the Alps in the second century BC, to invade the territories of the Romans, he is reputed to have used fire and vinegar to split rocks which barred his way. If stone is heated and suddenly cooled, it splits, but not along a predictable line. Hitting a heated rock with a hammer will also break it.

Todd (1859: 473-4) records an ingenious way of splitting stone with a fire along the top of a boulder, using soil to restrict the heat to the desired line of cut. He also tells us 'Small thin stones may be broken very readily by heating a large bar of iron and laying it on the stone where it is to be broken.'

Hitting a rock with a heavy hammer is the most basic way of shaping it. The results can be erratic unless the hammer is used with skill and precision. Natural fractures in rock can be exploited by inserting a steel wedge and hammering it in, until the rock splits. The first wedges were probably hard stones.

Wood swells when made wet. If wood wedges are hammered into a crack, and soaked with water, the swelling should eventually force the rock apart. In the 'olden days' wedges were often dried out by the quarryman's home fire.

Quicklime reacts explosively when water is added, especially in a confined space. Fairweather (1988: 7) records the use of quicklime to split Ballachulish slate. Quicklime was poured into fissures and '...left overnight to swell and split the rock.'

The 'chip and wedges' or 'pick and wedges' system, using iron or steel wedges, is an obvious improvement on wooden wedges. Slots, a couple of inches (5 cm) deep, are cut along a predetermined line with a pick or chisel. Iron wedges are driven into the slots with a hammer until the rock splits.

The 'plugs and feathers' method of splitting rock is probably the best known and the most efficient for the small scale required by wallers (Figure 24). Hill and Birch (1994: 38) confirm plug and feathers '…work in the same simple principle as the old traditional 'chip and wedges' by forcing open a row of holes until the whole block of stone cracks apart.'

A series of round holes are drilled, two metal 'shims', or 'shems', are put in each hole and a metal pin is inserted between the shims. The line of pins is struck with a hammer, in sequence. The rock will split along the line of pins. The process must not be hurried and will produce spectacular results, even curved cuts, if properly executed.

This method has many names, some international, some regional:

'Chip and wedges'
'Feathers and plugs' (Kentucky, USA)
'Feathers-n-quoins' (Quebec, Canada)
'Feather and tare' (in the granite areas of Dartmoor, Devon, southern England)
'Feather and tearers' (Plymouth Southern England)
'Peg and feather' (Donegal, Ireland)
'Pin and feathers'
'Plugs and feathers'

Figure 24: Striking a row of plugs in sequence splits the stone.
The holes are drilled along a line scribed by a stone saw. Image courtesy of Sunny Wieler.

'Plugs and wedges'
'Stook and coil' (Scots)
'Stook and feathers' (Scots)
'Wedges and shims'
In Kansas, according to Muilenburg and Swineford (1975: 39), this method is also called: 'Feather-and-plug; Feather-and-wedge; Feather-and-wing'.

Burgoyne (1895: 79) quotes 'feather and tearers' in a description of limestone quarrying in Plymouth, southern England. The 'feather' is the central wedge and the 'tearers' are the two outer shims.

The 'feathers' can be actual feathers. In Ireland, in the early 1900s, according to Conry (2000: 20), masons used '…any old strong feathers out of a hen's tail, or a turkey's tail or a goose's wings… to take the vibrations of the plug so that they wouldn't hop out of the hole when he hit it with the sledge.'

'Stook and coil' or 'stook and feathers' are, in effect, a larger version of 'plugs and feathers'. JSD describes the opening of ground to sink a coal pit from an 1812 account:

> …a borehole of two to three inches diameter, was put down several feet, by means of a steel augur; two long slips of iron named the feathers were placed down each side of the hole, and betwixt these a long tapering wedge called the stook, was inserted; this wedge was driven down with ponderous hammers, till the rock was wrenched asunder.

During a conversation with a mason from Worcester, England, he used the term 'cope' for the splitting of a large block of stone using pitchers and punches (types of chisel) around a line of fracture. In drystone terminology 'cope' is better known as a term for the top course of stonework on a drystone wall or the gable of a building.

Using wooden plugs, quicklime or steel pins takes time and are not ideal for large scale quarrying. Gunpowder and modern explosives make the task simpler and safer. A series of holes are drilled, often many feet deep, filled with explosives and detonated from a safe distance with modern fuses.

The quality of rock broken out in a blast depends on the type of explosive used. Modern explosives tend to shatter the rock because the blast produces a sharp shock. Older types of explosives produce more gasses which lift the rock, breaking it off in large sheets with a lot less fracturing. This is especially important with sedimentary rock. McMillan (1997: 32) writes:

> Large charges of explosive are rarely used when regularly shaped building stone is required. The use of excessive explosive produces minute cracks in the rock … coarse-grained black powder is preferred to other blasting agents (eg, dynamite) since it acts more slowly and less shattering is produced.

Extraction of fine quality rock, such as Carrara marble is a careful process, using methods, developed over centuries, to suit those quarries.

For more detail in stone splitting in the granite area of Carlow, Ireland, the reader is recommended to Conry (2000: 15-29). Another recommendation, Gage (2002), gives the history of splitting with wedges in greater detail, from a North American point of view.

The result of this splitting, and subsequent trimming, should be a 'purpose steen' which DSL defines as 'a stone suitable for some particular purpose in a wall – one that can be used for building, in contrast to rubble.' This stems from an original meaning of 'purpose' – 'neat and tidy'.

The following terms are associated with natural rock and how to split it

Awte (northeast Scotland) the natural plane of cleavage in wood or stone or a flaw in a stone.

Bed / bedding plane this is easily visible in sedimentary rocks as a horizontal discontinuation between the different layers, each with a slightly different colour or texture. The meeting point of these beds is a weakness which can be exploited. In geological terminology a 'bed' is the layer or stratum of a rock. The quarryman uses the term for the partings between the beds. Rock often splits easily along these planes. The beds may be level or inclined, depending on how the rock was laid down and affected by subsequent geological events. McMillan (1997: 20) calls these sandstone beds 'laminations' if they are below an arbitrary limit of 1 cm (0.4 inches).

'Bedding' is also a term associated with the foundations and laying stone. (See Sections 3.2 (i) (foundations) and 2.3 (laying a stone).)

Bait / bed (Scots) DSL defines this term, from northern England and Aberdeenshire as 'the grain of wood or stone' and adds 'bate' as an old English equivalent.

Blind seams Muilenburg and Swineford (1975: 44) describe these '…incipient cracks…' in the Kansas limestone which can be taken advantage of to break the limestone out to the proportions of thick fence posts.

Boord / bort (Caithness, Scotland) the grain of a stone, the natural plane of cleavage. As a verb it also means to split a stone along its grain, especially with reference to the flat Caithness flagstone. The term also appears as 'brod' in northeast Scotland. The men who split the flagstones in Caithness quarries are called 'boarders' or 'splitters'. Each splitter worked with a hammerman who strikes the steel wedge selected by the splitter. Caithness flagstone is usually easy to split along bedding planes which are consistent across the quarry floor. According to DSL – 'Stones are said to be *burdit*, when they split into lamina, perhaps from *burd*, a board; …like wood divided into thin planks.'

Cleavage / cleavage plane the line along which a rock easily splits, ideally along a smooth plane, parallel to the top edge. Sedimentary rocks are easiest split along the bedding planes. Metamorphic rock can also split or naturally delaminate along planes if the minerals in the original rock were stressed by heat and pressure to separate out and form layers (Figure 25).

Cutter (Scots) a crack or crevice in a stratum of rock.

Easy way (Aberdeenshire, Scotland) the easiest plane of splitting in granite. The Aberdeenshire granite quarries had many connections with those in Maine, USA. This provided ample opportunity for exchange of terminology and techniques.

Gaunwey / gaain-wey (Scots) literally translated as 'going way', the grain in stone.

Greek a Scots term used to describe the grain, the texture or particular quality of one stone compared to another. Stone with a 'small greek' had a small granular structure and took a good polish. 'Grete', a variation of 'greek', is mentioned in an extract from the Aberdeen *Press and Journal* newspaper for 28 November 1929. It confirms not much changed over time:

> With due allowance for the change wrought by the introduction of machinery and improved tools, the methods of the eighteenth century stone-cutter differed extraordinarily little from those of to-day. Your modern stone-cutter in splitting a big block of granite drills a row of holes along the grain, or run of the stone, and then splits the block by driving small steel wedges into these holes. The old stone-cutter practised the same method, only he used a steel pick instead of a drill and called the 'run' of the stone the 'greet'.

The line of 'greek' or 'grete' is therefore the easiest way to split a rock.

Arkell and Tomkeieff (1953: 51) also have 'greek' as a Scottish definition for 'Any coarse-grained hard rock, such as coarse sandstone.'

Grike / gryke fissures in the carboniferous limestone pavement in west Yorkshire and the English Lake District. They are also called 'gilles'. The original limestone was exposed by glaciation. Cracks and openings were further eroded by rainwater to produce wide, deep, slots – a sheltered habitat for plants. The cracks allow leverage of limestone slabs from the bedrock.

Rank (Scots) the grain of a stone, an indication of the best plane for splitting.

Reed (Scots) the direction of grain in stone or wood.

Rift Merrill (1903: 390) defines this concisely:

> The rift of a rock is the direction parallel to its foliation or bedding and along which it can usually be relied upon to split with greatest ease. It is best represented in mica schist, gneiss, and other rocks of sedimentary origin.

Rift is also a term for a fissure or cleft in limestone in Oxfordshire, England.

Rifter a heavy hammer from North America, used to break stone along the grain. It is laid against the rock and hit with a sledge hammer.

Rive / reeve (Scots) to burst or split. In Orkney and Shetland, a 'riva' is a split, crack, fracture or cleft in the rock. A 'river' (pronounced 'r-ee-ver') is one who splits slate in the English Lake District.

Run (Northamptonshire, England) the grain in stone.

Shakes Arkell and Tomkeieff (1953: 105) have this as an eighteenth and nineteenth century term for fractures in rock.

Skink DSL has this definition – 'splitting, the act of splitting or cleaving; cleavage, line of fracture.'

DSL also includes the phrase 'ilk stane has a skink o' its ain', which translated to 'every stone has its own line of cleavage.'

The word has an obscure origin, possibly derived from another sense of 'skink', to break up or reduce to smaller amounts.

Toe nails Arkell and Tomkeieff (1953: 119) define this as 'Curved joints between the sheet surfaces of granite.'

Vent a natural crack or fissure in stone.

2.3 Laying a stone

Laying one stone is simple. Laying others around it adds levels of complexity. Several factors must be considered to ensure each stone combines efficiently with those around it, to produce a strong structure.

Batter is mentioned frequently in this section. This is the narrowing of the wall with height, an important feature which adds stability and strength. Typically, a drystone wall narrows by one inch for each six inches of height. The topic is covered in more detail in section 3.4 and illustrated in Figure 40.

Each stone performs several functions

According to Sean Adcock's teaching notes each stone should perform eight functions, reproduced here with his permission. Each stone should:

Figure 25: Hard igneous rock with several opportunities for splitting.
Creag Beag, Badenoch, Scotland.

1. have its length into the centre of the structure.
2. be sitting level along its length and across its width.
3. cross the joints of the stones in the course immediately below with a good overlap.
4. follow the line and batter of the structure.
5. be in tight contact with the stones alongside, above and below and the minimum of gaps within the wall, and on the face.
6. sit solidly with good wedging and hearting to reduce the empty space to a minimum.
7. act as a strong stable base for building on top of.
8. be graded, with the biggest stones nearest the foundation and smaller ones higher up.

Adcock qualifies these points by adding:

> In any individual structure some or all of these points depend on the available stone and may be contradicted by individual craftsmen or local styles. Walling is about minimising your mistakes and compensating for them rather than compounding them, and not doing any of the eight very badly…

The combination of well-placed individual stones within tight courses produces good bonding and a strong wall. The experienced hand learns how to compensate for any 'broken rule' forced upon them by a highly variable material and can develop 'new' solutions. Therein lies the art of the dyking craft. An individual stoneworker's instinct, skill and experience are more valuable than any dogma, rule or textbook analysis.

Building a drystone wall, with successive courses of stone, is like writing a computer programme, with successive lines of code. On each successive course, or line, the waller, or programmer, uses their skill and experience to anticipate the route towards completion, allowing for any 'transgressions' in previous work.

Accept the stone with minimum of shaping

The stones used to build field walls and simple structures are used 'as they come' with hardly any shaping. The old cliché that a good dyker never picks up the same stone twice is confirmation the experienced craftsman has a well-trained eye and tends to only pick up stones he/she knows will be suitable.

Cairns (1975: 22) confirms this, referring to a dyker on piecework:

> I have heard a lot of masons talking about dykers never laying down a stone once it is lifted … The fact is that a dyker has trained his eye and his hand to work together to such a pitch that he simply doesn't lift a stone at all unless it will fit either one bed or another. He just can't afford it!

Mitchell (1992: 77) quotes a Yorkshire Dales waller — 'Every cobble's got its face, but it isn't any fool can find it.'

In Cook (2011: 168) Brendan Dunford refers to the field boundaries on Inis Meáin, Aran islands, Ireland, with '…angular, abrasive limestone rocks… locked into the fabric of the wall.' The location for each stone was carefully considered – 'Older farmers believed a stone had to be turned nine times before its perfect place was to be found in the wall.'

Corangamite (1995: 90) includes a dissertation by Rod McLellan, an Australian, who wrote 'Breaking or chipping stones to make them fit was seriously frowned upon although each stone is given a judicious tap with a small hammer to make it settle.'

Repairing a wall requires less hammer work than building a new wall. The stone has already been used and its best face already decided on. The stone 'knows' where to go.

Tighter joints are expected in modern drystone builds. More shaping is necessary; perhaps using chisels or a saw.

Lay stones — don't drop them

Dykers instinctively accept stones should be placed carefully on a wall; dropped stones disrupt the work they fall on. For the same reason a stone should be removed from the wall before it is 'punished' with a hammer.

Some writers offer a different opinion. John Vivian (1976: 101) quotes a conversation with a retired New England engineer who taught himself stonework:

> Another lesson learned painfully was not to set a rock down on the wall, but to drop it. 'If you set them down, sooner or later you'll pinch your fingers. So drop them about like this,' — and he dropped a rock about six inches. 'It will bed better too, that way.'

Shadmon (1989: 108) agrees — 'Stones should be dropped into place rather than put down so they will find their own level.' This advice may have some value when working with heavy New England boulders.

Flattest face up or flattest face down?

Borchgrevink (1982: 83-84) quotes Charlie Jardine, a Dumfriesshire dyker who advocates flat face down:

> If you prepare your bed for your stones with packing the centre and levelling the side of the dyke, always lay your stone on the flattest bed, no matter what the top o' the stone's like. You can always get another stone to sit in the top o' it, once you have built it up again ready for the next stone to go on.

Strength, not appearance, is the primary objective

It is a mistake to build a drystone wall with more thought to appearance rather than to strength. The completed structure is weak and short lived. Todd (1859: 462) recognised the importance of laying stone to produce a strong wall rather than a good looking one:

> If a stone has one good face side, and it will lie more substantially with that side down, let it be laid in that position… It is desirable to have a straight and smooth face to the stones in a wall, but when this object cannot be secured without detracting from the permanence of it, it is best to place the best sides down, even if the wall is not quite as even and smooth as we could desire.

Stephens (1853: 588) saw the need for good face stones, properly hearted:

> Very many dry-stone dykes in this country are constructed on erroneous principles, the stones being laid promiscuously, and more with a view to give a smooth face than a substantial hearting to the wall.

Todd (1859: 463) describes poor building:

> ...with merely a row of stones for each face, while the middle of the wall is filled with stones thrown in promiscuously, without being placed so as to lie firmly, and without having the face stones chocked on the inside.

Todd and Stephens are using the word 'promiscuously' for an archaic meaning, 'mingled together, confused', as defined in Bailey (1675).

A drystone dyke may look 'fair faced' with good line and batter and no running joints. This can hide a multitude of faults. The completed structure must stand up to critical questioning: Do stones appear tight on the outer face but lack hearting and pinning on the inside? Are the stones laid 'traced' with their length along the wall instead of into the middle? Are there enough throughbands? Was the stonework built to please the eye rather than for long term strength and utility?

Most of these points can only be answered if building work is checked as it proceeds. For this reason, certification schemes for drystone walling hold test days where candidates build a section, or a test piece, under the eye of a trained examiner.

Old stonework is not necessarily the best

Dawber (1905: 9) issues a warning about Cotswold limestone houses which can equally apply to old field walls:

> ...much of what they did we must perforce condemn now. Their walls, for instance, though thick and solid in appearance, were often merely an inner and outer shell, filled with rubbish and small stones, which had little or no cohesive properties and consequently could not withstand any settlements and suffered severely from the effects of wet and frosts.

Lay the longest edge of the stone into the middle of the wall

The 'correct' way to lay a stone is with its longest edge and narrower end (the tail) in towards the centre of the wall. For the best bond, with maximum friction, stones should ideally be touching along their full length. This is not always possible; hearting takes up the gap and tightens the stones. 'Tracing' is the opposite of 'length in'. Traced stones have their longest edge laid along the length of the wall. This often creates a weakness but, if the stone stretches a third or half way towards the centre, it should still be structurally strong.

In regions where the stone is small, for example the Cotswolds in southern England, traced walling is more acceptable; the build must be tighter to compensate. Williamson (2002: 107) also confirms this for limestone walls in the Mendips where the Carboniferous limestone is small and irregular and '...it is the custom to lay the stones with their length along, rather than set into, the wall.' Stone in the Cotswolds and Mendips is comparatively soft and easily shaped to get a tight fit.

GVA Dumfries (1794: 52) gives a good reason for laying the stone length in — to build a stronger structure and use the weight of the stones to hold themselves in place:

> ...the centre of gravity of each stone, is at as great a distance as possible, from the outside of the dyke. Hence, though the dyke should be put considerably from its horizontal to a sloping state, the stones would not lose their balance and tumble down. Placed in this way, every stone has a larger bed upon the one beneath it; and therefore, in case of a declivity, had proportionally more friction to overcome, before it can slide entirely from its bed.

This point once again emphasises the importance of maximising stone-on-stone contact and increasing friction.

The secure placement of stone is more certain if dykers work in pairs, one each side. Each will see the gaps behind the other's work and place hearting accordingly.

Tilt or not to tilt?

Ideally all stones should be laid flat and level. Level stones create the maximum value of friction between them and adds to the strength of the structure. There are exceptions and contrary advice. It is sometimes advantageous to lay stones 'on a cant' – with a tilt or on an edge.

In Cotswold walls the soft Jurassic limestone is best laid with a slight inclination towards the outer face, to shed water. This helps keep the inside of the wall dry and reduces the chance of waterlogged stone suffering frost damage. These tilted stones are 'watershot', a variation of 'watershed'. They 'shed' or divert the water away. The Duke of Buccleuch is said to have preferred dykers build his farmhouses in the Scottish Borders because they tended to lay the stone with a tilt to the outside. This reduced the chance of rain finding its way through the mortared walls.

In the Shetland Isles, Scotland, a 'watersten' (literally 'water stone') is a stone in a house or barn wall which has not been properly tilted towards the outside; it may even tilt to the inside. Moisture seeps back from the outer face of the stone and creates a damp patch on the inside wall. A 'weatherstane' is a Scots term for a projecting stone built, for example, at the foot of a chimney stack, to throw water away from the stonework.

Other sources advise a slight tilt to the stones. Stephens (1853: 590), writing in a Scottish context, repeats the need to shed water and the importance of good hearting to secure the larger stones:

> In good dry building, the stones are laid with a slight inclination downwards, from the centre of the dyke, towards each face, ... and to support their inclination, small stones should be wedged firmly under them in the heart of the dyke; whereas stones that are laid flat admit of no wedging to heart them, and receive none, to the risk of the dyke bulging out in both faces. The inclination causes the rain which may have found its way down through the top of the dyke to be thrown off both sides.

Borchgrevink (1982: 84) quotes Charlie Jardine's advice to bring the hearting '...up along wi' your sides.' This provides a bed for the next stone and, '...if you keep your 'heart' up, slightly up if anything' allows for the wall settling and sheds water to the outside. This is advice for the hard stone of Dumfriesshire, southwest Scotland.

Adair (1998: 27) also emphasises 'keep the heart up' in the middle so the stones tilt slightly to shed water away from the middle.

Water inside a wall can be advantageous. Garner (1984: 25) describes 'stone hedges', an earth bank faced with stone and topped with turf. Water finds its way to the middle of the structure and nourishes the roots of the plants and bushes growing on top of the bank:

> Stone hedging is the only branch of the craft where it is considered correct to lay stones tilting towards the centre. Where vegetation is relied on to bind the work rain must be encouraged into the centre.

In New England many of the walls are low and wide, built from flat stone laid inclined towards the centre. Thorson (2005: 61) captions an image as a '...classic inward-slanted, battered, well-laid double wall.' Thorson (2002: 125) refers to '...the double wall, in which slabby stones are slanted inward from two sides.'

How stones lie, whether they are flat and level or tilted in, or out, from the centre might affect the strength of the structure. Todd (1859: 467), referring to retaining walls, emphasises:

> ...the stones, especially the flat ones, instead of being laid level, as in other kinds of wall, should be laid *pitching inward* towards the bank or hill above, with their surfaces at a *right angle* to the face of the wall.

Todd advises this to allow for the frost / freeze cycle because '... if the frost should ever thrust it over any, it would settle back again to its original position as soon as the earth behind became all thawed out.'

Retaining walls in southern France are laid with the foundation stones slanting back down into the ground, at right angles to the batter of the wall. This is calculated to be more secure in countering the pressures from behind the retaining wall.

Lay stone flat on its natural bed

An experienced dyker can usually tell by looking at a stone how it should be laid. Bedding planes are especially noticeable in sedimentary rocks, the slight differences in colour and texture show how the rock was originally formed. These horizontal bandings are a potential weakness, as with poor quality or badly laid sandstone where, to quote London, (1988: 52), '...many buildings and graveyards are full of stones coming apart in layers.'

This quotation from (Powell 1889: 45-46), requires no comment:

> About thirty years ago the late Mr C H Smith, who had thoroughly studied the subject of lithology, said that the importance of laying stones in buildings upon their beds was generally over-rated, and that it signified little which way a stone was laid unless it

presented a decidedly laminated structure. We unhesitatingly maintain that soft, calcareous or limestone should be laid in the walls of a building upon its natural bed, and that the beds should not be exposed to inclement weather after they have been dressed.

The following terms relate to the placement of stones

Bedding a general term denoting how and in what orientation a stone is laid in a wall (Figure 26). The stratification in any rock type should always be considered to ensure it is 'natural bedded' or 'weel bedded'. These Scots terms mean the stone is laid with bedding planes, laid as found in the quarry, to best resist the forces of compression and weathering. This means the bedding planes are horizontal. 'Natural bedded' or 'weel bedded' should not be confused with 'bed' or 'lie-bed' which is the Scots terms for the surface of a stone upon which another is built, or the mortared bed upon which a stone is laid.

Cant-bedded a stone laid on its edge, or leaning to a side.

Closer a term for the last stone or brick in a course, perhaps cut to a non-standard size to fit the space.

Edge bedded / edgeset a masonry term describing a stone set on its edge, as a book sits on a bookshelf, with horizontal bedding planes lying vertically and exposed to the outer face of the structure. The Scots phrase 'stones laid on their hem' also describes those laid on their edge rather than with bedding planes laid horizontal. Another Scots term for this is 'cant-bedded'.

Sometimes stones are best laid on edge. Harbours and seawalls in Caithness in the north of Scotland are among places where large slabs are set vertically, with a long edge perpendicular to the face of the wall. (Figure 27). This allows air pressure, trapped by crashing waves, to disperse upwards.

Using flat stones, on edge, on riverbanks is, arguably, more secure than flat-laid stone because there is less chance of undermined. The slabs will tend to settle into the riverbed if disturbed by the water.

In the Scottish Galloway dyke or Irish *feidín* wall, and its variations, the large stones of the upper part are set vertically across the line of the wall for maximum height gain with minimum effort. In some Irish walls, particularly in the limestone areas of the west, where there are many long flat pieces, individual stones are often set vertically within the 'double'. This adds strength and visual interest.

Vertically laid stones also produce attractive walls on a smaller scale. In Welsh 'clawdd', a variation of a stone-faced earth bank, the stones are laid in vertical rows with their longest edge into the centre of the wall. Clawdd construction is detailed in Section 4.1.

Laying stones on their edge makes sense if the faces are smooth but the edges are rough. This produces more friction than if the stones are laid horizontally.

In the northeast England and County Durham there are walls with a course of short stones laid on edge for the foundation. It is a good way of building a foundation course on soft ground – when a series of edge-laid stone is likely more stable than one laid flat. This is also an efficient way of building a foundation course across sloping bedrock. Thin stone, laid side by side, on edge (with their long edge into the wall), are more stable than stone laid flat on top of each other. A course of vertical stones can be brought to uniform height by

Figure 26: Mortared seawall, showing edge bedded (examples in all the courses), face bedded (bottom course, second from left) and properly laid sandstone. Chanonry Point, Fortrose, Scotland.

trimming them to suit the irregularity of the bedrock.

After consideration we can understand why stones, set vertically in the foundation, or elsewhere, can often be more stable than the same stones lying horizontally on top of each other. Any series of vertical stones is best 'bookended' by large stones to limit sideways movement. (Figure 106).

The best example of vertically set stones is, of course, the Scottish 'locked top'. These copestones are laid vertically across the top of the dyke and pinned to create a solid top course.

Face bedding is a weak way to lay stone. The bedding planes in the stone are parallel to the face of the wall. Consider it like a loaf of sliced bread, laid length into the wall, so that only one slice can be seen at a time. If the rock type is a soft sandstone the planes are more likely to slough off with weathering.

Fat Tony (Canada) A stone laid on edge which is higher than it is long and is not a corner or quoin stone. It is a source of potential weakness because anything taller than it is wide is liable to topple, unless properly tied in.

Natural bedded a stone laid on its natural bedding planes laid horizontally, not on edge.

Roughback / rough back the end of a stone, hidden because it is built into the wall.

The end which faces out is chosen because it sits better with its colleagues, or better fulfils one of Sean Adcock's eight requirements, detailed at the beginning of this section.

Soldier / upside down soldier an English drystone term for a small stone set on its edge

Figure 27: Castlehill Harbour seawall, Caithness. The local flagstone laid on edge.

to fill the gap between two larger stones. It is taller than it is wide, and as tall as the larger stones it sits between. Presumably this reflects the stone 'standing to attention' rather than lying flat.

A general masonry term for this is 'closer' because the stone, or brick, 'closes' the gap. The 'soldier' is a potential weakness, it interferes with the bonding pattern and is visually 'disturbing'.

In brickwork a 'sailor' is a brick which sits on its edge with the broad face to the outside.

Shiner an individual stone laid on edge within the 'double' of a drystone wall, lying taller than it is wide, or with vertical sedimentary beds instead of horizontal. This is regarded as a building fault because it introduces a weakness and is visually disruptive. Individual examples on a wall face can be essential to, for example, fill in a narrow gap. One piece of information, traced to the Durham/Northumberland area of northern England, with no more detail, suggests they are named 'shiners' because 'on frosty mornings they shine with ice' – perhaps because of the way the grain of the stone lies.

Generally, as expressed by Mader and Zimmerman (2011: 56), 'Vertically placed stones cannot be used because they would destroy the harmonious appearance.'

By way of contrast Muilenburg and Swineford (1975: 117) promote the use of 'shiners' in Kansas limestone buildings 'for decoration or contrast' but go on to say these stones, laid on edge instead of on their bedding plane, have weathered more than others laid horizontally.

In regions where the drystone waller works with big stone the larger ones are often set on edge, and traced (laid with their long edge along the wall). They reach high up the wall, often to full height. These orthostats are secure because the stone has enough mass and size to be practically immovable.

Weel bedded (Scots) describing a stone as 'well bedded', laid properly.

2.4 Laying stone on stone to produce a structure

2.4 (i) Bonding

The place where stones, or anything, touches can be a point of weakness, or stress.

Laying one stone is simple; putting another next to it is only slightly more complex. Completion of a whole row requires care and attention. Setting a stone on the next, upper, course, securely on the two below, takes much more concentration to ensure the maximum surface contact between the stones. This produces a solid bond and a strong structure. It can also produce a good-looking structure, if properly executed.

'Bonding' generally describes the relationships between stacked stones or bricks. Brickwork has many bonding patterns. In drystone work good bonding can be difficult if the stone is not all the same size.

'Breaking the joint' is a basic principle of any masonry. One stone should lie across the joint between the two on the course below. This bonding goes a long way to ensuring a strong structure. If the upper stones do not cover the joints between the two lower stones, over two or more courses, the result is a 'running joint', visible as a distinct vertical line. It is an indication of weakness and potential structural failure (Figure 28). The running joint is defined in detail in Section 2.6.

A joint properly covered by an upper stone is variously termed a 'covered joint', 'crossed joint' or 'broken joint'. Murray-Wooley and Raitz (1992: 194) add 'break joint' from Kentucky and 'two-over-one joint' from New Hampshire, USA.

A 'butt joint' is the joint between two stones lying together on the same horizontal plane. Mader and Zimmermann (2011: 56) advise us to 'avoid long butt joints' – another term for a running joint.

A 1912 'formula for dry-dykers' gathered by The Rymour Club of Edinburgh, from Buchan (1984: 202) simply states — 'Pin weel, pack sma', Lay ae stane abune twa.' This translates to – 'Pin well, pack small, lay one stone above two.'

More detail is added in this verse:

A stane abin a stane,
Yin between twa,
Pack well and pin well
And make a strong wa'.

In Cumbria, northwest England, the advice is simpler — 'Yan on Twa, Twa on Yan' – 'one on two, two on one.'

Samuel Smith of Borgue in GVA Galloway (1813: 88-90) gives good advice, '…we submit the following observations as very proper to be attended to by every one who is desirous to have good stone dykes.' His recommendations are paraphrased below:

Figure 28: Bad coursing with running joints.

1. The dyker must use good material. Smith urges the use of large stone, 'no soft friable schistus', and no rubbish stone.
2. The stone should be laid on a proper bed, breaking the joint, with as many throughbands as possible. The propping up of face stones with small pinnings to give a smooth face must be avoided, this indicates strength has been sacrificed for appearance.
3. He stresses the importance of good hearting: '...the most prevalent and incorrigible fault of all, and which ought, therefore, to be chiefly guarded against, is neglecting to build the inside of the dyke. Here the stones ought to be of the same good quality and laid with the same care and attention as those of the outer faces. But instead of this, by the generality of dykers a parcel of rubbish is loosely huddled together, filling up imperfectly the intermediate space ...

Margaret Brooks (1977: 100) gives general guidance from a Yorkshire point of view:

In addition to digging the foundation, the four most important 'rules' of West Riding wallers seem to be: 'big stones at the bottom, cross the joints, keep the middle full and taper as you go up'.

Cairns (1975: 10-12), working in Scottish Border country, wrote:

> The life of a dry dyke, that is, its existence as a fence and not just a line of heaped stones, as a lot of them are nowadays for a lack of dykers, depends on four things… paraphrased below:
>
> 1. The quality of stones: the longer, and flatter, the stones the better the job.
> 2. The expert knowledge and care of the dykers who built it.
> 3. The stance of the dyke, the foundation. This should be built with care, allowing for hard, soft, wet or sloping ground.
> 4. Interference by gamekeepers, shepherds and others making holes to take out rabbits and set vermin traps, or farmers removing copestones for sitting on.

Dry Stone Conservancy in Kentucky, USA (2001: 46) provide these key points to remember when building basic drystone fences and retaining walls:

1. Lay strong foundations.
2. Hand pack the core with care.
3. Build an adequate batter.
4. Cover the joints.
5. Tie the faces well together.
6. Do not chink the face.
7. Construct a strong coping.

A more detailed set of principles and rules for drystone walling were devised by Neil Rippingale and the Dry Stone Conservancy in Kentucky, reproduced here with permission:

The Four Basic Principles:

1. Lay the stone with the length running into the wall.
2. Break the joints.
3. Keep the stones as level as possible.
4. Pack from the inside.

The Five Golden Rules:

1. As you lay each stone, pack it, pin it and make sure it is secure before laying another one.
2. Lay five stones then stand back and look at what you have done, that is the time for any alterations.
3. Move lines up according to the height of the stone.
4. Build to the line, not above it.
5. Don't hammer on the wall you have just built.

Robert J. (Bobby) Watt, President of RJW-Gem Campbell Stonemasons Inc. of Ottawa, Canada has written his 'Ten Commandments for snecked masonry'. This method of mortared

masonry, common in Scotland and Ireland, has been called 'Celtic Bond'. French Canadians know it as *travail ecossais.* The general principles of snecked masonry, with its strong bonding pattern, will be familiar to Scottish dykers used to working with random stone.

Watt (2007) provides a summary of the style:

> The principal characteristic of snecked work is the keying together of parallel courses of stone… achieved by interrupting the level height of a horizontal course by placing a taller stone that reaches up into the next course. This vertical interlacing gives a wall a tensile strength and enhances its compressive strength.

On the face of it the style is remarkably simple. According to Bobby Watt snecked masonry involves three kinds of stones:

> 'Risers or Jumpers' — These are the stones [which] extend up through the horizontal beds. They can be square, or almost square, or up to three times as long as they are high.
> 'Levellers' — These form the bulk of the wall. They are usually at least twice as long and up to five times as long as they are high.
> 'Snecks' — These are the smaller pieces that enable the mason to make up the differential in height between the top surfaces of the levellers and the risers.

These stones are blended to create a strong, good-looking wall, as described below and illustrated in Figure 29.

Bobby Watt's 'Ten Commandments' are:

1. Always lay sedimentary stones on their natural bed.
2. No stone should be higher than it is long (a Fat Tony), except at corners.
3. Avoid blocked or running joints. There must be only one stone on at least one side of any vertical joint.
4. Avoid setting any more than three stones against a riser.
5. Risers should be evenly distributed throughout the body of the wall and the grouping together of like-sized stones is to be avoided.
6. Avoid the 'stepping' effect of levellers of the same bed height and same approximate length set on top of each other.
7. Avoid lining up the vertical joints in alternate courses. This gives the wall too much regularity and detracts from the over-all appearance. Slightly offsetting the vertical joints imparts liveliness.
8. Generally, risers should never touch except at corners and openings. (Jambs)
9. Don't allow horizontal joints to run more than four or five feet (1.5m) and, if possible, break up the horizontals on short stretches between windows or doors.
10. Try to provide a substantial bonding lap. A minimum of a quarter, ideally a third, of the length of a stone being set should cross the joint between the two stones below it.

Figure 29 illustrates how important it is to lay stone, on succeeding courses, over the joint in the course below. The distance one stone extends over another is called the 'lap', 'overlap' or 'bite'. The term 'tied' also applies to the way one stone overlaps or covers the

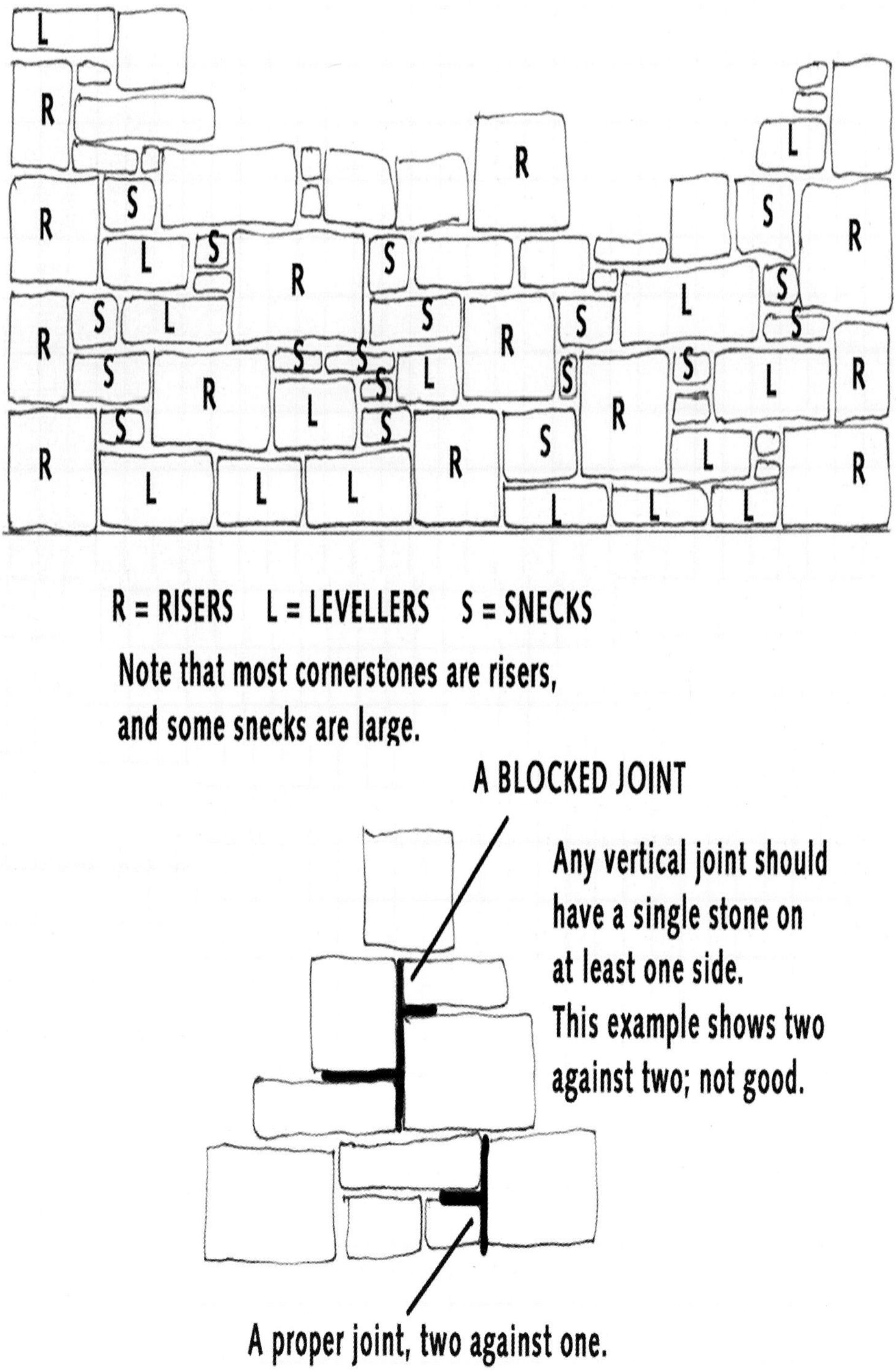

Figure 29: Snecked masonry. Diagrams courtesy of Bobby Watt.

Figure 30: The rider, or floater, is the top stone. Is it lying firmly on all three below? Note also the short 'bite' on the left side.

joint between the stones below.

When laying one stone over two it is easy to ensure solid contact between the stones. Laying one stone over three requires more care (Figure 30). Readers will note the shadow of a shim above the right-hand stone. This serves little purpose and will likely crush under pressure.

DSL records the term 'rider' from central Scotland in 1968, apparently a masonry term, defined thus:

> In building with undressed stone: a stone which completely covers a smaller stone beneath and partly overlaps the two adjacent to it…

In drystone work this corresponds to 'floater', a longer face stone which lies across three others. This is often described as 'one-over-three'. There is a fair chance the top stone covers, but is not gripping, the middle one. The dyker must ensure each stone has solid contact with all those below it. This 'one-over-three' is not necessarily a weakness; it is evidence which should be investigated. Figure 30 illustrates this point.

Nicholson (1838: 35) notes the limitations of long face stones and the need for solid contact with everything below:

> It is not much advantage in walls to use stones which have a greater length than three times their thickness; otherwise, from the difficulty of bedding them equally, they are liable to break in the wall; hence proper crossing of the joints with stones that are not so long as to have the risk of breaking, is a better method.

If long stones are not properly supported, they can break when weight is put on top of them.

Rainsford-Hannay (1957: 35) includes an eloquent description of drystone construction from a specification for a dyke near Gatehouse of Fleet, built for the Glasgow and South-Western Railway in 1865:

> The double to have both sides brought up together, having the stones properly blocked, laid close together, well hearted and packed in the centre, every stone doing its duty by its neighbour. The outer stones to lie with their ends generally inward, so as to stretch into the dyke as far as possible, for the better binding of the work.

Mader and Zimmermann (2011: 40-42), in an English translation of a German original, use the term 'dry-placed walls', and quote DIN (Deutsches Institut für Normung) 1053 for masonry standards:

> The rough stones have to be fitted in the correct masonry bond with little processing and without the use of mortar, so that there will be narrow joints and small cavities. The cavities between the stones must be filled with smaller stones in such a way that wedging creates tension between the wall stones.

Mader and Zimmermann (2011: 56-58) go on to discuss in detail the Stuttgart School style of building which '…became a tremendously influential model that influenced gardens, especially in southern Germany, until the 1960s.'

Extensive landscaping in the Killesberg area, near Stuttgart, Germany, for the 1939 German Garden show, built 54,000 square feet (5,000 square metres) of dry-placed red sandstone walls and steps and 200,000 square feet (18,000 square metres) of paving. The work was led by Adolf Haag (1903-1966). A consistent building methodology is essential for such an extensive scheme.

Mader and Zimmermann continue:

> The first design rule… is that the thickness of the stone courses tend to decrease from bottom to top. The second rule is that the stone walls must have a distinct shape with proportions of 1:2 to 1:5. Vertically placed stones cannot be used because they would destroy the harmonious appearance. The occasional installation of changers or changing stones is desirable because it enlivens the overall appearance. A changer is a stone with an approximately square face. Two stone courses are set on one or on both sides of the changer. The upper wall closure is done with heavy cover sheets or capstones that stabilize

the wall with their high dead weight.

The walls 'for structural and design reasons' have a 20% batter, about 1 in 3 – an indent of one unit for each three units of a rise in height. The wall gets narrower with height. 'Batter' is fully described in section 3.4 and illustrated in Figure 40.

A drystone wall is a handcrafted item. The craftsman's hand and eye are coordinated to make the best out of a highly variable material. Oblique light on stonework creates shadow

Figure 31: Well coursed mixed stone, Cumbria style. Note also the interlinking stones at the end. Dry Stone Walling Association collection, Crooklands, Cumbria.

lines between and along stone joints. This shadowing can create pleasing patterns and textures. In mortared work joints are normally pointed up, the surface looks flatter.

Pilipski (2002) considers the aesthetic appeal of stone walls and concludes:

> ...given two similar stone walls the one that exhibits the inclusion of some very large stones and more horizontal lines will appear more attractive than the one that has a uniform texture and more vertical lines.

2.4 (ii) Coursing

Good coursing is an essential part of bonding; with so many terms applied to coursing, the author feels it deserves its own section.

A course is a horizontal layer of individual stones along the face of a wall (Figure 31). It is also known as a 'tier' in USA or a 'rin', 'run' or 'ryn' in Scots. Coursing is fairly easy in regions with flat stone, and straightforward. Stones are graded into courses, the thickest at the bottom, graduating to smaller sizes in even courses towards the top. Smaller stones are used occasionally to bring a course up to height.

Figure 32: Herringbone coursing. Cornwall, England. Image courtesy of Rosemary Hardy.

According to Oxland (2011: 4) squared and coursed stone work is called *Kenchizumi* in Japan but 'Every mason I met in Japan thinks that this style is terribly boring and not necessary to create a good strong wall.'

Strict coursing could produce a horizontal line of weakness. Good interlocking and hearting in the internal structure is therefore very important.

Coursing is not so easy with rubble or irregular natural stone. There is typically no attempt at coursing beyond use of the largest stones at the foundation. Random rubble work, which means working with larger, irregular, material, can be brought to a level by including smaller stones. This allows the builder achieve a strict horizontal line to incorporate throughstones, bondstones or openings at specific heights, or prepare a bed for subsequent courses.

Standardisation of components influences, and dictates, the building method. It can be more difficult to build a solid structure with standard bricks than with irregular stone, unless mortar and a proven system is used, for example brickwork's English bond or Flemish bond.

A Working Man (1908: 101) tells a good yarn about coursing and bonding. An old stonemason was employed to lay bricks. The boss noticed he was not breaking the joint and told him so. The old mason replied – 'I know, and how the **** can I, when the bricks are all one length?'

Individual stones and courses in a drystone wall should always be truly horizontal; they should not follow every dip and rise in the ground. This helps guarantee stability and strength.

Williamson (2002: 102) considers a slope of 15 degrees (about 1 in 3.5) the maximum incline a wall should follow without stepping the foundation, in effect digging a stair up the slope for the foundation stones. Stepping straightens up the coursing, keeps the stones truly horizontal and reduces the chance of stones slipping on the incline.

The following terms apply to coursing in drystone and masonry work

Belt course a masonry term for a continuous horizontal course of flat stones; used to bring a rubble wall to a level.

Bite the extent to which a stone overlaps the joint between the two below it, or the length a lintel extends into the wall at the side of the opening.

Breaking the joint see the opening paragraphs of section 2.4 (i) for a detailed discussion.

Broken coursing (Kentucky) coursing where the stones do not form continuous horizontal lines.

Diagonal coursing Murray-Wooley and Raitz (1992: 191), with reference to Kentucky walls, define this as 'Fences laid with rock edges or most narrow faces towards the ground and coursed at an upright angle between 25 and 40 degrees from the vertical.'

These fences have several names, as described under **edge fences** in Section 4.1.

Herringbone coursing stone laid resembling the fish's bones, where each course is 45 degrees to the vertical with alternate courses leaning left and right. The most attractive examples of this style use small thin stone, the size of an average paperback book. Some good examples exist in Cornwall, England (Figure 32).

Brooks (1989: 16, 91) gives an example of a 'slate hedge', a style of stone-faced earth bank, like a Cornish hedge where the face of the structure is all laid in the herringbone style – 'Slate is best used upright or, if fragmented, in the herringbone ('Jack and Jill' or 'Darby and Joan') pattern.'The 'Cornish hedge' is described in detail, in Section 4.1.

Jumpers these stones extend up through the horizontal courses. They can be square, or almost square, or up to three times as long as they are high. Also called 'risers'.

Lacing course a course of stones put in occasionally to level up and strengthen rubble walls, an opportunity to re-establish a flat base line.

Lap the distance one stone extends over another is called the 'lap', 'overlap' or 'bite', illustrated in Figure 30. The term 'tied' also applies to the way one stone overlaps or covers the joint between the stones below.

Levellers face stones which form the bulk of the wall (see Figure 29). They are typically two to five times as long as they are high.

Figure: 33: A hasty dyke repair, with cement mortar thrown among the piled stones.

Nunozumi (Japan) literally 'cloth piling', as per Oxland (2011: 4). The coursing 'flows' like a pile of towels or cloth, it is not level, straight or of uniform thickness.

A western equivalent of this is 'wave walling', where courses flow like a wave. This can be the result of careless building with poor selection of stone or, maybe, careful work to produce an artistic effect.

Random coursing a style of drystone building with no clear coursing, associated with irregular shaped stones. Found in the upland areas of Scotland, England and Wales. There is little attempt at coursing, apart from when the wall is levelled off for the throughbands and the top of the 'double'. This contrasts with the exact coursing of the Cotswold walls, which use thin limestone with regular shapes and sizes.

'Disordered piling' (*ransekizumi*) is a Japanese term for a style using rough stone, closely packed together. This was used for building defensive castle walls and developed into a more sophisticated style, using larger stones, called 'Burdock piling' (*gobouzumi*), named for the stones' similarity to a cross section of the stems of the burdock plant.

Risers these are the stones which extend up through the horizontal beds. They can be square, or almost square, or up to three times as long as they are high. Also called 'jumpers'.

Snecks these are smaller stones which enable builders to make up the difference in height between the top surfaces of larger face stones. This helps produce a level course. The Scots verb 'to sneck' means to close an opening or gap. Figure 29 describes the use of 'snecks' in 'snecked masonry'.

In Scottish drystone work, 'sneck' has another meaning — a length of 'single' walling (one large stone wide) used to strengthen a wall or use up large stones.

Sreath is a Scots Gaelic term for a layer of stone, a course. This extract, from Dwelly's Scots Gaelic- English dictionary, might apply to any meeting of stoneworkers enjoying a social gathering after a hard, dusty week:

> On stated occasions of carousal it was customary among the Gael of old to sit in a circle, which they called *sreath*. The cupbearer filled the cup to the brim at every round and however potent the liquor might be, it was cleared off at a draught. These scenes of intemperance often lasted for three days and it was deemed effeminate in anyone to retire sober. At such drinking matches it was usual for two men to be in attendance at the door with a litter, to carry off to bed every individual as he fell senseless from his chair.

Tied a description of how a stone overlap those under it. A 'well tied' stone is 'well overlapped', 'well secured'.

Wave walling walls without strict coursing, appearing as haphazard. A 'wave' pattern appears, perhaps from careless building, maybe from artistic intent. This can result from building without a line or frame.

2.5 Structural failure and bad building practices

These terms apply to building faults or failing structures. Most drystone structures will eventually fall; damage is delayed if the structure is built properly and maintained regularly.

Blowout (New England) a collapse caused by stone shifting in a lower section of the wall. Most failures of a wall originate from the foundation; if base stones move, everything above them follows.

Breach (Scots) a hole, a collapsed part of a dyke.

Clabber (Scots) to roughly throw together. 'To clabber up' a dyke is to pile up fallen stones, with no method or hope the structure should last any time.

Clash up (Scots) to roughly build a wall with stones and mortar (Figure 33).

DSL confirms the phrase 'to clash a dyke' means to repair a dyke by throwing wet mortar into joints and crevices.

Clatched up thrown up or built in a careless clumsy manner. DSL defines the term – 'A house or wall is said to be clatched up when the workmen do it in such haste, and so carelessly, that there is little prospect of its standing long.'

DSL goes on to quote an entertaining reference from Roxburgh in the Scottish Borders – 'Clatchin' up a wasp-bike hole wi' glaur', which translates to 'stopping up the hole in a wasp nest with mud'.

Clooter o' stanes (Scots) a badly built stone wall. Probably a variation of 'clutter'.

Clutter defined in Mactaggart (1824: 139) as 'A piece of bad stone building, particularly if it be 'dry ware wark'.

Cobbled up a description of a badly built wall, one looking like a pile of stones.

Figure 34: A rowan tree (Sorbus aucuparia) established itself and burst the stonework with slow incremental growth.

Creel (Kentucky, USA) to lean, tilt, collapse or fall into a state of disrepair.

Creep if a field wall runs along a slope, ploughing and cultivation can create an accumulation of soil on the uphill side. Erosion on the downhill side can cause undermining. These factors, acting together, result in unequal pressures, and a wall which 'creeps' down the slope, before collapsing.

Cutting the found (Upper Annandale, Scottish borders) Prevost (1957: 98) gives this definition:

> Reducing the width of the foundation of a stone-dyke when rebuilding is known as cutting the found. This has the effect of thinning the dyke, a procedure viewed with disfavour by conscientious dykers.

Dit up (Scots) to close up a hole.

Expertise originally a meeting of master masons to inspect, discuss and advise on building faults, for example in Chartres, France, in 1316. Details may be consulted in Coldstream (2004: 60-62).

Fat Tony (Canada) a stone laid on edge which is higher than it is long and is not a corner or quoin stone. It is a source of potential weakness because, being taller than it is wide, it is liable to topple, unless properly tied in.

Front pinning this term specifically describes a building practice whereby small stones are used on the outer face of a wall to act as structural support for larger face stones.

These small stones are built into the wall *during* construction. They usually fall out, causing loose stones on the face of the wall, and inevitable collapse. Face stones should be pinned from behind, on the inside of the wall. That is much more secure.

'Front pinning' should not be confused with 'pinning', the old Scottish practice of tapping small stones into the gaps between the face stones after the wall is finished. Rainsford-Hannay (1957: 80) quotes 'Bristol pinners' as a derisory term used by Cotswold wallers for those who use pinnings in their walls. Cotswold wallers pride themselves in tight work with no need for pinning.

Gap a full height opening in a wall, caused by defect or damage. An 'Aran gap', from the Aran Islands, western Ireland, is a gap in a wall, usually between two distinct wallends, intended as an entrance. Piled stones take the place of a wooden gate. The stones are easily removed, and replaced, if access is required to the field.

Garadh griocaig (SG) a tottering wall.

Garadh tobhta (SG) a wall of an old house or ruin.

Garsty / gersty / gorsta / gorstie / gorsty (Orkney, Shetland) a ridge of earth and stones covered with grass. Perhaps an old ruined dyke left unploughed as a boundary.

Glip / glippik / glipek / glipik / glippik (Scots) a small opening or broken part in the top of a dyke. According to DSL the term derives from the Norwegian word *glip*, a small channel-shaped opening, a crack.

Hullerie (Caithness) a description of a stone wall which is loose, badly built or crumbling.

Ivy and trees plants can damage walls, especially if they grow unchecked and roots get between the stones (Figure 34). Roots can force their way through the smallest crack and may be all that is holding the stones together; the full extent of the damage is not visible until the plant is cut away.

A covering of ivy on drystone walls was once encouraged. Dutton (1808: 151) viewed ivy favourably in County Clare, Ireland — 'If ivy was planted to these dry walls, it would strengthen them greatly, and prevent cattle from throwing them down.'

Kyerro (Orkney) a heap of stones, as when a building falls into a ruin.

Larach / laroch / lairoch / lerroch (Scots) the site of a building or the remains of a ruined one, often just the outline of the foundation.

Letter box / post box a hole shaped thus, formed if face stones in a wall are badly butted together.

Lick and stick a derogatory term applied to stonework where thin stones are cemented on to a brick or block background, similar to tiling.

Lime bloom the whitish efflorescence found on stone or brick when mortar has been used too wet or with too much additive. The moisture finds its way to the surface and seeps down the face of the stone, carrying salts and impurities, which solidify as a white crust.

World War I concrete and cement work is reckoned better than World War II concrete because more care was taken with materials and less additives were used.

Offcasting (old Scots) the weathering or crumbling of soft building stone.

Poacher biggit (Scots) literally 'poacher built'. A roughly reconstructed piece of wall, a careless repair, as by a poacher after he has pulled a rabbit out. Robert Frost's poem

'Mending Wall' includes reference to this damage to his walls in Derry, New Hampshire:

The work of hunters is another thing:
I have come after them and made repair
Where they have left not one stone on a stone,
But they would have the rabbit out of hiding,
To please the yelping dogs.

Corangamite (1999: 78) refers to Australian instances: '…where walls have fallen down after people pulled out stones on the chase after rabbits. Once you pull a stone out, the wall's gone…'

Rabble (Scots) any kind of building fallen into decay.

Rack (Kentucky) to lean, tilt, collapse.

Rag (Scots) when a wall develops bulges and twists out of line. DSL includes reference to a legal case: 'Edb. 1767 Session Papers, Dick v. Tennent Proof 28' and the phrase 'These dikes were all rags and slaps since he remembers.' No more detail is given. Presumably this was a dispute between landlord and tenant regarding responsibility for dykes which had

Figure 35: An old dyke from southwest Scotland. It appears full of running joints and bad coursing but was solidly built by dykers who understood the stone and how best to use it.

degenerated into a series of bulges and collapsed sections.

Ramfeezle Cairns (1975: 10) describes this as the sabotaging of a dyke by levering stones forward an inch or two near the bottom of the 'double', on both sides, thus shortening the life of the wall. The damage might be the actions of a rival, perhaps the dykers themselves, to get revenge against a landowner who had cut their pay by inventing an offence such as poaching, or finding the wall not quite to the required specification. The word means 'to confuse or disrupt' and was reputedly coined by the Scottish poet Robert Burns (1759-1796).

Rapach (SG) rough building, untidy, slovenly, second rate.

Rauchle / raughle (Scots) a rough heap of stones ready to fall, a loose wall built without mortar.

Raxed (Scots) describing a wall which is strained and cracked.

Rickle / rickle dyke (Scots) a rough dyke, almost a pile of stones, a loosely built one-stone-thick drystone wall. 'Ricklie' or 'rickly' means shaky, loose, disjointed or dilapidated. A rickly dyke is therefore not always a thing of beauty. A 'rickler' is an unskilled stoneworker who produces poor work.

Rainsford-Hannay (1957: 95) includes part of the poem 'Davie the Dyker', originally published in *The Scottish Gallovidian Encyclopedia* by John Mactaggart (1824), wherein the dead dyker is praised for his good work:

A rickled rood ne'er left his hand,
His dykes for centuries will stan'.

See also **cape and rickle** and **rickle cope**, in Section 3.2 (iv), describing covers and copes.

Rumble down (Shetland) to demolish.

Rumbly dik / rumbly dyke (Shetland) a loosely built wall.

Rummle (Scots) any badly built piece of masonry, hurriedly put up, or in a rickety state. A ruin, a pile of collapsed masonry.

Running joint bad bonding, generally regarded as a building fault, detailed in Section 2.6.

Rush (Lake District, England) a collapsed part of wall or one in the process of collapsing.

Settlement 10% has been quoted as the amount a drystone wall will settle after construction; this means a five foot high wall sinks six inches. Even a fraction of that is not ideal; a good tight wall built on a solid foundation should hardly move before the moss grows. The author cannot trace the source of this estimate: it may be the north of England. To be generous we can assume the settlement is over the lifetime of a wall, perhaps 100 years, not immediately after construction.

Shard an English dialect word for a gap or broken place in a hedge or wall. 'Shard way' is a Derbyshire term for an opening in a wall to allow sheep through.

Shuit / shoot (Scots) the bulging or collapse of a building or wall.

Slap a gap in a wall, caused by collapse, or perhaps deliberately done to allow livestock through.

Slump (Cornwall, southwest England) to fall into disrepair.

Steeniger (Orkney) a mass or heap of ruins, a state of dilapidation. 'Steen' is an Orcadian variation of 'stone'.

Stengle (Shetland) to close up an opening temporarily, and roughly. This matches nicely with another Shetland word 'stigl' meaning to roughly and temporarily build up a dyke.

Step walling a disjointed effect where each course noticeably steps in from the one below, perhaps to correct an over-wide foundation. It should not be confused with a scarcement where only one course is stepped in, at ground level. A batter of two inches over twelve is smoothly achieved by using two or three thin stones. If the same rise is attempted with one thick stone, the face of the wall looks 'stepped' unless the larger stone has a face, natural or cut, which slopes to correspond to the batter.

Step walling can also occur if lines have not been properly used to ensure face stones follow the line and batter of the wall.

Tappiloorie JSD defines this as 'anything raised high on a slight or tottering foundation' with the suggestion it is a corruption of 'tapple o'er' — something likely to topple over.

Tender (West Lothian, Scotland) a description of a loosely built piece of wall which lacks strength and is likely to fail with little encouragement.

Thinning the dyke (Upper Annandale, Scottish borders) Prevost (1957: 98) defines this as:

> Reducing the width of the foundation of a stone-dyke when rebuilding is known as cutting the found. This has the effect of thinning the dyke, a procedure viewed with disfavour by conscientious dykers.

Throw the bond a stone is said to 'throw the bond' when, for example, a sill or lintel is introduced and the regularity of the coursing is interrupted.

Traced walling / trace laying a 'traced' stone is laid with its longest edge along the length of the wall. Normally the longest edge of a stone should lie into the middle of the wall, for maximum stability and strength. 'Tracing' may be unavoidable for an individual stone, and fairly acceptable if the stone stretches at least one third of the way across the width of the wall. This is equally important in drystone construction and mortared stonework.

Traced stone is quite common in the Mendips and Cotswold areas of England because their stone is smaller. Tracing is compensated for by tighter building.

Tumble a short length of collapsed stonework within a longer section of wall.

Tuutramuuny a very descriptive term from Shetland which DSL defines as 'apparently a ramshackle structure.'

The Shetland phrase 'A deak, caa's doo hit. I wid juist caa hit a tuutramuuny o' deevelry, fur doo doesna hae a richt stane atill him' questions whether the structure should be called a dyke because there's not a properly laid stone in it. DSL links the term with a Norwegian dialect word *tutra* – 'to tremble or quiver'.

Underputted (Scots) describing a structure which is underpitted or undermined, without a firm foundation or base.

Window walling loose drystone work with see-through gaps left between stones where they should tightly butt together.

2.6 Running joints and broken joints

Running joints are one of the most common faults in drystone walling, and one of the most obvious. With proper bonding the stones in each succeeding course should cover the joints (break the joints) between the stones in the course below. A running joint is a vertical joint up through two or more courses, visible as a dark vertical line (Figure 36).

In the context of a quarry a running joint is a vertical joint up through a series of beds in a rock face.

The French phrase *coup de sabre*, 'a sabre cut', a vertical cut as though made with a sword, and the Austrian term *schneiderfuge*, literally 'tailor's cut', are European language equivalents of 'running joint'.

Avoiding running joints is deceptively easy, according to the old rhyme:

A stane abin a stane,
Yin between twa,
Pack well and pin well
And make a strong wa'.

A dyker can defend the occasional transgression by claiming the joint is crossed behind the face of the wall, as when building with rough stone. Running joints are easily produced when working with standardised blocks or bricks. In brickwork this is avoided by having strict bonding patterns, for instance Flemish Bond or Double Stretcher Garden Wall Bond.

Some ancient structures, for example the brochs of northern Scotland, have vertical joints running up through many courses. The stone may have forced this upon the builders, the structure still stands after two thousand years because they understood what they were doing and compensated adequately.

The following terms apply to joints, good or bad

Aligned joints (Kentucky) running joints.

Angled joint a diagonal running joint.

Broken joint a joint between two stones, properly covered by an upper stone, is variously termed a 'covered joint', 'crossed joint' or 'broken joint'. Murray-Wooley and Raitz (1992: 194) add 'break joint' from Kentucky and 'two-over-one joint' from New Hampshire, USA.

Butt joint the joint between two stones lying together on the same horizontal plane. Mader and Zimmermann (2011: 56) advise to 'Avoid long butt joints.' This advises against stacking stones above each other to create a running joint.

Capginya (Mallorca, Spanish Balearic Islands) retaining walls in Mallorca, and round the Mediterranean, often include an occasional vertical construct of larger squared blocks set on top of each other, for the full height of the wall. This looks like a double running joint, one up each side of the 'stack'. This acts as a strengthening feature in the retaining wall and limits the extent of a collapse, like a vertical throughband.

Crossed joint (Yorkshire, England) a joint between two stones which is properly covered by the stone above. This should not be confused with 'cross joint' which is a defect. Mader and Zimmermann (2011: 56) describe 'cross-joint' as a building fault where stones in two courses butt together producing a vertical and horizontal line which resemble a cross. The Austrian equivalent is *kreuzfuge*, which translates to 'cross joint'.

Expansion joints long vertical joints found in mortared stonework, for the total height of the wall. This allows for seasonal expansion and contraction.

Jogging (USA) like a 'zipper joint', a running joint with a minimum of overlap up through the courses.

Lightning joint (County Mayo, Ireland) A very descriptive term for a running joint, gathered during a late-night discussion in a country pub.

Plumb-joins (North England) vertical running joints.

Rip joint a running joint, so named because it looks like a rip.

Riseband / risband (southwest Scotland) a vertical running joint.

Run (Vermont, USA) a running joint. The term is also used in Hawaii, recorded in Kona Historical Society (2002: 4).

Stack bonding (New Hampshire, USA) bad building with vertical running joints running up either side of similarly sized stones placed on top of each other. The equivalent French phrase *pile d'assiettes* (literally translated as 'a pile of plates') illustrates this perfectly.

Stacked joint (Kentucky, USA) a running joint.

Straight joint (Scotland and northern England) a vertical line of weakness caused by bad bonding on succeeding courses.

Zipper joint not quite a running joint, nevertheless a weakness, because the overlap of stones on each succeeding course is minimal, looking like the overlapping teeth on a zip fastener.

2.7 Wet and dry, mortared versus dry stone

Dictionaries of the Scots Language (DSL) includes the use of the word 'dry' in a 1540 instruction to build without mortar, if no lime was available to make mortar — '…and failyene of lyme to lay drywerk in all places necessary…' Another 1593 reference mentions a '…dyke of drye stanes…'

A drystone wall is obviously 'dry' because it contains no 'wet' mortar. However, that is a simple answer. The word 'dry' in this context may have more interesting origins. An old meaning of 'to dry' was 'to cheat', according to Bailey (1675). A 'dry' is also a flaw in a stone, usually a hairline crack.

Entries in DSL lead to an interesting proposition.

Consider these terms:

'A dry house cottar' is a householder who has only cottage and garden, no land.
'Dry handed' means without weapons — unarmed.
'Dry lodgings' are lodgings with no food provided.
'Dry talk' is an agreement settled without drinking.
'A dry letter' was '…in the language of an immigrant, one without a remittance.'

In these phrases the situation be 'dry' because each lack something. A vital component,

Figure 36: Running joints and general bad bonding. Caithness, Scotland.

drink, land or food, is missing. Could a 'dry stone wall' is 'dry' because it is a stone wall without mortar? It 'cheats' by not having mortar, something which was thought essential for a stone wall.

European language terms for 'dry stone' (for example German *trockenmauer,* French *pierre séche* and Italian *muretti a secco*), directly translate to the English language term 'dry stone'. Did the reference to 'dry' walls spread from the British Isles? When did these languages first differentiate between 'dry' stonework and mortared work? We shall leave it to others to do that research.

Stone walls were an obvious solution to the demand for reliable fences round agricultural land (see Section 3.1, a brief note on the historical development of drystone walls); the cost and inconvenience of mortar made it impractical for field walls.

The 'double' stone wall developed as the epitome of good walling design. Proper foundations, throughstones, capstones and hearting were already important features of mortared walls. It seems reasonable that mortared walls provided a basic pattern for the 'dry' stone walls.

This transfer of techniques and terminology from mortared to 'dry' work is further evidenced when Lavicka (1980: Section 7, 52-53) describes the construction of stone foundation walls in North America around 1899. These recommendations, for mortared work, will be familiar to any drystone waller and echo the instructions of many eighteenth and nineteenth century advocates of drystone fencing:

All stone walls, 24 inches or less in thickness, should have at least one header extending through the wall, every three feet in height, and every 4 feet of length… These headers serve to bind the stones of the wall together, and keep [the wall] from splitting apart crosswise when weight is placed upon it… the joints of the stonework should be broken… Where a long vertical joint occurs, the weight of the wall above may cause the wall to settle more on one side of the joint than on the other, and produce serious rupture… All wall angles should be well tied by long stones laid alternately in the wall, [therefore] the weight on the corners of the wall is more equally distributed, and the wall can be kept plumb and true… stone should always be laid on their natural or quarry beds.

The conscientious worker is urged to heart the wall properly, placing each piece carefully in a bed of mortar.

Rankine (1862: 389) provides another link between mortared and dry work:

Dry stone walls should be built according to the principles… already laid down for rubble masonry… with the single exception that the mortar is to be omitted.

Figure 37: One face of this mortared gable has fallen away and exposed the centre. Boulders, broken to produce a flat face, and smaller stones, assisted with mortar, combined to form a solid wall.

Rankine is a civil engineer and relegates dry stone walls to fence-walls, backing for retaining walls (to relieve pressure and provide drainage) and low retaining walls.

Clarification is often required regarding the use of the words 'cement', 'mortar' or 'concrete'. Cement is the powder. Mortar is a mix of the powder, water and sand. Concrete is a mix of the powder, water, sand and larger aggregates.

Lime mortars were the commonly available mortar when most of Britain's drystone walls were built. These mortars are made from burnt limestone, sand and water. They require more care and attention than Portland mortars. This makes them inconvenient and expensive for use in exposed and remote sites.

Portland cement is the most commonly known cement powder. It was patented and developed in the early nineteenth century and named so because of its similarity to Portland stone from Dorset, England. Mortar made from this cement sets under water. This was an important advantage to the Victorian engineers when they built the British infrastructure of docks, railways and sewers.

Mortared walls are generally built slimmer than a dry wall, and near vertical with less batter. This suggests mortar adds strength. McCutcheon (2000: 20), includes a tantalising phrase, without confirming the source — 'Some researchers claim that mortar may only add 10% extra strength to an expertly built dry stone wall.'

Friction between the surfaces of properly placed stones is vital for the strength of both wet and dry masonry (Figures 37 and 38). Mortar gives additional advantages, but the main structural strength of a mortared wall still comes from the proper placement of individual stones.

Arbennigol (2011: 11) tells us '[mortar] …doesn't actually glue the stones together, rather it ensures forces are transferred evenly across the joints and ensures a stable fit.'

Shadmon (1989: 116) gives more detail:

> Stone is usually stronger and more durable than mortar. Well-tied and stacked stonework without mortar is more stable, and often more attractive, than a wall held together with plenty of mortar. A major advantage of mortar is that, with its use, walls have to be less massive as it acts as bonding. Mortar seals the joints between stones, enables the use of different shapes and sizes and facilitates the use of rendering. Hard mortars are vulnerable to stress in walls. Mortar for stonework is less wet than mixtures used for other purposes to prevent dribbles and grime on the finished wall.

Course notes supplied by the Scottish Lime Centre, Charleston, Fife, entitled 'Masons Mortar Guide for the Use of Traditional Mortar for the Care and Repair of Historic Buildings' further confirm these points:

> Masonry buildings rely on their mass and the interlocking of individual units for their stability, and the mortar in masonry building serves in the main to provide a bedding medium for often very irregular components, filling the voids and maintaining the wind and watertight integrity of the building. There is normally no requirement for significant structural strength in the mortar of traditional masonry buildings, particularly in re-pointing work.

Stephens (1891: 230) saw the need to employ properly trained dykers because the 'dry' practitioners fully understood the need for good stone-on-stone contact:

> We suspect that many dry-stone dykes are built by ordinary masons, who being accustomed to the use of lime-mortar, are not acquainted with the proper method of bedding loose stones on a dry dyke as firmly as they should be, and are therefore unfitted to build such

a dyke. A builder of dry-stone dykes should be trained in the business; and, with skill, will build a substantial dyke, at a moderate cost, which will stand erect for many years.

It took some time before 'dry' walls were fully accepted. Agricultural writers in the eighteenth and early nineteenth centuries debated the use of mortar, its cost and effectiveness.

Dickson (1805: 112) defines a 'list wall' as a stone wall '...made partly with a cement, and partly dry...' They appear to be boundary walls; no more construction detail is given. He goes on with a full costing for building twenty feet of list wall in Somerset, southern England. The total cost, including all materials, was 11 shillings and six pence (57½ pence). The labour cost was 3 shillings and six pence (17½ pence).

The total cost of building the same length of drystone wall, with no cement, is given as 8 shillings and three pence (41¼ pence). The labour cost is reduced to 2 shillings (10 pence).

The cost is nearly 28% cheaper if built without mortar.

Dickson (1805: 112) notes the stone type can influence whether a wall is mortared. If the stone is flat bedded, they are usually built 'dry' — '...where a flat bed of stone can be had, they are generally constructed without the use of any cement whatever, and if the work be well executed, they will continue a long time.'

GVA Perth (1794: 115) comments on the unnecessary use of lime mortar because it was vulnerable to frost damage if water got into the wall and the mortar hides bad building. The writer also thought build quality had deteriorated since men were paid according to output:

> With regard to fences built with lime it may be observed that, except where the stones waste with the weather, lime is more ornamental than useful in all stone-fences, if the dry stones be laid on their fair bed; because lime enables the workman to conceal not only a penury of stones, but also the not laying them on the bed they ought. Besides, nothing can at any moderate expence be contrived to prevent water getting down into a stone wall and thereby destroying the effects of the lime, by the expansion of the water, owing to the alternate action of frost and thaw. It must be confessed that since the building of stone-walls by the piece had become so common, it is difficult to get the dyke-builders to bed the stones fairly.

Dutton (1808: 151) refers to the rough 'single' walls in County Clare, Ireland and goes on to describe a move away from using mortar — 'Gentlemen, and more substantial farmers, usually make double dry walls, sometimes dashed with mortar, but oftener without it... '

GVA Dumfries (1812: 151) notes:

> Stone and lime walls are not only very expensive, but, unless the top be well secured, they soon draw water and moulder. For garden fences, or inclosures about houses, these are undoubtedly the best; but in common field inclosures, they are by much too expensive.

Brewster (1830: 346) accepts tenant farmers will take the cheaper option when presented with the option of a mortared or dry built wall:

> A wall built of stone and lime is undoubtedly the preferable fence; but the expence (sic)

far exceeds the value of the interest which a tenant generally has in the premises.

Low (1834: 611) thought mortar added strength and was expensive:

> The cementing of the stones with mortar, adds, indeed, to the durability of the wall, but then the expense is generally too great in common cases. The wall, therefore, for the ordinary purposes of the farm, may generally be built of stones alone, though sometimes with a little mortar merely for cementing the coping, and occasionally for pinning or closing the interstices of the outside.

Many writers around the start of the nineteenth century still advocated pointing up drystone walls with lime mortar or applying a mortar coating; they thought it added strength and kept out the rain. They were slow to accept that a well built drystone wall was far superior to the flimsy barriers associated with agricultural systems and practices which existed before the Enclosure Acts. GVA Inverness (1813: 118) suggested a drystone wall coated with lime mortar will add to its strength and longevity — 'Where lime is easily procured, the harling of stone walls, would be a saving in the end…'

Dutton (1824: 164) gives an Irish example from Galway — 'When the wall is carefully built and dashed with good mortar, and in a proper season, it makes an excellent and permanent

Figure 38: Clay mortar resists the elements in a ruined cottage. Aultbea, Wester Ross, Scotland.

fence…'

Many modern drystone purists criticise the use of mortar; it has its place.

Mortaring copestones adds security and stability, especially on a public site. If mortar is missed between occasional copes this allows the top to flex with the body of the wall.

Mortar has been used in lieu of throughstones as a course of mortar along the length of the wall or, perhaps peculiar to Slaidburn, Lancashire (where some walls are built of brittle shaley limestone), a layer of mortar over the hearting which is not visible on the face of the wall.

In areas where long stone is rare, wallends are built of smaller stone mortared together, or, as seen in Ireland, the wall end is formed from a concrete pillar. The author has seen wall ends consisting of a 45-gallon oil drum filled with stone and concrete. The metal rusts away to leave a round pillar.

GVA North Wales (1810: 125) gives details of 36,000 yards (approximately 20.5 miles or 33 kilometres) of new stone wall on the Plas Gwynn estate. This wall was pointed up and had a twelve inch high mortared cope. 'The walls are five feet high, all entirely pointed; and the coping, of one foot depth, worked with mortar.' It cost 'about £4,000'.

One principle of mortar use is vital; the mortar should always be softer than the stone and sacrifice itself in favour of the stone. Repointing of mortared stonework is an inevitable and relatively cheap repair. If mortar is too hard it traps water at the joints and the stone will erode. This is especially noticeable when hard Portland cement is used with sandstone or soft Cotswold limestone.

In modern times there is a growing appreciation of well built drystone structures, often as part of an interior. Building control restrictions and realistic expectations for cleaning and draught proofing means mortar is incorporated into the work. These mortars may be traditional lime mortars or modern Portland cement mortars.

The following mortar-related terms have been found in descriptions of drystone work

Back-bedded a stone wall which looks dry-built but has mortar at the tail of the stones, fixing it to brick or concrete block. Metal ties between the block and the stone give added security. This imitation drystone disguises, or protects, a concrete load bearing structure. The stone veneer becomes a decorative rain screen.

Clay mortar the word 'mortar' can also apply to strong clay used as a binding agent. If the clay was mixed or 'gauged' with lime, straw, or other additives, this sped up the drying time and made it more workable.

DSL includes a 1795 Perthshire reference to '…that red clay substance called mortar… from its use as a cement in building.' North Americans builders still use the word 'mud' for mortar.

McAfee (2009: 34) confirms the use of **mud mortars** — 'Mud was commonly and successfully used as a mortar. It was often protected with an external lime render to prevent wash-out by rain.'

Examples of walls built of tight stonework and clay mortar survive in the west and north of Scotland (Figure 38). Lime pointing or a full coat of lime mortar on the outside kept out the weather. This coat was further weatherproofed by annual whitewashing. If weathering of the outer faces removes the protective coatings, the internal mortar is vulnerable. If the stonework is tightly built the walls keep most of their strength until time brings about collapse.

Dry fit (USA) a term for the setting of an individual stone in place and checking the fit before finally mortaring in.

Dry stack / drystacked (USA) general terms for stone work with no mortar — drystone as we know it in the British Isles. The same two terms are used in USA for work where the stone is discretely mortared — held back from the face of the stone to give the appearance of dry build. Other terms for this are 'dry face' (USA); 'dry joints' or 'dry seams' (Kentucky); 'dry pack' (Canada).

Kenlan (1997: 8) differentiates between 'dry-laid' (building without mortar) and 'dry-stack' (where mortar is used but is invisible) and adds:

> …because there is no mortar in dry-laid work, it has the disadvantage of being permeable by air and moisture. That's why its use is usually limited to retaining walls, where the passage of water is necessary, or to simple foundations where ventilation is desirable.

Lime / lime mortar limestone, when burned, undergoes a chemical change to become quicklime. Lime can be added to cement mortars to increase workability.

Lipped-dyke a Scots term for a drystone dyke pointed up with mortar. This practice was not restricted to Scotland and was common in the late eighteenth and early nineteenth centuries.

Loudon (1825: 444) describes drystone walls with '…about two or three inches on each side lipped with lime…'

He thinks this, along with harling, offers no real advantage over the common drystone wall beyond an improvement in appearance. Harling is a mix of thin mortar and small gravel, which is thrown (hurled) on to the wall to form a weatherproof outer coat.

There is a drystone wall in Apia, Western Samoa, with a coating of cement harling. Perhaps this 'sophistication' was introduced by Robert Louis Stevenson, a member of the Scottish lighthouse engineering family who lived on Samoa from 1890 to 1894.

Mortar a mix of the powder (a binding agent such as lime or cement), water and sand.

Mud / mud mortar mortars which use clay or 'sticky' earth as a binding agent.

Slobbered masonry Ribble Valley Borough Council (Slaidburn Conservation Area Management Guidance) public consultation document for December 2006 includes a definition of this Lancashire, England, style:

> Local traditions such as 'slobbering' ie, the uneven rendering of a rubblestone wall surface are part of the vernacular tradition of building and should be respected and therefore not painted or removed.

Sneck wi' lime the Scots verb 'sneck' means to close an opening or gap. The term 'to sneck wi' lime' is synonymous with 'to teeth with lime'.

GVA Aberdeenshire (1811: 199) includes instructions for the stone and lime walls of the houses to be '…sneck-pined with lime…' The gaps between the larger stones were pointed up and small stones pinned into the mortar. The small stones help keep the outer face of the stonework waterproof by reducing the amount of mortar in direct contact with the weather.

Teething with lime a Scots term for the process of pointing up a drystone wall with lime mortar. DSL, under 'teethe', gives a definition: 'to point a wall, slates, etc., with lime, giving an appearance like a row of teeth' — presumably because the stones stick out from the mortar in pointed rows.

DSL goes on to include a quotation from 1782 — 'All the walls have been built of mortar (i.e. mud), and only a very small teething or pointing of lime on the outside.'

Figure 39: Langdale, Cumbria, England. An area long associated with good agricultural practice.

3. THE STRUCTURE OF A DRYSTONE WALL

David Kerr Cameron (1978: 154) describes the building of the roads and roadside dykes in Scotland in the late nineteenth century. These dykes used a basic pattern, one which was straightforward to build, and flexible enough to allow for a wide variety of terrain and stone type:

> …dykes, mile upon mile, were built to march over hill and through haugh, lasting memorials to a hardy race of men who could translate stone in the singular into a plurality that was stock-proof, shock-proof and wind-proof. They would weather, those stone dykes with the farm steadings, till they achieved a great venerability, till you could think they had stood there from the first day of Creation.

3.1 A brief note on the historical development of drystone walls

Drystone structures have been built for thousands of years, all around the world. Skara Brae, in the Scottish Orkney Islands is a familiar example dating back 5,000 years. The Céide fields in County Mayo, Ireland, are part of an agricultural system dating back 3,000 – 5,000 years. Svizzero and Tisdell (2018) describe 'desert kites' in the Middle East. These long low walls (with associated enclosures, most likely used for hunting or gathering animals) may date back 6,000 to 7,000 years.

The drystone walls we see in the British Isles today are a relatively new addition to the landscape. The agricultural system in Britain did not require many fences prior to reform of agriculture in the seventeenth century. Stone walls, as we understand them, were comparatively expensive to build and restricted to protecting the extensive sheep and cattle pastures and hunting grounds associated with monastic or regal properties or manor houses. Boundaries round cultivated areas had less finesse, often simple lines of large stones set on edge and filled out with smaller stone or, a later development, tall, wide structures made of tight fitted small stone, and no mortar. Stone-faced earth banks were also common. These old boundaries were laissez-faire, they followed the lie of the land, seeking the easiest line and the most convenient source of building material.

In simple, general, terms, the ordinary tenant's land rights, prior to agricultural reform, were typically divided between a small area of cultivated ground and a share in the more extensive common grazing on rough pasture or moorland. A head wall or head dyke marked the boundary and kept the livestock out of the growing crops. Animals were allowed into the fields in the winter. These walls were ephemeral structures built of stone, turf or hedge in various combinations, with no expectation of them surviving more than a season or two.

The fences by the houses were very simple rows of earth and stone or woven hazel or willow.

McAfee (2011), Wildgoose (1991), Lord (2004) and Lord (2010) provide evidence of agricultural drystone walls in Britain and Ireland dating from before medieval times.

The walls associated with one Cistercian Abbey's sheep pastures were a definite improvement, as described by Lord (2010):

> It appears dry stone walls were being built widely in the Yorkshire Dales by the thirteenth century. The earliest walls are technically complex structures, and possibly represent an agricultural innovation brought in from continental Europe after the Norman Conquest. In the Yorkshire Dales dry stone wall construction continues more or less unchanged until the sixteenth century when new styles appear, and it is basically forms developed from those latter styles which have continued to be built to the present day.

Lord goes on to describe a typical medieval wall with a base width of 0.65 metres (26 inches), 1.6 metres (5 feet, 3 inches) high and 0.5 metres (19.68 inches) wide at the top with 'Projecting top stone forming [a] continuous lip: to prevent wolves jumping the wall.'

Thousands of miles of fencing were required to segregate livestock and crops following the Enclosure Acts (1604-1914) for England and Wales and subsequent improvements in agricultural practices throughout the British Isles. The main aim of enclosure was to reallocate the arable ground and bring moorland into production. Landowners made considerable fortunes from wool, grain and beef. These fences were built as efficiently, and cheaply, as possible. Hedges had their advocates though waiting for them to grow was not always practical on high moorland. The amount of lime mortar required, and the difficulty of hauling it to site, and managing it, meant mortaring the walls was unfeasible. Section 2.7 covers this topic in more detail.

Rainsford-Hannay (1957: 23) describes the development of drystone dyking in southwest Scotland:

> Dry stane dyking, in a comprehensive way, was first undertaken on the lands of Palgown in the west the lands of the Stewartry of Kirkcudbright. Somewhere about 1710, when the early Enclosure Acts were passed, the brothers McKie… leased pieces of land free to people who would work for them in the summer.

Doubtless similar innovations appeared in the north of England. Tenants often received rent concessions in exchange for clearing the land and building drystone dykes. After a few years, the rent for the improved land would be increased. If this was too high the tenants flitted to another piece of rough ground and improved that. The cycle repeated over generations; these tenants had little chance of prospering from their own hard work.

Alexander (1992: 85,86) describes how an Aberdeenshire tenant farmer substantially improved his land between 1805 and 1881 with little personal effort:

> …with the exception of limited help given in building dykes [twopence halfpenny per yard] – which added to value and became his own property – the laird … did not expend a single farthing on the farm from 1805 up to 1881; and yet the rent was twice doubled, and fifty per cent added in course of less than sixty years, largely through the skill, energy, and direct expenditure of an improving tenant, who reclaimed seventy acres, and built a dwelling-house and steading at his own charge.

Charles Fothergill, in Romney (1984: 113), describes drystone walling in 1805 Yorkshire:

> Stone walls without morter [sic] are the usual fences in this country: they are stronger and of longer duration than superficial observers would imagine… no landlord would be willing to let his land to any farmer who could not make up in a workmanlike manner any gap or breach that might accidentally occur in the fences [otherwise] there would be a necessity for new fences every few years.

The new boundary dykes were straighter; land was valuable. A declaration of ownership was essential to replace the traditional boundaries of an old tree, a ditch, a series of rocks, or a stream. Many older, rougher, walls were recycled into the improved field divisions. Demand for dykers' services increased. Their output was a highly visible manifestation of an agricultural revolution.

GVA Dumfries (1794) and GVA Galloway (1813) comment on the current state of enclosure, clearly a time of evolution and expansion; field boundaries existed where none existed before.

Wildgoose (1991), McAfee (2011), Robson (2004) and Winchester (2016) detail the progressive improvement of stone wall types in Derbyshire, Ireland, the Scottish Borders and northern England.

British agricultural writers in the eighteenth and nineteenth centuries were nearly unanimous in their praise of the new standardised drystone wall, with throughstones and cope, developed and finessed in the 1700s. These writers had an audience in North America, Australia and New Zealand. The oceans were no barrier to exchanges of information. This 'double dyke' was a more slender structure, easier, and quicker, to build than the older ridges of earth, stone-faced embankments or stone and hedge combinations. Drystone walls were immediately serviceable, unlike a hedge which took years to mature. Figure 40 illustrates the simple complexity of the 'double dyke'.

Winchester (2016; 45) accounts for the standardisation:

> By 1800, therefore, the evidence points towards an increasing uniformity in walling styles, as preferred practice (or fashion) spread, fostered by the specifications laid down in enclosure awards, the interchange of agricultural ideas among landowners, and the accumulated skill and knowledge of an increasing army of professional wallers.

The standardisation of the drystone wall (although the final structure ultimately depends on the available material) may have a parallel in the standardisation of sheep shearing in New Zealand in the 1940s and 1950s. Godfrey and Ivan Bowen developed the Bowen Technique by combining the best from individual shearers. They standardised a complex procedure; this increased output, using less effort. Their method was logical and teachable. Shearers no longer learned through trial and error. The quality of the shorn fleece improved; farmers and shearers prospered.

Compare this to drystone dyking. The early chroniclers of drystone field walls had little personal experience of manual labour (they came from the clergy or professional classes), but they repeated tried and tested methods and encouraged a standardised process, to produce a standardised outcome. Adaption of these practices elevated the dyker from labourer to skilled craftsman.

The New Zealand National Film Unit presentation 'Shearing Technique' (1957) refers to the need for strength, stamina, balance, grace, rhythm, a good eye and the need to shear '...with your brain as well as with your hands.' This has echoes in descriptions of drystone walling. Both jobs are physically demanding. Teachable techniques reduce the physical effort.

Shearers and dykers had a reputation for working hard and playing hard. Many legends were born on the shearing floor, and on steep hillsides.

Drystone walls spread throughout the British Isles remarkably quickly (Figure 137). Rainsford-Hannay (1957; 20) imagines an educated man who has travelled from southern England to Perth in central Scotland and how that person would have seen '...a bewildering pattern of hedges, stone walls, roads, lanes and woods... How many of us know that this irregular patchwork of hedges and walls did not exist before 1760 north of the Thames, and that the work was fairly complete only by 1850.'

According to Rainsford-Hannay, southern England, south of the river Thames, had been fenced with walls and hedges before 1760 but '...most of England and Scotland was open country.'

Incidentally, Murray-Wooley and Raitz (1992: 106) reference this with the more definite: 'Stone walls did not exist before 1760 north of the Thames...'

Winchester (2016: 39) writes: 'The majority of dry stone walls in Britain date from the 'age of improvement', the century from c.1750 to c.1860...'

He further advises that the improvement of land and consolidation into larger holdings had '...begun in south-west Scotland by the 1720.' This improvement continued throughout Scotland until it reached northwest Sutherland in the 1870s.

In Europe there was a comparable era of drystone activity. Löbbecke (2012: 10, 35) tells us the '... technique of corbelled dome buildings dates back to the fifth millennium BC...' but application to smaller drystone construction is a lot more recent:

> It has been consistently shown in the various countries of southern Europe that the corbelled dome huts still visible today were not built until the eighteenth century; only in very isolated cases can one be dated a little earlier. The height of their dissemination was in the nineteenth century and they were seldom built in the twentieth century.

Christian Lassure (2004) gives a French perspective, with a similar timescale, in an article which, in its full form, challenges the accepted importance of drystone structures in cultural heritage:

> Prior to the incentives to land clearing offered by the Monarchy in the 2nd half of the 18th century and the division of common land in the first third of the 19th century, dry stone structures were unknown in many a territory, and forest prevailed in uncultivated margins... The emergence, life and desertion of dry stone structures cover an unsignificant (sic) and recent portion — approximately a century— of a rural history that spans several... In the rural society that bore them, dry stone agricultural structures were considered as nothing more than production tools... Once a plot had been deserted for lack of profitability, its dry stone fixtures stopped being maintained as they had reached the end of their useful life. This explains why today so many areas with dry stone vestiges can be seen falling into

disuse and being overrun by vegetation, with their rightful owners refusing to lift a finger.

The six states of New England, USA, are famous for their stone walls. Like Europe, the bulk of their drystone network is a relatively recent addition to the landscape. Russell (1976) and Thorson (2002) detail the development of agriculture, and enclosure, in New England. This is a fascinating cycle from colonial forest to eighteenth/nineteenth century farmland and back to twentieth century forest cover, as the fortunes of agriculture flourished and waned.

The early colonists found a forested land, with rocky ground, especially inland from the deep soils near the coast. Timber was the obvious choice for fences and buildings. After deforestation the stone was pushed to the surface by frost and erosion and cleared to the edges of cultivation. The landscape became a patchwork of small fields surrounded by piles of stone, some built into rough fences incorporating timber posts and rails.

It is reasonable to assume that the landscape also included stone structures built by native peoples, going back many centuries. Ancient walls, associated with the control of livestock and hunting, are found in other continents.

European-style agriculture had been practiced since the earliest colonists to North America. Russell (1976: 9) includes a contemporary reference to 'Rootes and Garden Hearbes and some old Walls', at the ruins of the failed Popham colony (1607-08) at the mouth of the Kennebec River in Maine. The extent of the walling is not given. It may have been a small garden area or a more extensive field system.

It is reasonable to assume the Popham walls were built, at least partly, of stone. These early settlers brought social structures, farming practices and seed from Britain. They might have been familiar with the medieval drystone walls protecting monastery grounds in England and may have used this knowledge to build stone walls to protect crops and control livestock.

Thorson (2002: 8), gives a timescale for enclosure in the period between the 1780s and the 1830s:

> ...there is a mistaken impression that stone walls are primarily a colonial phenomenon. They are not. Although walls were being built from the time of the first settlement to the end of the pioneering stage, most were built in the half century between the end of the American Revolution and the construction of the first railroads.

In the early 1800s New England farming modernised, including revision of the fences, as described by Thorson (2002: 141):

> ... in general, the primitive tossed walls and stone accumulations of the preceding epoch were either built up and levelled off or disassembled and rebuilt into their well-ordered early- to mid-nineteenth-century counterparts.

These were low, wide walls, built with two outer faces, with little or no batter, infilled with smaller stone, and flat capstones. Figure 50 shows an example from Rhode Island. They did not have the vertical copestones associated with the standardised 'double' drystone wall (Figure 40), developed in the British Isles during the eighteenth and nineteenth centuries.

The boom in sheep farming in New England in the first half of the 1800s promoted pasture improvement and the building of stone walls, just as it did, in the same period, in the British Isles, particularly Scotland.

Murray-Wooley and Raitz (1992: 84-85) illustrate the flow of 1700s and 1800s migration from the British Isles to northeast USA and onwards. The migrants carried their skills to Kentucky, where the British style walls were adopted, and adapted, by landowners who had a '...desire to emulate English estates...'

Interestingly, the Kentucky walls were, very generally, built from quarried limestone or stone from creek beds, along highways and as new field boundaries. The New England walls were, very generally, built from glacial field clearings already laid out along the edges of cultivation.

Most regions of the world have stone walls, in some form or other, for agriculture, ceremony or shelter. Indigenous populations had built stone structures for ages before European colonisation.

One of these ancient landscapes is a 3,900 square mile sector of the Mpumalanga Escarpment in South Africa. This stonework is entirely separate from the famous ruins at Great Zimbabwe. Remains of buildings and walls indicate complex systems which retained slopes, protected crops and improved irrigation. According to Delius et al (2001), this was at its peak between 1500 and 1820 — well before the arrival of colonists in the nineteenth century – and predating the main thrust of the Agricultural Revolution in Britain. Delius et al (2001: 144) goes further:

> The walls stand in mute but eloquent reproach to the host of commentators who have suggested that prior to the arrival of settler farming African agriculture was rudimentary, subsistence oriented, transient and barely capitalised.

These African walls, predating any European influence, confirm drystone construction is instinctive, flexible, and universal.

Drystone walls exist on every continent, including Antarctica. Shackleton (1919) records how, in April 1916, he left the men of his wrecked Imperial Trans-Antarctic Expedition on Elephant Island when he sailed to South Georgia for help. The men built stone walls, four feet high, as a base for their upturned boats, for shelter, until they were rescued four months later.

3.2 The structural components of a drystone wall

The components described in the following sections refer mainly to the basic 'double dyke' (Figure 40) but apply to most variations because, as pointed out by Winchester (2016: 8), '... walls which appear very different at first sight may be structurally almost identical.'

A freestanding drystone dyke is a complex structure. Not all stones are the same; some have been carefully selected to perform a specific function. A dyke is a cooperation between different components, eloquently summed up by Rainsford-Hannay (1957: 35) in a quote from an 1865 specification as '...every stone doing its duty by its neighbour.'

This cooperation is equally important whether it is a mortared wall or a drystone wall. There must be as much stone-on-stone contact as possible, that's where the main strength comes from. Many old cottage ruins stand long after the external roughcast coats and the mortar has been weathered away (Figure 38). Some rebuilding, repointing and a new roof would take such a ruin well down the road to restoration.

Patrick McAfee listed seven components in the 'International Freestanding Wall' at his training centre in Dublin:

1. Foundations
2. Quoin stones for wall ends, openings and corners
3. Face stones, those which form the outer faces of the wall
4. Hearting
5. Through stones
6. Cover stones
7. Coping stones

Each component has a specific task. Some walls do not have every component because walling styles are dictated by the local stone and how it can be used. A skilled waller understands the material and can make something out of any rock type. It is almost certain individual dykers perfected a style of wall which was restricted to one or two parishes.

Nineteen regional styles of wall are detailed in the booklet 'Dry Stone Walls, the National Collection', published by the Dry Stone Walling Association of Great Britain (DSWA) in 2002. Only one (from southeast Scotland) has a coverband. Only seven have obvious protruding throughbands. Long bond stones and throughstones, flush with the face of the wall, may be in others.

All 'double' walls in the booklet have copestones; a variation from the Isle of Skye has a turf cope. Doubtless other local styles exist, variations of those detailed in the DSWA booklet.

Local geology influences local walling styles. Upland ground is generally full of shattered boulders and rocky crags; lowland ground, by

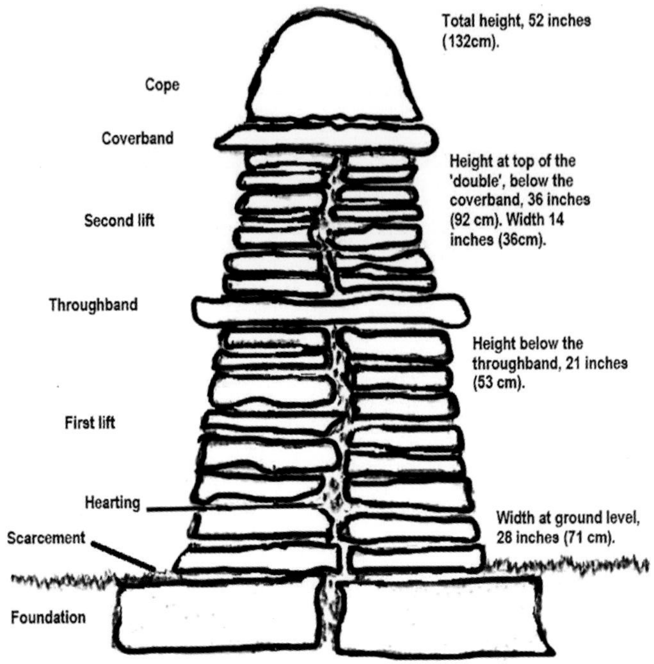

Figure 40: A cross-section of a typical freestanding 'double' dyke. Figure 82 is a good example of a wall end, and wall, showing all these features.

contrast, has more sedimentary rock and rounded glacial deposits. The hill farms of the Pennines and southwest Scotland developed a style of walling more rugged than that in the pastures of southern England where low limestone walls or hedges are the most common field boundary.

A freestanding 'double' wall consists of two outer faces with smaller stones in the middle and occasional stones laid across the width to add strength. A freestanding wall is typically a boundary between two fields, so it has two sides. A retaining wall is typically built against an embankment and can be represented by Figure 40 split down the middle — one face to the outside and a rougher face against the embankment.

In areas where the stone is too rough or too big, 'single' walls prevail. These are one stone wide, made from large boulders, with smaller boulders placed on each successive course.

Garner (1984: 1) is not condemning the walls in the Mendip area of Somerset, England when he describes them as '…irregular stone, unrefined technique and no separate cope.'

He recognises the skill of the local wallers who managed to produce a stockproof wall, despite the poor quality of the local stone.

Collapses are frequently caused by a slipped foundation; the site must be cleared to get at the original fault. 'Stripping out' is the controlled dismantling of a fallen length of wall in preparation for the rebuild. Copestones, coverbands, throughstones, face stones, foundation stones and hearting are carefully separated and set aside for their specific place in the rebuilt wall. If there is a shortage of any component the modern dyker can bring it to site with little effort.

To 'redd out' or 'rid out' is a Scots term for the preparation of the day's work. In a dyker's case this might be the dismantling of a collapsed section and sorting the stone ready to rebuild. At the end of his ten-hour day a Scots dyker will 'redd up' — tidy up the site and put tools safely away.

3.2 (i) Foundation and scarcement

A strong foundation is essential to support the wall (Figures 41 and 42). Most structural failure in drystone walls is caused by a badly built foundation which has slipped or sunk.

After deciding the line of a new wall, the next stage is to gather the stone and lay it out. The biggest stones form the base of the wall, they are best laid nearest to the proposed foundation, leaving a couple of yards clear, for workspace, on each side. Setting the lines and frames confirms the edges of the stonework (Figure 19, 43 and 61).

Preparing the ground for a foundation is usually straightforward – simply strip off the grass and dig down a few inches to firm ground.

Nineteenth-century wallers recognised the importance of a good foundation and, according to Raistrick (1973: 17), talked of 'setting a wall on its feet' and 'giving it summat to stand on.'

McAfee (2009: 201) tells us a building should have a 'good hat and a good pair of boots.' It should have a sound roof, a good foundation and a proper system for management of rainwater. This applies equally to a drystone wall. A good cope and a solid foundation go a long way to guaranteeing a strong structure.

Foundation stones are normally the largest, laid flat with their longest edge into the centre of the wall, tightly together with maximum contact between individual stones. Inevitably there are variations, human inventiveness and geology being what they are.

Vertically set foundation stones are recorded in Northumberland, near Hadrian's Wall. The stones are thin slabs twelve inches high, laid on edge to provide a stable base where the wall crosses soft ground.

Shadmon (1989: 106) includes an illustration with the caption: 'Well-built foundations… should slope towards the centre.'

This does make structural sense. Flat foundation stones, or rounded boulders, slightly sloping down towards the centre of a wall, are less likely to slip outwards once pressure is put on them from above.

Stephens (1853: 589) recommends men working in pairs, one each side of the wall, because they are '…the best number to make the best work, and they assist each other with stones which one would not be able to manage.'

He goes on to advise, for a two-man team on undulating ground, 15 yards (13.716 metres) between the frames is a sufficient stretch of building to take on at one time. On level ground this may be stretched to 30 yards (27.432 metres). This makes sure the wall stays a consistent height and reasonably parallel with ground.

The following terms apply to the foundations of a drystone wall

Aggar an embankment of earth or stone, upon which the Romans built their roads, typically four to five feet high and 45 feet (13.716 metres) wide.

Apron rock (Kentucky) the below-ground part of the foundation.

Base (Kentucky) foundation.

Basecourse Pride (1996: 14) defines this as '…the lowest (exposed) course of a stone wall (usually projecting), a plinth.' This fits the description of a 'scarcement'.

Bearing medium the ground surface which bears, or supports, the foundation.

Bedding (Kentucky) the foundation.

Bedrock walls (New England) Hubbell (2006: 41) identifies these as walls which cross solid bedrock. There is no danger of disruption from frost heave though the wall could slip, if the bedrock is on a slope. Iron pegs in the bedrock created something the foundation stones could grip on. This helped stabilise the structure.

Bôn clawdd (W) the bottom of a wall.

Box an old Scots term for the area dug out for the foundation of a road. This trench was filled with stone.

Butters MacWeeney & Conniff (1998: 60) describe the construction of massive 'single' walls in Ireland and how 'The bottom row of butt-stones, or 'butters', may be a yard thick and up to two yards long…' These large foundation stones were the footings for walls ten or eleven feet high, built with hydraulic assistance from a JCB tractor.

Cerrig sylfaen (W) foundation stones.

Clach steidhe (SG) foundation stone.

Cutting the found (Upper Annandale, Scottish borders) Prevost (1957: 98) considers this dubious practice:

> Reducing the width of the foundation of a stone-dyke when rebuilding is known as 'cutting the found.' This has the effect of 'thinning the dyke', a procedure viewed with disfavour by conscientious dykers.

Trimming the dimensions of a dyke speeds up the rebuild, but the structure is likely weaker.

Foot stones (Dumfriesshire, southwest Scotland) large stones for the foundation, from Borchgrevink (1982: 89).

Footings this term is used as a general description of the foundation. Callander (1986: 45-46), is more specific. He defines these as long flat stones used to even up the height of a boulder foundation before starting to place the face stones. Callander writes about the granite drystone dykes of Deeside, Aberdeenshire, Scotland, where squared stones are built in level courses. It is important to start by levelling off an uneven foundation to create a reasonably flat base for the squared stones.

Found / foond the foundation, the stance of the dyke. Cairns (1975: 11) writes:

> If it is built on a very hard rocky hill the foundation ('foond') is inclined to scatter in places if the slightest fault is made in laying the said foond. Of course, this doesn't happen for many years. If on the other hand the foond is soft… then the dyke will swing from side to side, depending on the soft ground.

Found / foon (northeast Scotland) a ring of stones upon which haystacks were built, the foundation of the stack. This kept the hay off the ground, up from the damp.

Found stane (Scots) a foundation stone.

Foundation the first layer of large stones laid directly on the ground. No great depth is necessary for a normal dyke foundation; cutting through the turf and a few inches of soil to get down to firm material is usually enough. Strictly speaking, the foundation will be the portion of the wall below ground level. The wall is that part built on top of the foundation. The roughest stone can be used in the foundation and the earth dug out to accommodate any awkward shapes.

Founding pint (Orkney, Scotland) a drink, with bread and cheese, traditionally given to workman after the laying of the foundations of a building, to bring good luck. This ceremony, with local variations, is still performed in many parts of the world.

Frost heave water expands by approximately 9% when it turns to ice. Anything sitting on, or in, wet ground could be affected by that expansion when the water freezes and pushes up and/or sideways. Heave is especially notable in the climates of Canada, New England and northern Europe. The effects could be one big event or, more likely, a series of smaller freeze/thaw events over several winters.

As far as drystone walls are concerned, one remedy is to dig foundations down, to below the frost line. This will still leave the wall vulnerable to heave pressures from the side. A less expensive option is to build the wall in a shallow trench with a few inches of well compacted free-draining gravel to take water away from below the foundation stones.

Frank Lloyd Wright (1970: 141) successfully developed a method of building frost-resistant foundations for his houses, a system he used from the early 1900s:

> One of the best foundations I know of, suitable to many places (particularly to frost regions), was devised by the old Welsh stonemason who put the foundations in for buildings now used by Taliesin North. Instead of digging down three and a half feet or four feet below the frost line, as was standard practice in Wisconsin, not only terribly expensive but rendering capillary attraction a threat to the upper wall, he dug shallow trenches about sixteen inches deep and slightly pitched them to a drain. These trenches he filled with broken stone about the size of your fist. Broken stone does not clog up, and provides the drainage beneath the wall that saves it from being lifted by the frost. I have called it the 'dry wall footing,' because if the wall stayed dry the frost could not affect it. In a region of deep cold, to keep a building from moving it is necessary to get all water (or moisture) from underneath it. If there is no water there to freeze, the foundation cannot be lifted.

Mader and Zimmermann (2011: 56) describe a German foundation:

> Like traditional vineyard walls, the dry-placed walls of the Stuttgart School followed certain rules: a foundation depth of 8 – 16 inches (20 – 40 cm) when they are built on undisturbed subsoil. The foundation is built on a gravel frost barrier with large-size foundation stones.

The Stuttgart School was a style of stonework, developed in Killesberg, near Stuttgart, Germany, which influenced landscaping until the 1960s.

Todd (1859: 462) describes the use of 'bolders' in wall foundations in New York State and the need for consistency in the building material to avoid frost heave damage:

> And when the foundation is a part small bolders and another part large flat stones, and another portion something else, the wall will not all settle alike, and the frost will not lift it all alike, and by constant lifting and letting down, here and there, the wall will begin to spread out in places, and in a few years, will tumble down.

Deane (1797: 106) considers the different soil types and precautions for frost when building in New England:

> On a hard, sandy, or gravelly bottom, if built with good stones, a wall will stand well on any soil, clay and mire only excepted. On a clay soil it will stand if the foundation be laid in a trench, near as low as the earth commonly freezes in winter.

This appears to be advice for boulder walls because Deane goes on to recommend less frost heave precautions for flatter stone — 'But a wall of flat or square shaped stones, will stand tolerably well on any soil, laid only on the surface.'

British wallers and dykers express surprise at the North American emphasis on frost-proof foundations; early accounts of British walling also mention the dangers of freezing.

GVA Fife, Scotland, (1800: 136) recommends 'The dike should be founded so far below the surface, as to be beyond the reach of the frost…'

Loudon (1825: 443), also writes from a British viewpoint:

> In the construction of walls, it is essential... that they have as dry and deep a foundation as possible, in order to guard against frosts... from nine to twelve inches is the smallest depth that it should be below the common surface, especially if the soil is open and porous...

In GVA Dumfries (1794: 48-50) Bryce Johnston recognised the freeze and thaw cycle was the reason many dykes failed, and thought damage could be avoided — 'These dykes, well built with a sufficient quantity of throughbands, stand for many years, if they are founded on a soil that lies on a sandy or sandy-gravel bed.'

Dykes built on sandy soil or gravel are free draining, the ground will be drier and therefore less affected when water freezes. Johnston proposes how to solve the problem of frost heave:

1. The line of the longest dykes of the inclosure must run from south to north thus giving maximum exposure to the sun with no build-up of frozen ground on a sheltered north side.
2. The foundations should be dug into the ground, but only nine inches deep to keep the expense moderate.
3. The earth should be laid up for nine inches on each side of the wall. This gives the foundation stones a maximum depth of 18 inches (9 inches into the ground and 9 inches soil banked against the stonework). Johnston notes frost in the county very seldom gets below 18 inches.
4. Stones should be laid length into the wall, to put their centre of balance inside the wall. This reduces the risk of stones moving. If they do heave up, they will easily drop back with the thaw.

Todd (1859: 466), referring to walling in New York State, repeats concern about frost and how to avoid damage and advocates not banking earth up against the wall because the soil will find its way into the wall and cause damage with years of freeze/ thaw cycle.

Hubbell (2006: 96) also recognises the alignment of New England walls will affect their resistance to frost heave damage. He proposes that walls which run east to west suffered damage because the north side of the wall is sheltered from the sun and remains frozen. The south side of the wall gets more heat from the sun and is subject to more freeze-thaw cycles. Frozen solid ground on one side and soft on the other eventually causes instability and leads to collapse.

'Frost polygons' are a phenomenon, recorded in Arctic regions, whereby frost heave produces a landscape full of rounded shapes with small stones on the outer edge. These patterns can look like a series of collapsed walls, if viewed from high above. On the ground the landscape looks like a scree.

Fundation a Scots form of the English word 'foundation'.

Girt muckle (Shetland) a large stone, set upright in the foundation. This is often the easiest way to use large boulders.

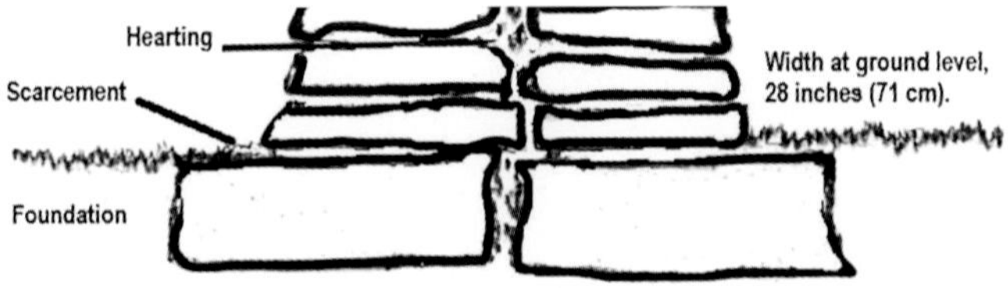

Figure 41: These foundation stones extend beyond the first course of the 'double' to form a 'scarcement'.

Ground line the line on the ground the wall will follow.

Ground-wa-stane/grund-wa-stane (Scots) this translates to 'ground-wall-stone' — a foundation stone.

Grounder (southwest England) a large stone used in the bottom course of a Cornish hedge, a type of stone-faced earth bank.

Grundstane (Shetland) literally 'ground stone', a foundation stone.

Lift the point where a dyke comes above ground, where the first course of face stones is laid on the foundation. See Section 3.2 (ii), 'the double', for a fuller description of this term.

Niho (Hawaii) literally translated as 'tooth', a base or foundation stone.

Nose bleed Patrick McAfee, Irish mason and writer, uses this term for working low down at boot level where a lot of bending is required. Any builder is happy to get away from the foundation, up where the work is drier and cleaner.

Pressure the footprint of a stone wall is remarkably light. If granite weighs a pound for every piece one square inch by 12 inches high, a wall four foot high (48 inches) exerts four pounds of pressure per square inch. This is 576 pounds per square foot (12 inches x 12 inches = 144 square inches x 4 pounds per square inch). In summary — a four foot high granite wall will exert 576 pounds per square foot pressure on the ground, equivalent to 0.28 Bar or 2812 kilogram-force per square metre.

This is a lot less than a stationary adult human male though the comparison is not a fair one. The human is standing on two legs, with a smaller surface area touching the ground than the foundation of a drystone wall. The person is also taller. Readers can amuse themselves by calculating ground pressure, using the simple formula – ground pressure equals weight divided by area of the base, whether it is a boot or flat-laid stone.

The point of the calculation is to prove that a good foundation spreads the load.

Sacrament this is a variation of 'scarcement', included by Callander (1986: 86) in his glossary. He defines this as 'extra width at the base of a dyke, which is the difference between the width of the foundations and the width above ground, normally about 2 inches.'

Scarcement / scarsement the measurement between the outer edge of the foundation stones and the outer edge of the first course of building stones sitting on the foundation, typically two or three inches (50-75 millimetres). This 'platform' strengthens the foundation and provides a more stable base for the wall. Scarcements are not common where the rock is mostly rough rubble or boulders because such a finessed edge is difficult to produce with that material.

In the inside of many sheep pens the foundation has a scarcement of six inches (100 mm) to prevent the undermining of the wall by sheep as they mill around in the confined space. Corangamite (1995: 59) details how a deep foundation and a 'rock apron' below ground prevented rabbits from burrowing under a wall.

This protection for the lower sections can go up the wall as well as along the ground. Munday (2012: 23) includes an illustration of a sheep yard in South Australia '...with large

Figure 42: This Kansas limestone foundation, after more hearting, will be a solid base for 'the double'. Note the two guide lines.

flagstones on end placed along the inside of the wall at ground level, presumably to prevent sheep from dislodging the lower structural stones when large numbers [of sheep] were penned up.'

If a dyke runs across a slope some early writers advocated the scarcement should be mainly on the downhill side, at least 4 inches (100 millimetres), with less need for it on the uphill side. This should provide a more secure base for the dyke. GVA Galloway (1813: 86) covers this point — 'If they are built on a declivity, most of the scarcement, or the whole of it is on the lower side of the dyke…'

Scraw (Mourne, Ireland) the first layer of grass and soil, removed before the foundation is laid for a 'single' wall. See Figure 80 for a fuller description of this Mourne style.

Sidlins / sidlens (Scottish Borders) a slope, a hillside. If a dyke foundation is built up or down a steep slope it should be stepped along the length, like a stairway, by cutting into the ground. This holds the large foundation stones more securely and allows stones to lie truly horizontal.

Skin to pare off the surface layer of vegetation and soil, the initial site preparation for a foundation.

Murray-Wooley and Raitz (1992: 26) describe the construction of 'a plantation fence', an older style drystone fences in Kentucky, built 1770 to 1850, which:

> …do not have a foundation but commence at ground level with only the sod layer removed. One mason cautions that these fences should only be built during a 'light [new] moon'; if laid 'during a dark [full] moon it will sink in the ground and fall down.' …Here, a new moon is waxing or becoming lighter, and a full moon is waning or becoming darker.

Soft ground the wise man never builds his house on sand and the dyker tries to avoid

Figure 43: Huge boulders form a base for this drystone wall in the Langdales. Cumbria, England, 1953. © Illustrated London News Ltd/ Mary Evans.

building on soft ground. Some written sources see no trouble with this, accepting there will be additional maintenance in the future, if the wall sinks.

Stephens (1853: 588-589) recommends removing loose soil and vegetable matter on soft ground but thick turf can be built directly on top of:

> When the surface consists of old firm thick sward, the dyke may be founded upon it; but in forming foundations, it should be bourne in mind that dykes are apt to sink in *soft* earth of every kind, to its injury – not merely by curtailing its height as a fence, but by twisting its structure and causing it ultimately to fall. When the soil consists of vegetable *mould*, it should therefore be removed altogether, and its intrinsic value as a compost will amply repay the trouble of removing it.

GVA Hebrides (1811: 170) had a similar opinion. Dry ground is considered less stable than wet ground because wet ground is level and compacted:

> …we have found that, contrary to general expectation, the inclosures built upon moss or moist ground, hold out very well, while those which were founded upon dry ground frequently give way. The reason, however, was obvious to any accurate inquirer. The moist or mossy ground is generally level, and the pressure and weight of the materials are consequently equitable and even. On dry land, not infrequently too of a loose sandy texture, the dyke is often built upon a slope…

In more recent times Borchgrevink (1982: 87-88) quotes Charlie Jardine, dyker, who recommends not breaking the strong fibrous surface of damp mossy ground on moorland, because the vegetation will support the dyke, though it will sink over years. When building in lowland fields Jardine recommends removing the turf to a depth of four or five inches (100-125 millimetres) for a solid foundation. He also recommends rebuilding existing dykes on old foundations, even if they have moved, because the foundation stones will have sunk as far as they can go. There is no need to haul out and reset them unless they are lying at a very awkward angle or totally out of line.

Goulder (2018: 24) tells how he filled a boggy trench with brushwood to provide the firm base for a dyke's foundation. He based this solution on a recollection that sections of the Settle to Carlisle railway line were built on top of tons of brushwood pressed into the peat.

The wood will not rot while submerged in the acidic bog.

Trautwine (1937: 582) gives detail of a stone bridge foundation 'built upon a bed of gravel only about two feet thick, overlying deep mud so soft that an iron bar 40 feet long sank to the head by its own weight.' One of the bridge piers sank and was restored. Trautwine notes this 'illustrates the bearing power of a thick layer of well-compacted gravel.'

Steethe (Caithness and Orkney, Scotland) the foundation for a building hence 'a steethe o' stanes'. A 'steethe-stane' is a foundation stone. The Scots Gaelic word for foundation is *steidh*. Note the close connection between 'steethe' and *steidh* which sound similar, despite different spellings. Caithness has some Gaelic heritage but, like Orkney and Shetland, Norse influences predominate.

Stonner (Caithness, Scotland) according to DSL this is 'A stone foundation; a hard pan below the top soil ...prob[ably] variant of Stanner, altered to conform to *stone*.'

'Stanner' is defined as '...shingle, the rounded water-worn stones, pebbles and gravel in the bed or on the banks of a river or on a sea beach...' This material is naturally compacted, an ideal base for a dyke.

Underputted / underpitted (Ayrshire, Scotland) a description of a wall which is undermined, without a firm foundation or base.

Width of the foundation a typical drystone wall is 24 to 32 inches wide at grass level, depending on the stone and the height of the wall. Section 3.3 covers the dimensions and proportions of a drystone wall.

Callander (1986: 48) gives advice on foundation width when building with Aberdeenshire granite:

> The batter of a dyke, is in general 2 inches per foot and that is the extra width that should be added to the base of a dyke for every extra foot in height than the normal four and a half foot dyke.

The 'batter' of a wall is the gradual narrowing of the stonework from foundation to the underside of the cope. The dyke is, in simple terms, built like a truncated letter 'A'. The batter adds strength and stability to the structure. In the example above the dyke narrows, on each side, by two inches for every foot of height.

Cairns (1975: 8) reckons for every additional 6 inches in height, add 2 inches width to the foundation. He is dyking in the Scottish Borders where the stone is more varied, smaller and easier to work than granite. The Borders dykes have steeper sides than the Aberdeenshire dykes; the batter, the inward inclination from the vertical, is less.

If there are a lot of large stones the foundation can be widened to accommodate them. Todd (1859: 464) covers this point:

> The width of the wall on the ground must be determined, in part, by the size of the foundation stones. Should there be a good number of large bolders [sic], from four to five feet in diameter, it will be best to have the wall about that width.

3.2 (ii) The double

The basic principles of drystone construction and how to lay stone have been covered in

Sections 2.3 and 2.4. Some components within the 'double' merit individual description.

Block a large stone or boulder in a wall face. 'Blocker' or 'block-in-course' is a masonry term for a stone whose height rises through two or more course levels. This adds strength and establishes a bonding pattern. They are also known as 'jumpers'.

Blonk (central Scotland) a big stone in the face of a dyke.

Cerrig bras (W) literally 'fat stones' — big face stones.

Changer / changer stone like a 'jumper', they 'jump' up through two or three courses.

Closer the final stone or brick in a course, cut to size if the space is not a standard size.

Double / dooble / doubling the part of a drystone wall which consists of two 'skins' of stonework with smaller hearting stones packed between (Figure 45). A 'double dyke' is now accepted as the standard style of dyke, often described in North America as a 'double-walled fence'. 'Single walls', by contrast, are only one stone wide; the stones are larger and would be difficult to use in a standard 'double' wall unless they were broken up.

Face / face stone / facing stone / facet stone the exposed outer face of a wall. A stone with at least one face exposed on the outside of the wall. Ideally a stone's best face should be visible on the outside but the strength of a wall should not be compromised for the sake of good appearance. Hazard (2006: 5) used the term 'facet stone', a variation of 'face stone' which he confirmed, in personal correspondence, was learned from an Irish mason in Glasgow in the 1970s.

Fair faced a dyke with a good external appearance, a good bonding pattern and no bulges.

Gull an old Scots word used by the Reverend Samuel Smith in GVA Galloway (1813: 87). He tells how the 'gulls' are placed on top of the throughbands when the 'double' has reached two feet five inches high – 'The stones immediately above the bands, (termed *gulls*) are so dressed as to lie solid upon them [the bands] without being supported by small stones or pinnings.'

In Pride (1996: 39) the definition is extended, a 'gull' is:

> …one of the stones between the bands and the coping in a drystone wall; these are dressed so as to lie solid on the bands without being supported by pinnings and to provide a solid bed for copestones.

These descriptions fit a style of walling found in southwest Scotland where the top half of the dyke consists of large stones sitting on a coverband laid at approximately half height. This style is nowadays called a 'Galloway dyke', fully described in Section 4.1.

Gall is a Scots Gaelic word for a boundary stone and, incidentally, a Suffolk, England, word for a large egg-shaped flint.

Jump rock / jumper (Kentucky, USA) a tall face stone which reaches up through two or more courses. This is potentially unstable and, to counteract this, the stone should reach well into the wall and be in solid contact with those around it. They are also known as 'blockers' in general masonry, following the meaning of 'block', a large stone in a wall face.

Mader and Zimmermann (2011: 56) describe jumpers as 'changers' or 'changer stones', they are judged desirable '…because it enlivens the overall appearance.'

Lifts there are potentially three 'lifts' in a drystone wall, depending on the local style

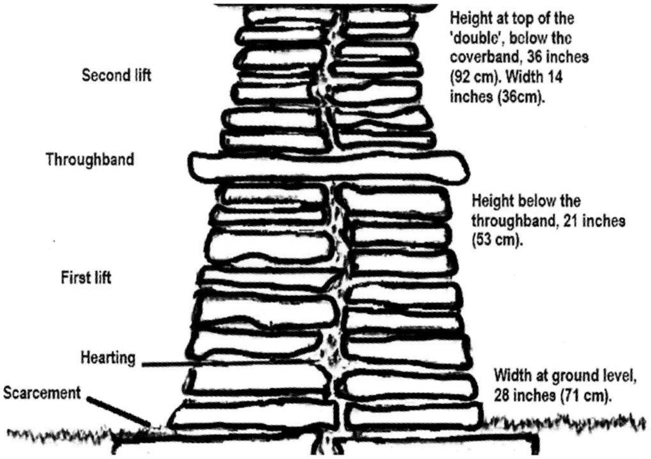

Figure 44: The 'double' consists of the 'shank' of the dyke, the two faces of stonework between the foundation and the cope, with throughbands at approximately halfway.

(Figure 44).

The 'first lift' is the section of a 'double' wall from the foundation to the first row of throughstones. The 'second lift' is that part of 'double' between the throughstones and the covers and copes. If there are two rows of throughstones the 'final lift' lies between the second row of throughstones and the cover and cope.

Prevost (1957: 100) defines the term 'lift' thus — 'The point of contraction, where the dyker begins to build the base proper, is called in Galloway the LIFT.' He is describing where the first course is laid above ground level, on top of any scarcement.

Instead of using the terms 'first lift and 'second lift' Cairns (1975: 2), in an illustration for a dyke in the Scottish Border country, has 'under course' as the 'double' between the foundation and the throughstone and 'top course' as the 'double' between the throughstone and the cover and cope. His term for the throughstone is 'band'.

In another variation Hart (1980: 41), with frequent references to southern Scotland and northern England, includes a cross-section through a typical 'double' dyke. The stonework between the foundation and the throughband is called the 'first course'. The stonework between the throughband and the top of the 'double', under the coverband, is called the 'second course'.

'Lift' can also refer to the individual stones. Individual pieces of thin slate do not have much 'lift'; there can be two dozen courses between the foundation and the copestones. A stone with 'good lift' is thicker than the others and takes the height up faster.

In mortared work 'lift' can refer to a definite amount of building work. According to McAfee (2009: 203):

> When a lift is completed, scaffolding is erected so that the next lift can be built. A lift in stonemasonry is generally not more than 1.35 metres. The slower set of lime mortars may restrict the vertical lift height that can be completed in any given time period.

Drystone walls are seldom higher than six feet (1.83 metres). There is no need to wait for mortar to cure; the whole height can be built in one go.

Racking if a length of wall is dismantled for repair it will not normally have vertical stonework at each end of the gap. 'Racking' is the stepping back of successive courses of stonework to produce a step-like end which can be built to. This produces a smoother link between existing and new work. An exhibition wall is often 'racked back' to expose a cross-

Figure 45: A collapsing dyke, showing 'double' and hearting.

section of features in one short length, from foundation to the copestones.

Rib / ribs (Scots) the external walls of a dyke.

Seconder this term is used in the 'single' walls of the Mourne Mountains in Ireland for those large boulders built above the foundation, on the second course, as described in DARDNI (the Department of Agriculture and Rural Development Northern Ireland) (2013: 27). See Figure 80, for more detail. The term is also used in southwest Scotland for a smaller stone used to bring a copestone to height.

Shank Prevost (1957: 100) includes this Upper Annandale (Scottish Borders) term for 'The double of a stone-dyke when completed', illustrated by the expression 'This dyke has a grand shank.'

Skin a term used by Scottish dykers for the outer face of a dyke.

Sleek / sleekit / sleekit skin / sleekie-skin / slee'd skin (southwest Scotland) Prevost (1957: 100) defines the first two of these terms as 'Smooth. Used to describe the *double* of a stone-dyke which has been built neatly, to please the eye.'

Such a dyke is said to be 'weel skinned' (literally 'well skinned'), according to Rainsford-Hannay (1944: 28-29). He warns against judging the dyke by its outside appearance. 'It may have a smooth countenance but its heart *may* be full of corruption.'

Rainsford-Hannay (1957: 95) includes the poem 'Elegy to Davie the Dyker', from Mactaggart (1824),and praises the dyker's work – 'His dykes had ne'er the sleekit skin, A' fair without and fause within…' In this case 'sleek'd' means 'sleekit', in the sense of 'sly, calculated to deceive'. Davie's dykes were never deceptively tidy on the outside and 'false' on the inside.

The Reverend Samuel Smith, in GVA Galloway (1813: 89), warns against dykers who '… make their work *well skinned,* [but] they give themselves no concern about its firmness and durability.'

Stephens (1853: 558) repeats the warning:

Figure 46: The throughband is an occasional long stone lying across the width of the dyke, typically at half way up the 'double'. It adds strength and provides a solid base for the upper part of the dyke.

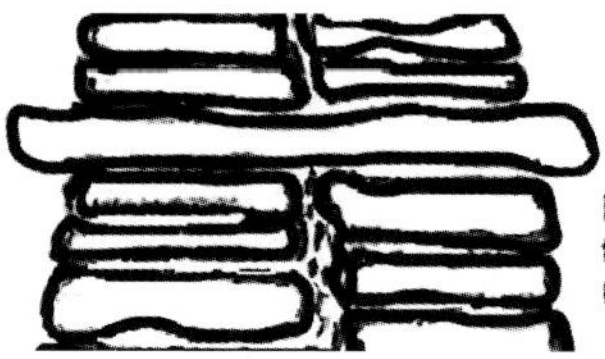

> Very many dry-stone dykes in this country are constructed on erroneous principles, the stones being laid promiscuously, and more with a view to give a smooth face than a substantial hearting to the wall.

He uses the word 'promiscuously' for its original meaning, 'mingled together, confused', as per Bailey (1675).

'Show walling' is walling work which emphasis a well finished face rather than strength. The term is associated with walling competitions.

Watershot / watershed stones laid to 'shed' or divert the water away. The Duke of Buccleuch is said to have preferred dykers build his farmhouses in the Scottish Borders because they tended to lay the stone with a tilt to the outside. This reduced the chance of rain finding its way through the mortared walls.

In the Shetland Isles, Scotland, a 'watersten' (literally 'water stone') is a stone in a house or barn wall which has not been properly tilted towards the outside; it may even tilt to the inside. Moisture seeps back from the outer face of the stone and creates a damp patch on the inside wall.

Weatherstane a Scots term for a projecting stone built, for example, at the foot of a chimney stack, to throw water away from the stonework.

3.2 (iii) Throughbands and thruffs

The early dykers understood the importance of laying long stones across the total width of the wall to add strength, and limit the extent of any collapse. This was already well known to masons who incorporated bondstones in their mortared walls.

Margaret Brooks (1977: 102) confirms throughstones are essential, especially where the walling stone is flat, smooth and evenly sized:

> Limestone walls, with their extremely irregular stones, settle into a stable pyramidal heap, ...whereas the flatter and more regular sandstone and shale, built in more vertical walls, are more likely to bind together as separate faces which could fall apart from each other. Throughs are thus more crucial in non-limestone country, which is fortunate, since they are harder for a waller to obtain in limestone country, but they are a valuable and important element in the durability of any dry stone wall.

In the nineteenth century the expanding railway network meant heavy materials were more easily moved across country. Transporting stone for houses or masonry walls makes economic sense, less so for drystone field enclosures. Nevertheless, wallers made the effort to get suitable material. Margaret Brooks (1977: 98) illustrates this point, for Yorkshire at least — 'In areas where flat stones large enough to go through the width of a wall are rare, wallers used to carry good throughstones over long distances on their backs.'

Throughstones add strength by reaching across a wall. Long stones can also add strength by reaching vertically (Figure 47). They can reach right across the wall or only half way. These uprights are key features of the Aran style of *feidín* wall (Figure 87).

The following terms describe throughstones or arrangements of stones which perform that function

Anchors / anchorman (USA) the retaining wall equivalent of throughbands, long stones which provide strength by reaching from the face of the wall and back into the earth embankment behind.

Band / bandstone / band-stane / bandstane / band-stane / baand-stane / band stone / bandstanis (Galloway, Scotland) a binding stone, a throughband, extending across the width of a dyke. The ends can be flush with, or extend beyond, the faces of the dyke. The verb 'to band' means to make a dyke secure by inserting band-stones. A dyke is 'well

Figure 47: Occasional vertical stones perform the same function as horizontal throughstones. They do not always stretch all the way through to the other side. Inis Oírr, Aran Islands, Ireland.

banded' when it has sufficient throughstones.

Band height (Scots) the level above ground at which it is intended to lay the throughbands. Normally this is half way up the dyke and at regular intervals along its length. If the wall is high enough a second set can be set approximately half way between the first set and the top of the 'double', each throughband midway between the two below.

Bind stone / binder (USA) a throughstone, a long stone stretching into the wall to strengthen it. Mader and Zimmermann (2011: 56) give a German definition – 'To guarantee the stability of the dry-placed wall, a third of the stones are binders, stones that span the entire wall depth.'

'Binder' is also used in the Mendips, southern England, where the carboniferous limestone is small and irregular. Garner (1984: 1), includes an old photograph where the '...waller is laying a 'binder' (a long stone stretching into the wall)...' The stone reaches from one edge to half way across the wall.

Williamson (2002: 107) also refers to limestone walls in the Mendips:

> ...the walls tend to be less broad and less tapering than in the rest of England. To overcome the paucity of stones suitable for throughs, stones are used which extend some two-thirds of the way across the wall.

Bond stone / bondstone large stones used in mortared walling to provide a strong structural link well into the width of the wall, or across the face of an opening. Bond stones can act in pairs, one from each side, overlapping in the middle to act as a two-part throughstone.

Bonder a form of 'binder', a tie stone, a rock extending part way across a wall in areas where rock is not long enough for a total stretch, for example in the Mendips, southern England.

Carreg drwodd (W) a through stone.

Cross stone Murray-Wooley and Raitz (1992: 191) have this as a synonym of 'tie-rock', described as — 'A rock laid with its longest dimension running entirely through the fence as a header, having the effect of tying the two faces of the fence together.'

Deadman / dead man / dead man's extension / dead head (USA) the equivalent of a throughband, used in retaining walls to provide strength and stability. They reach from the face of the wall, back through the body of the wall, and make solid contact with the ground behind.

Doubling stone listed in Alan Brooks (1989: 18) without specifically mentioning the function, it might be a face stone or a throughstone. Murray-Wooley and Raitz (1992: 197) quote Brooks and consider it a type of throughstone.

False band (Annandale, Scottish Borders) also known as a 'split band', using two stones, one from each side, if there are none available for a proper throughband. According to Prevost (1957: 101) 'The insertion of split bands is a deceitful practice if a dyke is being built to specifications at a contract price.' The use of two stones is structurally weaker than using one stone, unless they are properly placed, and packed securely with hearting.

Gaffs are three-quarter throughs in the Mendip region of England. This could might originate from an old slang variation of 'gaff', meaning 'to cheat'.

Galf stones (North England) a term for three-quarter-length throughstones, used in pairs, one from each side of the wall, overlapping in the middle. This is not as strong as a single stone.

'Galf stones' should not be confused with 'half stones' or 'half throughs'. They are separate terms.

'Galf' may derive from the Scots Gaelic word *gulfa* (meaning 'narrow'), finding its way into Scots and northern English dialect. This is evidenced by a 1575 entry in DSL from Haddington, East Lothian, Scotland, under 'scaffald': 'For making of skaffattis to the eist galf wyndoke of the kyrk…' This translates to 'For making scaffolding to the narrow east window of the church…' A 'galf' stone is therefore narrow — too short to reach across the wall.

Half through / three quarter through a throughbond which consist of two pieces, one reaching in from each side, overlapping in the centre of the wall, a compromise when there is a lack of long stones. The stones must be approximately equal length and well butted together for the maximum friction between them. This aspect is well covered by Adcock (2012: 24-26).

Headers (North America) a specialised throughstone running *across* the whole width of a wall, as opposed to a 'stretcher' which runs *along* the length of the wall. Headers and stretchers are important components at wall ends, corners and openings. 'Interlocking headers' according to Adcock (2012: 25), are a pair of throughstones, one from each side, each reaching three-quarters of the way across the width of a wall and whose tails interlock.

Leac a Scots Gaelic word for a gravestone or flagstone, a term also used for a throughstone.

Locking stone (USA) a throughband, a stone which locks, or connects, both sides of the wall.

Mortar assisted throughs mortaring smaller stones is an option in areas where the stone is too small to reach across the total width. Wall ends have been mortared for strength if suitable stone is not available locally, or cannot be easily brought in from another site.

Garner (1984: 24) includes an image with the middle four courses of an Oxfordshire wall pointed with mortar to compensate for a lack of throughstones.

Personal correspondence with a drystone waller in Dorset, southern England, confirmed '…some stretches where a layer about a foot high at the level of the throughs is mortared.'

Parpen / parpin / parpane / perpend / perpender (Scots, with variations recorded elsewhere) a bond stone which reaches across the entire width of a wall. Good faces are required on each end because both are visible, on opposite faces of the wall. The term is more often associated with masonry work. *Parpaigne* is a French equivalent of 'throughstone'.

Pierre de liaison (French) literally 'a linking stone', a throughstone.

Placing of throughs typically throughs are placed every metre or every yard along the length of a wall, depending on the availability of suitable stones and the local style. GVA Northumberland (1813: 62) gives details — '…about half way up a row of through-stones are put, at the rate of nine or ten in a rood of seven yards…' This means up to ten throughs in a length of seven yards (6.4 metres or 21 feet) — an almost continuous course of throughs if the individual stones are more than a foot wide.

Brown (1869: 147) describes placing two sets of throughstones in Yorkshire:

Figure 48: This section of dyke has suffered storm damage. The throughstone on the left side of the opening remains on duty – limiting the extent of the collapse, and supporting the stone.

> The first layer of through-band stones should be laid one in every four feet in length at least, and at about eighteen or twenty inches from the ground. The next layer should be put in at similar distances, but alternating with the others, at a height of three feet six inches from the surface.

Williamson (2002: 51) records standardised specifications for construction during the eighteenth and nineteenth centuries. One, for Grassington in Yorkshire, specifies:

> …there shall be laid in a Workman-like Manner 21 good Throughs in every Rood of Fence, and the first 12 to be laid on at the height of 2 Feet broad, and the second 9 to be laid on at the height of 4 Feet from the Ground…

The wall is in Yorkshire, so the rood was probably 7 yards or 21 feet; the specification was for a wall six foot high plus a flat cope which was not to exceed four inches.

Raistrick (1973: 19) describes Pennine walls with two sets of throughs, at two feet and four feet from the footings and, in some walls, '…a third set of thinner throughs just under the copings.' The third set fulfils the function of a coverband.

Currently, drystone dyking courses almost universally teach setting throughstones at one yard or one metre centres, halfway up the 'double', unless this is contradicted by the local style.

Projecting throughbands according to Raistrick (1973: 19) ragged, projecting, ends are left on the throughs because it is too laborious to cut each to the right length before building them in. He advises against cutting or breaking them flush with the face after building is complete because this can cause damage to the throughstone, inside the wall.

Stephens (1853: 590) acknowledges through-band stones are frequently left projecting from one, or both sides, of the dyke to prove there are through-bands '...but the practice is objectionable, inasmuch as any projections serve only as stepping-stones for trespassers to climb over the dyke.'

Projecting throughs are used by livestock as rubbing posts and, in sheep pens, can be a hazard to men and sheep in a confined space.

Projecting throughs allow the top part of a wall to spread and still be supported. Many old dykes are saved from a final ruinous state by throughstones which continue to link the two faces of the dyke and restrict the extent of a collapse (Figure 48).

Conversation with a Yorkshire waller suggested throughstones protruding to one side of a wall is confirmation of ownership. The stone may protrude on the owner's side or be flush on the owner's side. This varies from one part of Yorkshire to another, an example of local styling – particular to one county, one parish or one group of wallers.

Skintled work a Kentucky description of tie rocks (throughstones) which project. This may derive from the old Scots term 'skint' meaning 'to hasten or to hurry'. 'Skintled work' could therefore be hurried work; no time was taken to trim the stone flush with the face of the wall.

Sneck (Scots) a term applied to a throughstone, one set across the wall for strength and to inhibit collapse. In mortared masonry work a 'sneck' is a small squared stone used with larger stones to create a good bonding pattern. The term also applies to a short section of 'single' dyke constructed within a longer 'double' dyke, to use up big stone, or to add strength, perhaps on a slope.

Split band 'a deceitful practice', fully described under **false band**.

Tail binder a stone which reaches two thirds or three quarters into the wall and lies on the tail of a similar stone which reaches in from the other side. In combination they constitute a conventional throughband.

Three-quarter through a stone which does not reach all the way over a wall, like a **half through**.

Throo stean (Cumberland, northern England) a through stone, as pronounced in the local dialect.

Thorough bands / thorough-band an alternative variation of 'throughband' from Wilson (1849: 606), Stephens (1853: 590) and Stephens (1857: 355), all writing on agricultural improvement in the British Isles.

Through a short form of the words 'throughstone' or 'throughband'.

Charles Fothergill, in Romney (1984: 113, 115), records the use of 'throughs' in six feet high walls during his tour of Yorkshire in 1805. There are three courses of throughs over the height of the wall:

> When the wall is raised... about 20 Inches from the ground... a large flat stone that will

> reach thro' the wall and is about 3 Inches thick is then laid on... [t]hree of them are used in the perpendicular of one of these walls and are put in rows 20 Inches above each other... they strengthen the wall and render it firm...

Drawings with the description suggest the throughstones are laid in a continuous course, not spaced at regular intervals. Fothergill says copestones were sometimes laid on to the top course of throughs. They, in effect, act as a **coverband**, as detailed in Section 3.2 (iv).

Through bond the 'bond' achieved by a throughband as it extends through the wall. The individual throughband stone overlaps the joint or joints between the stones below it, and links and binds together the two outer faces of the wall.

Throughbands / throughband / through-ban the throughstone, a term from Galloway, southwest Scotland. Rainsford-Hannay (1957: 34) advises, 'Outside Scotland they are called 'throughs'.'

Throughbanes (Scots) throughstones.

Throughstones longer stones placed at vertical and horizontal intervals along a mortared or drystone wall to tie both sides together and stop them from 'bellying out' and collapsing.

Throwband stane a variation of 'throughband'.

Mundell (1981: 122) includes the poem 'Dry Stane Dykes' by Helen Burness Cruickshank (1886-1975). The poem's last verse uses the throwband stane as a metaphor for Scotland's potential as a political weight to hold a shaky world together:

> And lang may Scotland haud her ain traditions,
> Yet claim her richtfu place in League o Nations.
> Wha kens? Her wecht, the throwband stane micht be
> Tae bind a shoogly warld in Amity.

A footnote to the poem defines 'Throwband-stane' as 'The long balancing stone thrust out on either side without which the dyke would fall, since it contains no mortar.'

Thruch-stane / through stone /throughs / thrughs / thrucks / trough-stane / throch-stane southern Scottish variations describing a flat gravestone. According to the DSL entry for 'thruch stane', they are 'strictly one resting directly on the ground but also applied to one lying on a pedestal or on four feet.' Most of the DSL references are concurrent with the era of drystone dyking; two date back to the fifteenth and sixteenth centuries. Perhaps the term was adopted by dykers because a throughstone, lying across a half-built dyke, resembled a flat gravestone.

Thruffs (Yorkshire) throughstones. Also known as 'thruffers', in north Yorkshire. Raistrick (1973: 18) describes their purpose:

> The throughs are necessary to bind the two faces together and prevent the wall 'bellying out' and bursting. A 'weel thruffed wall' rarely fails unless the foundations slip, and even then the gap is never very serious.

Thrufter (New England, USA) a throughstone, a long stone placed to bind the two sides of a wall together.

Thrusts Adair (1998: 36) mentions this as another term for throughstones, one he had heard in the north of England.

Tie (British) a throughstone used in the 'wall head' or 'cheekend', detailed in Section 4.5. A 'broken tie' is a two-piece tie stone, used where stones are too short to reach right across the face of the cheekend. A tie stone literally 'ties' the two sides of the wall together. The term 'tied' also describes the way one stone overlaps or covers the joint between the stones below it, thus creating a secure bond.

Tie rock (North America) a throughstone, a rock laid entirely over the fence, tying the two faces together. In a Kentucky context Murray-Wooley and Raitz (1992: 197) tells us —'Coping-rocks and cap courses that cover the full width of the fence also act as tie-rocks. A course composing entirely of tie-rocks is a tie course.'

Tie stane (Scots) a stone which links the wall of one house with its semi-detached or terraced neighbour. Synonyms are 'tusks', 'tuskstones' or 'toothing'. These are masonry terms for projecting stones in the gables of a terraced house wall; they link with corresponding stones in the walls of the adjoining houses. They are usually left projecting when the neighbouring house is demolished because these tuskstones are an integral part of the remaining house.

In County Kerry in Ireland 'spirit stones' are those which project from the gable. They have no practical purpose (for thatching etc) and are believed to be a resting place for the spirit of ancestors. In the Channel Islands, between England and France, similar stones are called 'witch stones' because witches are supposed to rest there.

Tieband Adair (1998: 36) quotes this as a Scottish term for a throughstone.

Tie-in rock (Kentucky) a throughband.

Tie-stone (New England) a throughstone.

Toothing stones which link adjoining house gables. A **tie stane**.

Trough-stones a term used by Lambert (1829: 152):

> …if the stones are indifferent, have two or three carts of a good long binding kind, dropped among the bad… in order that they may be placed occasionally across it, to bind the bad stones together. Manage to have at least nine or ten, of what are called trough-stones, to every six or seven yards at least, in every course of work.

This is a lot of throughstones for such a short length, suggesting the building stone is very bad or the 'trough-stones' are simply longer stones which reach well into, though not entirely across, the width of the wall.

Trow-baand / trowband a variation of 'throughband' from Orkney, Shetland and Caithness, reflecting the local dialect and pronunciation in those parts of northern Scotland.

Tusks / tuskstones projecting stones, a variation of **tie stane**, the stones which link adjoining houses.

Wooden throughs Prevost (1957: 101) writes:

> Dykers have been known to use wooden through bands as a substitute for stone, and oak bands were used in the construction of a dyke on Poldean in Wamphray parish. Red cedar shingles were used as *covers* in a 100-year-old dyke at Galt in Canada.

Wamphray lies 10 miles (16km) north of Lockerbie in southern Scotland. Galt is now part of the city of Cambridge in Ontario, Canada.

Todd (1859: 465) notes the use of wood in New York State walls:

> When stone fences are made of small stones alone, where there are no flat stones to bind the wall together, small strips of wood, called binders, about an inch wide, and one-fourth of an inch thick, which are usually split out of cedar or some other durable wood, are laid between all the courses of stone, as thickly as thought proper, but one binder extending entirely across the wall to every six or twelve inches will be sufficient.

Todd recommends the binders are sawn off flush with the face of the wall.

Thomas (1864: 42), published in Albany, New York, also mentions the use of thin wooden throughs:

> Where the stones are not good for building, the wall may be bound together by strong cross ties of durable wood, placed about half way up, or at two or three different distances up — cut to the right length with a saw, and split thin, and two or three inches wide.

Wood within stone walls is nothing new. Marcus Vitruvius Pollio the Roman architect and engineer (c80 BCE — c15 BCE), recommended using charred olive wood ties in the city walls to bind the two faces of the stone wall together. They were the steel reinforcing bar of the day.

3.2 (iv) Covers and copes

The top of a drystone wall can be a regimented line of shaped stone or a ragged collection of irregulars carefully marshalled into a unit.

The coverband is a continuous course of flattish stone laid across the top of the 'double'.

The copestones are laid on top of that. Not all walls have coverbands. Most walls have some kind of cope.

If there are insufficient large stones, the top of the dyke may taper off to a rickle of small stones, be topped with turf, or incorporate supports for wooden rails or wire. Cutting symmetrical copestones might be worthwhile for a wall round a garden but expensive, in terms of time and material, for a field wall.

Copestones lie across the top of the wall and frequently overhang by at least an inch or two.

They secure the stonework beneath and, if vertical, are an efficient way of achieving height. The minimum height for a vertical copestone is generally accepted as 10 to 12 inches. A typical coverband is 4 or 5 inches thick. Wilson (1849: 606) estimates constructing a cover and cope combination 15 inches (380 mm) high is equal, in labour, to building an additional nine inches (230 mm) on to the 'double'.

Good copestones, if properly set, also do the work of coverstones and act like another set of throughstones at the top of the wall.

Figure 49: The coverstones and copestones secure the top, and help make a dyke stockproof.

Charles Fothergill, in Romney (1984: 115), details the construction of a six feet high (72 inches) wall during his tour of Yorkshire in 1805. He describes three sets of three inch thick throughstones, set at vertical intervals of 20 inches. The highest set of thoughstones are near the top of the wall, at 60 inches or five feet, what we would consider as a coverband. Romney continues — '…and upon that perhaps a row of stones; when the top *through* is not used, a row of stones set obliquely is thought to be the best.'

From the description it seems the top set of throughstones are laid touching, side to side, and not at regular intervals. This suggests a cope of flat slabs. Sometimes copestones were laid on the top of the flat slabs. If there was not a top course of throughstones the Yorkshire wallers laid a course of copestones, on edge, set at an angle.

Nicol (1803: 297-298) stresses the importance of a good cope on walls built of stone and lime mortar, his point is equally important for walls built 'dry':

> But certainly, after being at the expence [sic] of building a *good* wall, the further expence of placing upon it a *good coping*, is ill withheld. This should be considered before the work is set about; for, provided a wall is set on a *fair foundation*, if *well coped*, although built of inferior materials, it will frequently be found to last longer than one built of the *best* materials, and *badly coped.*

A solid cope with a good regular top line takes the eye off rougher work in the body of the wall. If there is a good selection of stone the cope is a chance for a craftsman to add their own individual style. The parallel lines of a well built cope and projecting throughstones look like a living thing as it snakes over the landscape.

There is some regional, anecdotal, evidence of copestones placed in a way that indicates ownership of the wall, to one side or the other. The stone may be more finely cut on the owner's side or overhang. This is an interesting study topic for local walling groups.

If the top of a wall is wide, for example over two or three feet, it may be too wide for a formal cope. The effort for cutting stone and placing them is out of proportion to the benefit. The top course of face stones can take the place of the copes by reaching well into the centre of the wall. Hearting fills, and levels, the space between the face stones.

The following terms describe how to top off a stone wall

Buck and doe a type of cope with alternating low and high stones, a castellated cope.

Bunkering Prevost (1957: 97) has this as a term from Upper Annandale, Scottish Borders, defined as 'the use of two small stones instead of one long stone when building the single.'

Figure 50: Flat laid capstones. Rhode Island, USA.

Bunkering is a compromise, making the best use of available material.

The 'single' in this case consists of long stones reaching across the width of the wall in the upper part of the regional style we now know as a 'Galloway dyke'. This style is described in Section 4.1, stone structures and wall types.

Caip / Caip-stane / caip-steen / caipin' stane / cape stane / capee stane / cope (Scots) a coping stone, or the whole cope of a dyke.

Cams / cam stones (English Lake District) cope stones. According to LDSPB (1977: 5) 'Cam stones often slope downhill and protrude towards the neighbour's land.'

The overhang helps prevent animals climbing over from the neighbour's ground.

The word probably shares origin with the Norwegian word *kam* which can translate to 'comb' or 'crest'. Many Lake District place names have a Norse origin.

Cap (New England, USA) the cope. Thorson (2005: 71) defines 'hard cap' as a top course of slabs and 'puddle cap' as an informally piled layer of small stones.

Cap course / cover course (Kentucky, USA) the coverband. The horizontal course of continuous flat stones between the top of the 'double' and the copestones.

Murray-Wooley and Raitz (1992: 189-190) give 'bell', 'flat cap', 'lip cap', 'overhang', 'overlay belt', 'set out course' and 'shelf cap' as Kentucky terms for a 'projecting cap course', one which overhangs the face of the wall by two or three inches and is the course upon which the coping rests. The small overhang will have some deterrent effect and, along with its duties as a base for the copestones, creates a definite line which adds to the attraction of the structure.

Cape and rickle (Scots) a style of coping where the 'cape' is a course of flat stones covering the 'double' and the 'rickle' is a tapered pile of small stones above that, total height around twelve inches, 30 centimetres. This style is found wherever there is a scarcity of good solid stones to make a vertical cope.

Capes (North Riding of Yorkshire, England) copestones.

Cape-stane (Scots) the cope stone. The term has been used as a metaphor for a remediless calamity as in the Scots expression 'the last sad cape-stane of his woes'. The cape-stane is the ultimate, whether it be the best or the worst. It is the crown, the finishing touch, the apex.

Capping base (Kentucky) the coverband, literally the base for the copestones.

Capstone (USA) flat stones laid on top of dry or mortared stonework, overlapping the edges by an inch or two, to finish off the wall, keep the interior dry and help protect the stonework below (Figure 50).

USA sources, including Thorson (2002: 147), distinguish between 'capstones' which are slabs laid flat across the top of the wall and 'copestones' which are laid vertically. Vertical coping stones are common in Britain but seldom seen in New England apart from '… wealthy, old, thoroughly English communities like Newport, Rhode Island.'

Reed (2003: 44), writing from North America, refers to capstones which he also calls the 'cover course'. He continues — 'Traditionally, this type of wall is completed with a row of coping stones is set on top of the capstones.' These stones are set vertically or at a slight angle.

McRaven (1989: 37) tells us 'I like to cap with thick, flat stones laid ledge (flat), to keep out as much water as possible…'

McRaven also includes an illustration, labelled 'vertical capstones' with a caption – 'Rural drystone walls were often capped with a vertical course, probably to discourage climbing. I

Figure 51: Dyke with copes, no coverbands or throughstones.
Post and wire add to the height.

prefer heavy stones laid flat.'

Corangamite (1995: 90), detailing drystone walling in the western district of the state of Victoria, Australia, also differentiates between flat and vertical copes – 'Top stones laid flat were called capping stones, or coping stones if they were placed vertically.'

In some particularly wide New England walls there is no formal cope; the wall tops off with final courses of large stones on each side. The space between is filled with smaller stone.

Carreg copa (W) a coping stone.

Carreg glo (W) a smaller stone used as a wedge to lock the larger vertical copestones.

Carreg grib (W) literally translated as 'ridge stone', a coping stone.

Castellated / crenellated / chevaldefrise a vertical cope with alternating high and low stones. Variously known as 'buck and doe', 'cock and hen', 'cow and calf', 'Yorkshire tops', 'Jack and Jill' and 'one up, one down' (Figure 52).

Cement coverbands Hart (1980: 46) notes their use:

> A practical innovation is the use of cement coverbands. These are used by some county councils, and leave a neat and workman-like finish. They have the advantage of standard width and thickness, so provided the wall is brought up properly, they form a level base for the final row of coping stones.

Cerrig cant (W) literally 'rim stones', coping stones.

Clach-chinn (SG) copestone.

Clonk Rainsford-Hannay (1948: 81) defines these as 'Big stones in the single, above the cover band and below the cope.'

This is confirmed by Prevost (1957: 97) who defines 'clonk' as a Galloway (southwest Scotland) word for 'the big stones in the single, below the cope stones.'

Callander (1986: 84) says 'any large stone taking up the width of the dyke below the cope.'

From these descriptions it is clear 'clonk' refers to stones in the upper part of the regional style we know as a Galloway dyke (Figure 58), in the section of the dyke built 'single', which can be more than half of the total height of the dyke.

In an earlier piece, Rainsford-Hannay (1944: 28), appears to confirm 'clonk' as the cover-band though it is not entirely clear whether the copestones or the cover-band stones are 'clonks':

> The cover-band stones fulfil three purposes: they bind the dyke together; they protect the inside from nearly all the rainwater; they form a bed for the copestones, often called 'clonks.' Copestones or 'clonks' should be of a good shape, 12 to 14 in. high. Skill is wanted to get them to bind and lock together.

Rainsford-Hannay (1957: 34) tells us "Clonk' is a good expressive word but it cannot be found in a Scottish dictionary.'

DSL quotes only Rainsford-Hannay's use of 'clonk' and advises the word may stem from the Norwegian word *klunk*, a lump.

'Clonk' was included in a list compiled by the Dry Stone Walling Association membership; defined as a 'larger than normal pinning, usually triangular to fit a hole exactly.' This fits the description of a large stone filling a gap in the copestones.

Cock and hen (Cotswolds, England) a style of castellated coping with alternating high and low stones (Figure 52). Also expressed as 'cocks and hens copin' in North America. This contrasts with 'flat copin' where a flat laid capstone tops off the wall.

Ceiliog a iâr is the Welsh term for this type of cope, which translates as 'cock and hen'.

Codgers the top rows of small stones in a Cornish hedge, per Alan Brooks (1989: 94). This may derive from a slang term of 'codger' for an old man, often in dire straits. 'Cornish hedges' are detailed in Section 4.1.

Comb / combers (Cotswolds, England) the top stones on a drystone wall, probably

Figure 52: A variation of 'cock and hen' coping.
Newtyle, Tayside, Scotland.

derived from the Old English *comb* — 'a crest'.

Comble-stane / comble stone (Angus, Scotland) the top stone of a pile. DSL includes, in its definition, this term's origin from the Old Scots *combill* — the highest point, and Old French *comble*, 'top or summit'.

Cope / coping / cope stones / coping stones the top course of stones on a drystone wall, masonry pillar or chimney, which covers and protects the work below, particularly from water penetration. An overhang also acts as a drip edge to help keep water off the face of the stonework below. In drystone walls the cope stones can be set vertically, slightly angled or laid flat across the top of the 'double', all depending on the local style.

'Cope' is also an English term describing the splitting of a large block of stone along a line of fracture using hammer and chisels.

Corbie steps / corbie stanes a Scots form of the English term 'crowstep gables' or 'crawstep gable'. The step-like, corbelled, cope at the top edge of a house gable, used by crows as a lookout.

Cover / covers / cover stones a course of long, flattish, stones reaching across the top of the 'double' to level off the wall, strengthen the work beneath and act as a secure base for the copestones. Prevost (1957: 98) notes 'covers' as a term from Annandale in the Scottish Borders and defines it as 'flat stones of a minimum specified length which are laid on top of and cover the double.'

Stephens (1853: 590) insists the final course of the 'double', under the cover stones, should be substantial and secure:

> The cover also acts as a thorough-band at the top of the dyke; but in laying the cover, the *levelling* of the dyke to form its bed should not be made of very small and very thin stones, as is too often the case, as neither have stability, are easily broken, and are constantly in danger of slipping out from under the cover and cope.

Cover band Prevost (1957: 98) notes 'cover bands' as a term from Galloway in southwest Scotland and refers the reader to 'covers', the term used in Annandale, a district to the east of Galloway. This term can apply to the height at which covers are placed or the whole course of coverstones, seen as a unit along the length of the wall.

Hart (1980: 46) writes:

> Coverbands are used in Scotland, but not always elsewhere. Their advocates stress their dual virtues; they bind the wall together again — it [the wall] has been built in two faces since the throughbands — and they keep rainwater from driving straight down into the wall centre.

Most walling styles rely on a good wide copestone to bind the wall together at the top. The copes can sit on a coverband or a layer of thin stones which bring the 'double' up to a level. Alan Brooks (1989: 72) details these two variations:

> In parts of Scotland, a heavy coverband of throughstones is placed at the top of the wall just below the coping. Pennine wallers don't do this but, where suitable stones are available, they level off the top course with small, thin flags laid side by side to make a good bed for the topstones.

Cover course (Kentucky, USA) also known as the 'cap course'. Defined in Dry Stone Conservancy (2001: 28) as:

> ...a row of rocks laid on top of the face courses that binds the faces together. The cover course also gives more height for less effort — it is much quicker to place a layer of cover rocks than to lay an equivalent height of face rocks. A cover course also makes the coping easier to build.

Cow and calf coping a type of castellated cope with alternating high and low vertical stones.

Crawstane / crowstone (Scots) the top stone in a gable. Probably derived from the Scots word for crows, birds of the *Corvidae* family, who use these points as lookout posts.

Crowstep gable / crawstep gable the step-like cope at the top edge of a house gable. Named for the crows who use these steps as lookouts.

Crow-stones D'Arcy (2006: 30) notes how '…'crow stones' are left on the wall tops rather than in the gaps between the stones' on the walls in the Burren, western Ireland.

These stones have been lifted off the field and put safely on top of the wall, away from livestock and machinery. They sit solitary, and conspicuous, like the bird.

Crown Hart (1980: 46) uses the phrase '…the desired line of the crown…' when referring to 'setting the cope' on a drystone wall and in his glossary defines it as 'the top of a bank or hedge.'

Crownstone (Scots) the saddle stone or top stone in a gable, a term more appropriate for mortared work.

Dike tip the top of a dyke.

Double copes if individual stones are not wide enough two may be used, one reaching in from each side, overlapping in the middle. One good stone, reaching across the total width, if available, is set at intervals to strengthen the cope. This style is common in Kentucky, USA.

Neil Rippingale, Master Craftsman Dyker, Scotland and USA, uses the phrase '50:50 or 60:40' as a reminder that double copes should extend equally from each side. In Kentucky the depression in the middle of the wall, where the double copes interlock, is sometimes filled with mortar to strengthen the top of the wall, and deter the rascals who steal stone.

Flat cap (Kentucky, USA) the coverband

Frost stones (New England, USA) long heavy flat stones on top of a wall. Their weight helps keep the wall together in frost heave and when it settles in the thaw. These are particularly seen in Maine where the local masons are skilled in the fabrication of granite.

Galloway cape a style of rough stone coping developed in southwest Scotland, an early use of a cover and cope combination to secure the top (Figure 53). A distinguishing feature of the style of drystone dyking developed in Dumfries and Galloway in southwest Scotland, the one most remarked upon by early writers, was the top section of rough stones built 'single' across the top of the 'double'. Sheep farming was expanding in the nineteenth century, and the Galloway cape was a radical improvement, with more mass and presence than a turf top or tapered pile of small stones. This cope makes up the top quarter, or top half, of the dyke.

DSL quotes from page 374 of *Farmer's Magazine* for August 1811:

> The Galloway cape upon a well built stone wall is known to be unassailable by stock of any kind, if the wall is built five, or even four and a half feet in height before the cape is constructed.

Under 'Caip, Cape, Cap', which DSL defines as the Scots form of the English 'cope' and the '…highest part of anything', the definition goes on to include a 1799 reference:

> A stone dike… four feet in height, with a cape of stone, will defend your plantation from cows and horses; but a foot more, with a cape of turf… will be necessary to defend it from sheep.

In GVA Galloway (1813: 85) the Reverend Smith does not refer to the Galloway cape by that name but emphasises the need to build the top of a dyke carefully, not as a careless mass. The first stones of the cope (which correspond to the coverband) are carefully laid across the wall. Good sized copestones are carefully set on top, wedged to form a level top line:

> When the stones are good, one-half of the dyke is sometimes built single: but it is a practice which ought rarely to be admitted. It is too much the practice to huddle together the stones in a careless manner in the upper part of the dyke. It ought never to be so. Much of the durability of a dyke depends on the levelling of double accurately, laying the first coping stones in such a manner that they may rest firmly and press equally upon it, and making these above to wedge nicely and balance exactly, so that the centre of gravity may always be in the middle of the dyke. On the top none but stones of good size ought to be admitted, and these laid with so much care that the workman may boast, as he sometimes does, that a man may wheel a barrow along the top of his dyke without shaking a stone.

Figure 53: The Galloway cape, as described in early writings. Note the tight build, despite the shattered rock type. Southwest Scotland.

Hard cap (New England, USA) Thorson (2005: 71) describes this as a coping style with flat slabs spanning both sides of a 'double wall' – one built with two outer faces and smaller stones in the middle.

Heightening a cope this has been achieved with stone, furze, wood rails or rope.

If a drystone dyke has a coverband, and the top is wide enough, it is straightforward to remove the copestones and add a few courses of stone. A cope could also be heightened by adding another course on top of the existing cope. This must be stable and have 'presence', to deter any ambitious livestock.

Wilson (1849: 605-606) declares a five foot high wall would not hold in mountain sheep:

> …nor, indeed, can a much higher fence do it; for they have a clambering way of scaling, not leaping, high walls.… There is a device, however, which has been practiced with some success on these walls… and that is, to erect long stones, like pavement stones, about a foot above the ordinary coping, at short distances from one another, among the coping stones. These stones have a notch cut into their upper edge, which can support a small rail of wood, or a rope of twisted straw. This rail or rope appears to them so frail, as not to be able to bear their weight at such a height, seems to deter them from attempting to leap the wall. Sometimes small branches or furze are stuck in between the cope-stones as objects of terror to them, but which are objectionable, in as far as when they begin to shrink in withering, they lose hold on the wall, and are blown about among the sheep by the wind: and, as they generally get entangled among their wool, they become sources of serious annoyance to the sheep…

Furze is the common gorse (*Ulex europaeus*), also known in Scotland as 'whin'. At one time gorse was a popular hedging plant and was crushed and fed to livestock.

Macdonald (1887) advocates fencing of hill pasture because it restricts the movement of animals and the introduction of disease and adds:

> Where there is an abundance of stones at hand, dyke-fencing is, if anything, preferable to wire-fencing… it is less liable to suffer from heavy snowstorms or frost than wire … [and]… affords more shelter to sheep. Dyke-fencing is specially (sic) commendable on exposed ground, and if it cannot be raised to a sufficient height conveniently, a wire or two…[is]… run along the top of it in order to prevent the sheep from leaping over.

GVA Galloway (1813: 84) warns against building a dyke too low:

> A few inches deducted from the height of a dike is most injudicious economy… Nothing can place the mistaken economy of such a practice in more striking light than the means employed to remedy the evil. The farmer is obliged every season to devote many days labour to cope his dyke with whins, or to strew whins along the sides of them, to the great injury both of the dykes and of his pastures.

The writer recognised the thorny branches of the whin took time to gather and put on top of the wall and, if they fell off, were a hindrance to grazing animals. Thorns also became

Figure 54: A thin flagstone holds fence wires above the cope. Latheron, Caithness. See also Figure 55.

entangled in the sheep's wool and reduced its value.

GVA Dorset (1812: 174) also quotes the use of 'bushes' to increase the deterrent value of a wall:

> …where stone is much more plentiful than wood, the fences consist of walls of loose stones, and sometimes they have a few bushes under the topmost stones, to prevent sheep from climbing over them.

Another way of adding to the height of a wall to deter livestock was to erect a 'sheep syme', a rope set along the top of a stone dyke, as in Cumberland, northern England. Fixing willow and twigs to the cope also added to the height.

A post and wire or post and rail fence, running beside the wall and reaching above it, also add height. This is easier than taking off the top courses of stone and rebuilding higher, a situation which may become necessary if the use of a field changes from arable to sheep or cattle grazing.

In medieval times prickly additions were also used to inhibit trespassing by people.

Figure 55: Another view of the cope in Figure 54.

Coldstream (2004: 6) tells us:

> …for all its inherent dangers, the building site was open to the curious. At Troyes Cathedral in France, for example, the authorities were obliged to cover unfinished masonry with briars to prevent children from climbing on it…

'Berd' and 'berdio' are Welsh terms for the fixing of thorns on top of mountain walls to increase the effectiveness of the barrier.

Dutton (1808: 152) uses the term 'bearding' for walls in County Clare, Ireland:

> Walls are sometimes so badly built, and so low, and the breed of the sheep so active, that it is necessary to put a bearding of furze, briars, or thorns, which soon fall off, and leave gaps, that require constant mending.

The 'saddle stones' in Figure 57 shows an arrangement of copestones where poles were laid, in notches, between occasional uprights, to raise the total height of a dyke.

Hogsback a long copestone, semi-circular in cross-section, maybe three feet long. Typically, these are fabricated from sandstone or limestone, are fairly expensive and limited to use on special walls. 'Half barrel' is another description of these copes.

Jack and Jill Brooks (1989: 16, 91) describes fragmented slate in a herringbone pattern in a 'slate hedge' (a style of stone-faced earth bank) as: 'Jack and Jill or Darby and Joan'. Brooks thought the use of a herringbone pattern (Figure 32), for at least the top two courses of a slate hedge, provides '...a good rooting medium for plants growing along the hedge's crown or 'comb'.'

Kaipit (Scots) caped, covered, coped or topped.

Keap-stone / keaping-stane / kepstone (Scots) a copestone. Terms more associated with buildings than drystone walls. The words include elements of 'cape' and confirm the stones are the final course on top of a structure.

Large stone coping (Ireland) a description of coping with large stones, butted tight against each other along the top of the wall, with the minimum of cutting.

Line / to line the dyke (Upper Annandale, Scottish Borders) Prevost (1957: 99) defines this as:

> The process of using a line or string when finishing the top of a stone-dyke and when laying the *cope stones*. This ensures a neat finish, a straight top and 'lines the dyke with the ground.'

The dyke therefore 'flows' with the land, allowing for occasional dips and rises. Long sections, coped at one time, ensure a 'sweet' top line, not a disjointed irregular appearance.

Locked top this arrangement of wedged vertical copestones secures the top of the 'double' (Figure 56). It is a quick way to achieve the top twelve inches or so of the dyke.

In GVA Galloway (1813: 86-87) The Reverend Samuel Smith confirms the origin of this style of coping in southwest Scotland:

> In pursuance of this idea of making the top as firm as possible, the late John McAdam, Esq. of Craigenguillen, about sixty years ago, invented a mode of putting on the top of a dyke, which had obtained the name of a *locked top*. This is made by chusing [sic] out a set of broad and flattish stones, which are laid edge-ways on the top of the dyke as close as they can be put together; and when a considerable extent has been thus laid, thin stones are driven in, like wedges, at small intervals, which bind the whole so firmly together, that, when well built, a stone can hardly be taken out of the top without an iron-crow.

Smith continues with more detail:

> The locked top is formed of stones not less than ten inches long, so dressed and laid that one cannot be removed without injuring the rest. Where stones will not admit of dressing, the locked top is dispensed with.

Smith means the stones should be ten inches tall when he says 'long'. He notes that, where the stones are too rough to shape, the locked top is not used.

Mundell (1981: 120) represents the Scottish Borders by quoting a well-known dyker:

> The top should be set straight up. If you have them lying about you cannot key them

> tight together,' said Charlie [Jardine]. 'If the topstone is a bit loose a pinstone hammered in will tighten it.'

McRaven (1989: 37) describes a locked top:

> Flat stones, if they're thin, tend to get pushed off. Flat stones standing on edge and wedged in tightly last longer and have been popular in the past.

His illustration of this style is entitled 'vertical capstones'. McRaven is based in Virginia and will have seen the locked top in Virginia and Kentucky; it is less common in New England.

Incidentally, Rainsford-Hannay (1957: 45) credits the invention of the locked top to 'John MacAdam of Craigengullich as far back as 1753.' The Reverend Smith, above, refers to 'John McAdam, Esq. of Craigenguillen' – variations in spelling were less important in the eighteenth century.

Mortared copes these give additional security, especially where the wall is in a public space or vulnerable to disruption by livestock, especially cattle, using the top of the wall as a rubbing post. The mortaring can be discrete, almost invisible. There is a danger the wall may settle beneath a rigid mortared cope and leave it hanging; this is less likely if the wall is built strong and tight and the occasional copestone is left unmortared to add some flexibility.

Low (1834: 612) contrasts turf and mortared coping and recommends a coverband with a mortared cope:

> Different kinds of coping may be placed upon the wall, to defend it. One of these consists merely of turf, two sods being laid upon the wall, with the earthy sides placed towards each other. Another species of coping consists of large stones, which, being closely built and wedged together, are cemented by mortar. This is a complete and durable species of coping; but when it is used, a row of flat stones should be laid on the top of the wall immediately beneath the coping, and made to project a little on each side of it.

One up, one down a description of a **castellated** cope.

Overhanging copes and covers these provide a deterrent, inhibiting livestock from climbing in, or out. Margaret Brooks (1977: 103) details how Myers Johnson, a waller from the West Riding of Yorkshire, built a course of coverstones, projecting a few inches, with copestones on top to secure them:

> Sometimes he builds a plain flat top of smallish throughs placed side by side. This is more susceptible to dislodging so actually he often combines this with the weight and stability of upright stones. He thinks this is a particularly effective hindrance to sheep…

GVA Perth (1794: 116) advises how to sheep-proof a wall: 'Even a common wall of five feet high may be rendered fencible against sheep… by laying on the top a tire [sic] of flags set on edge, standing in a perpendicular position and of such a height and length, as to project over the wall 6 or 8 inches on both sides.' If flat stones for a vertical cope are scarce,

Figure 56: A 'locked top' using rough boulders. No coverstones.
The tight cope has survived a collapse in the 'double'. Inverness-shire, Scotland.

they should be supplemented '...by putting land stones between them, so as to keep them at six inches distance...' This combination of round fieldstones and flat flagstones produces a kind of castellated cope. If flagstones are not available for a vertical cope he considers '... long stones of any form may be used for the same purpose, taking care always that they shall project over the wall, and that a coping of small field stones be piled above, tapering to a proper ridge.'

GVA Perth (1794; 116) also recommends '...short truncheons of any kind of timber be driven, leaning somewhat to the side of the wall on which the sheep are...' These timbers would then be bound with plaited withes or straw ropes.

If a wall is built along a slope, one side may be taller than the other. This can be compensated for if the copes and covers on the topmost, uphill side, overhang. The uphill side can be equally stockproof though its effective height is less than the downhill side.

Winskill Farm, near Settle in Yorkshire, England has medieval walls which include flat copestones overhanging to one side, reputedly to keep out wolves. Lord (2004) and (2010) details these walls and uses the overhanging cope to determine how the walls were originally built and which side of the wall was inside or outside the pasture.

In the English Lake District 'cams', the local term for copestones, often overhang to the neighbour's side of the wall to deter livestock.

Pin gwyddel, literally translated as 'Irishman's pin', is a Welsh term for a stone poking out to stop sheep jumping the wall.

Corangamite (1995: 22, 59) has several references to rabbit-proof drystone walls in Australia. Many were built by professional wallers from Scotland in the 1880's. The 'double' wall was preferable for repelling rabbits. A tightly built 'double' wall was more of an obstacle than the open stonework of a 'single' wall. Foundations were dug down to bedrock with a rock apron below ground (like an exaggerated scarcement) to prevent rabbits burrowing under. Overhanging copes with additional timber slats under the cope, to increase the overhang, added to the rabbit proofing. Wire netting, supported by timber slats under the copes, was also used to create an overhang.

One attempt at rabbit-proofing was a wall running between Lake Corangamite and Lake Purrumbete in the Western District of Victoria, Australia, apparently built around 1880. Corangamite (1995: 22, 59) does not state a length but, from the map, a direct line between the two lakes is about 5.5 miles. The outside of the wall was built with a smooth vertical face and cope stones projected 10 cm (4 inches), or more, to keeping rabbits out. The inside face had a slight slope, no projecting cope and two sets of throughstones to act as a ladder; rabbits inside the wall could get out, and be kept out, by the more vertical wall face on the outside. A man at a rabbit-proof gate guarded the road. According to the account 'the efficacy of the wall is not known.'

Trias and Arrom (2002: XLIII) write about stock proofing in the Spanish Balearic Islands:

> As walls became higher they were given a finish to reinforce them and in any case, improve their appearance. One of the most interesting finishes is on goat walls, whose last course projects out from its face to prevent livestock from passing.

Overlay belt (Kentucky, USA) a coverband, the longer flat stone between the 'double' and the 'copestones'.

Peen (Scots) a peak or apex, as in the coping of a dyke.

Puddle cap (New England, USA) Thorson (2005: 71) describes this as a coping style where there are no formal copestones, only a layer of rubble or smaller stones along the middle of the top. This use of 'puddle' probably comes from an old meaning of the word — 'a state of disorder, a muddle, mess, confusion.' 'Cape and rickle' is a similar Scots style.

Rag (Orkney, Scotland) the cope on a dyke, named for its ragged appearance.

Rickle cope DSL advises the word 'rickle' derives from a Norwegian dialect term *roykla* meaning 'to throw (stones or peats) together to form a heap; to build (up) quickly or unsystematically.'

This is a good description of a rickle cope, a coping style where a course of small stones is piled on top of the 'double' or coverstones sitting on the 'double'. This style is ideal for areas where there is a shortage of suitable stones to build a **locked top**. 'Rickle' is also defined in DSL as a loose, carelessly thrown together pile, often applied to a rough built stone wall.

Rigbone (Shetland) a local description of a style of cope using large stones, probably derived from 'rig-bane' or 'rig-bone', old Scots words for the backbone or spine.

Saddle stones (New England, USA) Hubbell (2006: 44) has an image of a very low drystone wall with vertical slabs (the saddle stones) laid at regular intervals across the wall, each with a notch cut into the top edge to support long poles. This raises the height of the wall to about two feet. This coping feature exists in Scotland, but the wall is a lot higher with taller notched stones built solidly into a normal vertical cope (Figure 57).

Wilson (1849: 606) describes these vertical slabs with notches, designed to deter sheep:

> …erect long stones, like pavement stones, about a foot above the ordinary coping, at short distances from one another, among the coping stones. These stones have a notch cut into their upper edge, which can support a small rail of wood, or a rope of twisted straw.

Wilson thought the frailty of the rope deterred the sheep. In Cumberland, northern England, this rope was called a 'sheep syme'.

Scotched coping according to Cridland (1922: 140), 'The irregular or so-called scotched coping is not at all satisfactory for a garden enclosure, as it is a line of agitation and most unrestful.' These copestones are roughly dressed, or 'scutched'.

Seconder (Dumfriesshire, southwest Scotland) a smaller stone placed under a short copestone to raise it to the correct height.

Figure 57: Saddle stone. Poles, or a rope, would have been laid, from notch to notch, to add height and scare livestock. Dalcross, Inverness, Scotland.

Sheeters (Dumfries and Galloway, southwest Scotland) coverbands.

Shelf cap (Kentucky, USA) the coverband.

Sickle-coping Nicol (1803: 299) mentions this in a footnote, apparently referring to the rough cope on top of a Galloway dyke — 'The sickle-coping seems to intimidate sheep and cattle more than any other…' No further description of the cope is given. A sickle is a small

Figure 58: The regional style known as a 'Galloway Dyke'. Note 'single' stonework on top of 'double'. A good use of large round boulders. Contrast this with Figure 59 where the 'single' consists of flatter stone. Southwest Scotland.

reaping hook, a type of scythe. 'Sickle' may refer to the roughly crescent shaped copestones. The words 'intimidate' and 'terrify' were often used to describe their effect on livestock.

Single a 'single wall' is built one stone thick with the largest boulders as foundation graduating to smaller stones in the cope to give a level top line. The term is also used by Prevost (1957: 100) and Rainsford-Hannay (1957: 33) for the upper part of a drystone wall in southwest Scotland. Large stones are laid across the full width of the wall, immediately on top of a continuous set of throughstones (Figure 59). This could be the top sixteen inches or so of cover and cope of a normal 'double' wall or the top half of the regional style now known as a 'Galloway dyke' (Figure 58). This top section of the dyke has also been described as 'single-work' or 'singling'

Slopes and inclines when building or repairing a dyke on level ground the angle of the copestones should follow those on nearby dykes. When coping a wall on an incline the setting of stones usually starts from the bottom. Should the individual stones lean uphill or downhill or be vertical? There are several opinions and many local styles.

It is fairly straightforward to lay copes from the bottom of a slope when they lean downhill. First lay a large stone, to act as a 'bookend'; lay the first copestone against that one, and so on, up the slope. Each copestone reaches across the width of the wall and secures the horizontal stones below it. They also sit snugly, side by side, with maximum surface contact. Gravity will tighten rather than loosen them. Copestones can be tightened by tapping small

Figure 59: A nineteenth-century boundary dyke with 'single' top, nearly half of the total height of the dyke, set on a continuous line of throughbands. A variation of the 'Galloway dyke', using flatter stone.

wedges between them.

Laying copes from the bottom of a slope, with their tops leaning uphill, is more awkward; each copestone must be laid *under* the last one. The effect is more easily achievable by starting at the top and working downhill. This style of cope is more likely to be insecure.

Rainsford-Hannay (1957: 76-78) is adamant the copes should be vertical, maybe lean slightly uphill, not leaning downhill — 'If it is thought necessary that stones should lean a little, they should be set to lean uphill, against the hill as the wallers say.' He notes the high incidence of downhill sloping copestones in the north of England.

Drake (2008: 77) refers to copestones leaning uphill but '...in more than 25 years of walling I have only come across one single example of that technique...'

His experience, in Cumbria, northern England, is of downhill sloping cams (the local term for copes). He accepts uphill sloping cams were appropriate in the exceptional case where cams are few and the slope is so steep '...that to lay them flat against the wall-top would mean they were tilted downwards, and liable to slip out of place.'

Mundell (1981: 120) confirms the standard for the Scottish Borders — 'On a steep brae topping is always against the lie of the hill.' The copes are laid sloping downhill.

In Kentucky, USA, copestones are traditionally sloped downhill, as per Dry Stone Conservancy (2001: 30). In practice copestones are generally laid as near vertical as possible, each one tightening against its neighbour as the coping climbs up the slope. Sloping copestones is easier with thin stone. Boulder copes tend to sit solidly rather than lean in any direction.

The top courses of the 'double', or the covers, if present, are often stepped to allow for the gradient and provide a more secure base for the copestones.

Anecdotal evidence suggests copestones on the walls of the North Yorkshire moors deliberately slope towards the farm, to act as a guide on misty days. It is more likely the farmers are familiar with the details of each wall and can calculate the direction home.

Section 3.5 also discusses general points regarding building drystone walls on a slope.

Snab (Scots) a variation of 'snap', a roughly triangular coping for a drystone wall with the point upwards. A derivation from the Scots word 'snab', a projecting part of a rock or hill, a rough point.

Snap a vertical stone cope. DSL has the word 'snap' possibly originating from the old English word 'snape', to taper.

GVA Selkirk (1794: 17) has one of the earliest descriptions of dyking and notes turf tops are still the most common in his area:

> [The dykes] are sometimes built with a snap, that is, the upper part, for a foot or two, of loose stones set on edge, leaving a small space between each, and tapering to a point at top: but dikes built without a snap are most common, and are five feet high, including a row of feal or sod on top, to prevent the stones from falling off.

The term 'snap' originated in southwest Scotland. It spread nationwide as this type of cope gained popularity. This development in coping secured the top of the dyke and created a formidable barrier to livestock, a definite improvement on earlier attempts, such as turf and thorns.

A 'snap fence' or 'snap dyke' is a wall with a 'snap' cope. Walker (1812: 105) stresses the importance of a good top to a 'snap dyke':

> To obtain these cop-stones of a proper size and form, is the most important article in building a dyke of this sort. It is on the length, strength, and weight of these flat stones that the strength and stability of this dyke must depend.

Soft capping defined by Morton (2011: vol 1, xv) as '…using living plants and soil as a thermal and moisture-buffering layer, thereby impeding the decay of masonry structures.'

A layer of soil or turf over an old building or a ruined gable's top edge looks attractive and helps protect the stonework. Many old structures survive when buried and deteriorate if exposed to the elements and curious mammals.

The vigorous rooting systems of some plants will cause damage but Dutton (1808: 151) thought ivy (*Hedera helix*) strengthened a wall — 'If ivy was planted to these dry walls, it would strengthen them greatly, and prevent cattle from throwing them down.'

Nowadays anyone seeing the damage caused to Irish walls by ivy would disagree.

Lassure (2009: insert 6) tells of the planting of irises on the top of drystone huts in southern France. The roots help contain the earth on top of the roof and add to the waterproofing.

Soldier cap / soldier row / soldier course (Kentucky) a cope with stones set with their longest edge upright. Elsewhere in brickwork and drystone walling, 'soldiers' are small stones set on edge, literally standing to attention, within the 'double', to fill a gap between two larger stones.

Soldier coping (Galway, Ireland) vertical coping using stones which are all the same height.

Stobbie (Scots) describing the top of a dyke as rough, spiky, prickly or bristly.

Tablet a masonry term for a coping stone set flat on top of a wall.

Tabling / table / table stone (Scots) the stone coping of a building's wall or gable. In Caithness, northern Scotland, this is a term for the coverband, below the cope, which helps keep the inside of the wall dry. This will have been especially important for older walls in that county, built of softer stone, or stone which tended to delaminate. Nowadays stone, from deeper in the quarry, is harder.

Tie rock (USA) a throughstone, a rock within the body of the wall which lies across the wall, linking the two faces of the wall and therefore strengthening it. Copestones and coverbands are also types of tie rocks because they sit across the wall and tie the two faces together. For a fuller description of a tie as a throughstone see Section 3.2 (iii).

Tie course (Kentucky) Murray-Wooley and Raitz (1992: 197) include this definition —'Coping-rocks and cap courses that cover the full width of the fence also act as tie-rocks. A course composing entirely of tie-rocks is a tie course.'

Tobbing off a Cornish hedging term for the topping off of a Cornish hedge. The domed earth top is typically covered in turf. Cornish hedges are described in detail in Section 4.1.

Top (Annandale, southwest Scotland) Prevost (1957: 101) defines the term as — 'To top, to lay the top stones on a dry-stone-dyke. The stones are set from one end only and that the lower end.'

The copestones are laid from the bottom end of a slope, and up the hill.

Top-heavy according to Alan Brooks (1989: 18), '…fully one third of the wall's weight is supposed to be in the coverband and coping.' This seems a high proportion for normal 'double' walls but could be the case in some of the big Galloway style boundary dykes where the coverband and copestones can be approximately half the height. The upper half of a dyke can become unstable if too many large stones are used. In GVA Galloway (1813: 85), The Rev. Samuel Smith advises the builder to take care:

> … and by no means to overload the coping, as is often done, with large projecting stones, which, if not made to project with the greatest exactness, or if the dyke, from any cause, should sink unequally, must infallibly occasion terrible breaches in it.

Cairns (1975: 23) saw nothing romantic about dyking, especially the heavy lifting for the top courses:

> … when people watch the covers and cope being 'clinked' into position, they just can't believe that it's brute force combined with art! Some of the heavy cope can make the blue sparks shoot out of the dyker's eyes and his upper lip curl up. It's a very funny feeling and it makes one a bit dizzy when one stoops for the next one.

Top range (Scots) the highest course or coping, on a wall.

Top stones Prevost (1957: 101) defines this as a term from Annandale — 'Stones of certain dimensions and shape selected and used for the coping of a stone-dyke. These were sometimes stipulated to be 10 inches or more high.' The specifications for drystone dykes in the eighteenth and nineteenth centuries were quite detailed and include exact specifications for cope and throughstones.

Borchgrevink (1982: 84, 90) quotes Charlie Jardine describing the fitting of 'copestones' which match the description of coverbands. On top of these 'copestones' the upright 'top stones' bring the dyke to finished height. This top combination adds 18 inches (460 mm) to the total height of the dyke.

Prevost and Borchgrevink are both writing about dyking in southwest Scotland.

Topping out / topping off a ceremony to mark the fixing of the topmost stone in a structure. Scandinavian custom says a branch of a tree is fixed to the highest point. This ceremony is more associated with buildings than stone walls.

Figure 60: Turf cope. Sutherland, Scotland.

Toppings (Yorkshire) copes.

Tops / toppers (England) copestones.

The term 'toppers' is also used to describe the top course, in effect the cope, on the 'single' walls built in the Mourne Mountains, Ireland. The stones are large, possibly 175 kg (385 lb); they are placed carefully, with consistent batter from side to side, and lying across the joint of the two beneath. See Figure 80 for a full description of this Mourne style.

Turf coping ideally two layers of turf sods are laid on top of the stonework, the first layer is grass side down, the top one grass side up (Figure 60). This soil-to-soil contact promotes the continued growth of the grass, especially in regions with high rainfall. Raistrick (1973: 17) refers to '...occasional parts of Weardale, where a row of sods is built in at the top, and the matted roots bond the top stones together.'

GVA Dumfries (1812: 151) thinks sod-coping is inferior to the Galloway cope (a solid stone top) because sheep and wind cause damaged to the turf.

In areas where the copestones are relatively small, for example the Durham Dales in central England, they are sometimes laid on a layer of turf which acts as the coverband. This releases any long stones for use as throughbands.

GVA Caithness (1815: 75) recommends a stone cope (a snap cope) though he prefers a turf cope if the stone weathers badly:

> This rough, or snap coping, was introduced into this county, about 20 years ago, and is found to answer well, where the stone is weather-proof; but in case the stone yields to the weather, the old mode of covering the top with sods, is preferable.

Some of the older dykes in Caithness are made from stone which was quarried from nearer the surface; it was softer and more inclined to deteriorate compared to stone extracted from lower levels in more recent years.

Dutton (1808: 151) mentions the local coping style in County Clare, Ireland and warns against use of turf:

> Some few put a coping of mortar, but the usual one is sods, which last but a short time, and greatly injure the ground; from which they are stripped: the wall is usually finished by throwing all the small broken stones on top...

Dickson (1805: 111) also advises against a turf cope:

> They [the walls] should always be coped at the top, either by means of stones laid edgeways, or with turf about six inches in thickness, ...the former is... by far the most advisable method, as being the most durable, and affording much difficulties to animals in their attempts to get over.

Nicol (1803: 298) warns of the danger of frost:

> ...*coping* with *sod*, is by no means advisable. It may tend to keep the uppermost stones together for a time, until water penetrate the turf; but afterwards, it is of mischievous consequence to the wall.

Vertical copes details of, and justification for, this style of coping is included in several of the definitions in this section. Vertical copes are ideal for the narrower patterns of wall which developed after the 1700s.

Older New England walls seldom have vertical copes. The build style is an older pattern compared to those found in the British Isles, the walls are wider and shorter. Vertical coping is an inconvenience because the wider walls would need longer copes which would have to be shaped. Large heavy slabs, laid horizontally, do the job.

Rankine (1862: 388) describes the use of thin stone set on edge:

> Random rubble coping forms an exception to the general rule that laminated stones should be laid with their layers parallel to the beds of the courses.... stones are very often set on edge with their layers vertical, and perpendicular to the length of the wall, so that the edges of the layers alone are exposed to the air at the top, as well as at the sides of the cope.

This exposure means that vertical copestones should be carefully chosen, no use of friable or 'rotten' stone which will weather badly.

The **slopes and inclines** entry in this section deals with instances where the copes are not vertical.

Wooden covers according to Prevost (1957: 101), 'Red cedar shingles were used as covers in a 100-year-old dyke at Galt in Canada.' Galt is now part of the city of Cambridge in Ontario, Canada.

Todd (1859: 465) advocates laying two strips of wood, three inches wide, the longer the better, along the top of the wall with their edges even with the face. The copestones were set on these timber strips.

Corangamite (1995: 22) includes an image of 120 year old timber slats placed under the copestones to create an overhang. These were primarily placed to deter rabbits and also acted as a top course for the 'double', upon which the copestones sit.

Yorkshire tops a coping system with alternate high and low stones. A type of castellated cope.

3.2 (v) Hearting, pins and trigs

Hearting is the smaller stone inside a wall. It secures the larger stones and acts as the bed for subsequent courses. Hearting should be placed carefully by hand, not shovelled or poured from a bucket. Some narrow drystone walls have hardly any hearting. If the face stones reach in from both sides, and touch in the middle, or there are good throughstones, this can compensate, in the short term, for a lack of hearting.

Early commentators on drystone walling stressed the importance of good hearting. In GVA Galloway (1813: 89-90) The Reverend Samuel Smith of Borgue laments:

> ...the most prevalent and incorrigible fault of all... is neglecting to build the inside of the dyke. Here the stones ought to be of the same good quality, and laid with the same care and attention as those of the outer faces. But instead of this, by the generality of dykers, a parcel of rubbish is loosely huddled together, filling up very imperfectly the intermediate space, and affording no firm resting place for the throughbands and stones that overlap; so that, in a short time, the two sides either fall together, or large spaces are left in the

middle of the dyke.

The hearting material should be of good quality. GVA Galloway (1813: 88) warns the reader:

> ...the dyke-builder... has many temptations and various ways of deceiving... The quarries abound with soft friable schistus, which falls to pieces by the influence of the sun or air.... the cunning workman, if he dare not risk putting them in the outsides, throws them in the middle of the dyke, where they remain concealed till the falling in, or bulging out, shews of what stuff it is composed.

Stephens (1853: 588) emphasised the need for good face stones, properly hearted:

> Very many dry-stone dykes in this country are constructed on erroneous principles, the stones being laid promiscuously, and more with a view to give a smooth face than a substantial hearting to the wall.

Todd (1859: 463) also criticises poor hearting:

> ...with merely a row of stones for each face, while the middle of the wall is filled with stones thrown in promiscuously, without being placed so as to lie firmly, and without having the face stones chocked on the inside.

Todd and Stephens are using the word 'promiscuously' in an archaic sense, 'mingled together, confused', as defined in Bailey (1675).

Adair (1998: 27) emphasises 'keep the heart up' in the middle of the wall so the face stones lie solidly on it and tilt slightly to shed water. When an old wall is dismantled for repair the inside is often remarkably empty. This usually means the builders did not put in enough hearting to stabilise the face stones. When they flexed, the loose hearting fell to the base of the wall and the outer faces drifted apart.

Hearting should be angular. Railway ballast is a good size; it consists of hard angular crushed stone of around 50 mm (2 inches) diameter. It is termed 'track ballast' in North America. Gravel or pea gravel is not a wise choice; it is too rounded, too small, like filling the wall with ball bearings.

Dykers working in pairs, one each side, should see the gaps behind the other's work and place hearting accordingly.

Dykers on piecework (paid by the yard), were known to neglect the hearting because it took up time.

This is a suitable place to repeat the tale (however mythical) of the dykers who asked for immediate cash payment. The farmer checked their work; it seemed satisfactory, so he obliged the men. A few days later he passed the dyke and noticed a bad smell. The 'dykers' had chopped up turnips to use as hearting and been anxious for payment before the vegetables rotted.

The following terms apply to hearting and its use

Back filling rough masonry built behind a facing or between two faces. This is the mortared equivalent of hearting between two drystone faces. The fine face of ashlar work, large cut stones with very fine joints, often has a rough back face with rubble infill because solid ashlar work is too expensive and time consuming to build.

Backing up (New Hampshire, USA) laying stones in the heart of the wall, behind the face stones.

Backup that part of the wall which is behind the face stones.

Figure 61: Building a drystone wall in Swindale Valley, near Haweswater, Cumbria, in the early 1950s. © Illustrated London News Ltd/ Mary Evans.

Bee's feet Cramb (1992: 163) has this as 'Slang for small stone used in random rubble.'

These are the small stones, stacked between larger blocks in mortared walling, probably named for their similarity to the jointed segments of a bee's leg.

Bugdalin / bugdealling DSL defines this as 'loose material to fill in between stones in building a wall.' The word is derived from Dutch or German terminology applying to 'Anything used to line the hold of a ship before putting in the cargo; hence packing of any kind.' The word was possibly introduced to Scotland by Dutch sailors and, in a stoneworking context, was applicable to mortared walling rather than drystone work.

Bullets (Northern Ireland) small stones, gathered for hearting.

Bulletstane (Scots) a round stone.

Button a small lead slip put between blocks of stone by masons to ensure the mortar is not squeezed out, and the courses are kept level while the mortar is curing. Oyster shells were used in Durham Cathedral, England (twelfth century). The modern plastic version will not swell, compress or corrode.

Carreg glo (W) a smaller stone used as a wedge to lock larger, vertical copestones.

Cementum (Latin) stone chips, quarry stone for walls, small stones or rubble.

Cerrig bol clawdd (W) literally 'wall belly stones' — hearting or filling.

Cerrig llanw (W) literally 'filling', the hearting.

Cerrig sofl (W) literally 'stubble stones', pinnings.

Cherry-cocking / cherry caulking / cherrycaulking/ cherry pointing / cherry stones / cherry cocks (Scots) the small stones pushed into the mortared beds and joints of masonry to reduce the amount of mortar used, and protect the joint from the elements. These small stones are also known as 'pinnings or 'gallets'. The act of placing the small stones is known as 'galleting' in England.

Chinking (New Hampshire, USA) jamming or hammering small stones into cracks after the wall is completed. Chinking adds little to the structural strength. It produces a tighter looking face with fewer gaps. It also reduces the openings available to rabbits or vermin. Care is necessary to make sure the hammering does not disrupt the larger stones. This must not be confused with 'front pinning'.

Chocking the support of larger stones by smaller pieces, to bring the stone's face into alignment or allow for an irregular shape. Todd (1859: 463) emphasises the need to chock the face stones from the inside and not throw in the hearting stones. Chocking on the outside face is insecure and inevitably falls out. Too much chocking can cause unstable building. Ideally each stone should sit directly on the one or two below, with a minimum of small stones between.

Chocking stones smaller stones which 'chock' or assist in the support of larger stones.

Clinkers (Scots) broken pieces of rock.

Clots (Yorkshire Dales, England) small broken pieces of stone.

Clover stones (Perthshire, Scotland) small stones gathered off the fields after ploughing, used for hearting.

Cog a smaller stone used to secure the large boulders in the 'single' walls built in the Mourne Mountains, Ireland. These walls are built of large stones, one stone wide, with largest at the foundation, decreasing in size towards a level cope. See 'ditch' in Section 4.1 and Figure 80 for a full description of this Mourne style.

Corefill a mix of stone and lime mortar used in masonry walls to fill the space between the two outer stone faces. When a section of stone and lime mortar wall is completed, a thin hot lime mortar mix is poured down into the wall to fill up the remaining voids. This adds considerably to the wall's solidity and weatherproofing.

Crottle / crotle / crottil (Scots) a small fragment of rock or any hard material, a crumbly substance in its raw state. These words also apply to lichens, used to dye wool for tweed.

Fill / filler / fillers / filling / fillings this term, with slight variations, is widely used throughout Scotland, England and North America. Rainsford-Hannay (1957: 76) refers to the use of 'filling' in Yorkshire but reckons:

> ...'hearting' is the better word, because every hearting stone should help its neighbours and be set to bind and grip with adjoining stones. The word 'filling' conjures up the spectacle of a careless man taking shovels full of sand and rubbish and tipping them into the wall. An autumn's rain will wash this stuff into the bottom of the work and leave the wall hollow.

Flitter / fluther (Scots) a splinter of stone.

Foggin' the wa's (Scots) as late as the nineteenth century this was an autumn activity; stuffing the gaps in house walls with moss to keep out the worst of the weather. It was a crude version of internal plastering or rendering. JSD quotes 'fog' or 'fouge' as a generic name for moss in Scotland and refers to '…the most miserable hovels built with stone and turf, without mortar, and stopped with *fog*, or straw, to keep the wind from blowing in upon them.' JSD also mentions 'A rowing stane gathers nae fog' from a 1776 collection of Scottish proverbs. 'A rolling stone gathers no moss' is a modern translation.

'Yorkshire fog' (*Holcus lanatus*) is a type of grass, commonly known as 'velvet grass'.

'Fog cups' is an eighteenth-century Scots term for a type of lichen, described in DSL as being 'of a greenish mixt colour.'

Front pinning there is a difference between 'pinning' and 'front pinning'. Pinning is the act of tapping or placing small stones into the voids between face stones for the sake of appearance, after construction is complete. Front pinning is the wedging of face stones, on the outside of the wall, during construction. Front pinning has no structural value and is a symptom of bad building practice. These insecure pins will fall out. For maximum security any wedging or pinning of a face stone should be from behind, from the inside of the wall.

Gallets chips or spalls of stone. Galleting is the insertion of small chips or spalls of stone into the mortar joints of rough masonry to tighten the wall, protect the mortar joint, reduce the amount of mortar required or add detail to the appearance. Also called 'garretting' in the USA.

JSD suggests 'gallet' may come from the Scots Gaelic word *gallad*, a little girl, which in turn derives from *galla* meaning brightness or beauty. The link is tenuous, though these little stone chips do add to the charm of properly mortared walls.

Garreting (USA) the insertion of small splinters of stone into mortar joints before the mortar has firmly set. A variation of 'galleting'.

Glut-lionaidh / glutaran (SG) the packing in a wall, the hearting. Dwelly defines this term as the packing of gravel in the middle of house walls. According to Walker and McGregor (1996: 4) the drystone walls of Hebridean Blackhouses can be five to seven feet (1.5 to 2.1 metres) thick. That requires a lot of infill.

Grout (New England, USA) Wood (1977: 364, 368, 373) uses this term for small, loose stones. They are put into the bottom of dry walls, when built on heavy clay soil, to assist with drainage. They are also used for chinking which '…pick up the weight of the stone, and make that rock good and solid.'

Hakahaka (Hawaii, USA) the loose filling between the two outer faces of the wall.

Haut a term from southern Scotland and the Scottish Borders, defined in DSL as a variation of 'hot', meaning to heap up or heap together or '…to gather with the fingers, as one collects stones with a garden rake.' This activity is familiar to any dyker who has gathered the last small pieces of hearting from the grass.

Hearting probably the most widely used term for this important component of any drystone wall. Prevost (1957: 99) lists 'hearting' as a Galloway word, comparable to 'packing' which is referred to as an Annandale term. The test for adequate hearting was a kick. If the face stones moved this suggested there was not enough hearting. Annandale and Galloway are neighbouring districts in southwest Scotland, each proud of their heritage and traditions.

Hert a Scots variation of 'heart', to fill the centre of the dyke with smaller stones.

Infill (Mendips, England) hearting.

Knogging Arkell and Tomkeieff (1953: 67) define this as 'small refuse stones used for the inside of walls', a term from Northamptonshire and Worcestershire in England. This will refer to wet masonry where the finer cut stones are used for the outside face and small rubble is used as infill. 'Refuse' is used in the sense of 'waste' or 'rubbish'; the stone is unsuitable for use as a face stone.

Lacing course a horizontal course of cut or shaped stones constructed to level up, and strengthen, walls made from irregular rubble or lumps of flint.

Middles / middlings (Yorkshire, England) hearting.

Nickers (Scots) small wedges used to bring the stones in a course up to a level.

Packing (Annandale, Scottish Borders) Prevost (1957: 99) quotes this as comparable to 'hearting' which he lists as a term from nearby Galloway.

Peggings (Dumfriesshire, southwest Scotland) Borchgrevink (1982: 90) includes an illustration of the top section of a dyke with the caption — 'Note the little stones, 'pinnings' or 'peggings', put in to finish off the face of the dyke.'

Peinneagan (SG) broken fragments of stone used for filling crevices in the wall, as packing and for levelling the building stones.

Pin / pinn (Scots) as a noun this is a small stone, a 'pin stane', used to wedge or secure a larger one or fill up a cavity in a wall. As a verb it means the act of pinning. 'To pin' can also mean to make a small hole in, say, window glass, by throwing a stone.

Pinnings / pinings small stones wedged into the front face of a wall, supposedly to strengthen the wall, mainly to please the eye. Margaret Brooks (1977: 104) gives the opinion of Mr Johnson, a waller in the West Riding of Yorkshire:

> After the wall is built, Scottish wallers in particular are quite keen on tapping or wedging stone chips, called pinnings, into the interstices to add to the strength and security of the wall. Over-zealous filling of cracks can force other stones out of alignment and, anyway, Mr Johnson considers pinning a symptom of laziness in building.'

Inevitably there are contrary opinions. Loudon (1825: 444) thought pinning every vacant space added considerably to the durability of the building, and harling afterwards with a coat of lime mortar gives '…the whole a finished substantial appearance, which renders them at once agreeable to the eye…'

Such pinning can be especially disruptive when the face stones are small. Several writers in the late eighteenth and early nineteenth centuries still thought there was an advantage to mortaring the outer face of a 'dry' built wall.

GVA Inverness (1813: 118) suggests pinning should only be cosmetic, large stones should not sit on smaller ones:

> It is also a good rule, though not generally known, never to rest the weight of the larger stones in the second or third tier upon small pinnings put in below to support them; but upon the other larger stones in the tier next under them. If any part of the weight is supported by the pinnings, the wall will soon come down. Their use ought only to be to please the eye, by filling up the interstices betwixt the large stones, after these are built.

Rainsford-Hannay (1944: 28) also advises against separating face stones with a layer of smaller material, and when pinning is justified:

> 'Pinning' must not be used to steady large stones on those beneath by inserting small flat stones and pins. This is very often done, but it violates the principle of making the centre of gravity pass through the bedding planes. For this reason dykers should have enough stone from which to choose. When that is the case, good dykers go ahead fast and 'never luik at a stane twice' as the saying is, and need not spend much time in dressing stones with the hammer. This practice of pinning to make a bed has crept in because men have been given gaps to repair without additional stone.

Pinning proper comes when the dyke as such is finished. Then it can be improved in appearance, made rabbit proof, and perhaps very slightly strengthened by driving in small stones into the interstices with firm but gentle taps with the hammer.

Rainsford-Hannay (1957: 80) gives 'Bristol pinners' as a derisory term used by Cotswold wallers for those who use pinnings in their walls. Cotswold wallers pride themselves in tight work with no need for pinning.

'Pinnings' may have had very specific meanings at one time. Nowadays the term tends to be a generic term for any instance where a larger stone is stabilised by a smaller one.

Pinstone a small stone used in pinning, for example to secure and tighten a copestone.

Plugging Corangamite (1995: 91) defines 'pluggers' as those who followed the wallers, filling in small holes in the face of the wall by tapping in chips of stone. In nineteenth century Australia it was important to make walls rabbit proof. The procedure is like Scottish 'pinning', when small stones are carefully tapped into any spaces between larger face stones.

Pointing the mortaring of joints in stonework to fill up the space between the stones, to provide a weatherproof seal and for decorative purposes. There are various styles, many of them specific to a region. The mortar can be flush with the stones' surface, raised in some way to define the outline of the stone or scored with a sharp point to produce faux ashlar lines.

Pointing adds little or no structural strength.

See Section 2.7 for a discussion of mortared versus 'dry' masonry.

Riddlings (Scots) the small stones retained in a riddle, or sieve, which can be used for hearting.

Scablings / scabblings / scapplings / scafflings the small stone fragments produced when dressing stone. They are useful for hearting.

Scalpings a mix of stone and quarry dust which is sometimes, unwisely, used for hearting walls. The fine material soon washes out and leaves the interior of the wall hollow, and weakened.

Sclaffert (Scots) any thin solid substance, a piece of laminated rock which can be equally useful for the inside or face of a wall.

Sclift a small sharp fragment, broken off a stone when hit by a hammer.

Scone (Scots) a small stone, cut to fill a space, a masonry term. In carpentry a 'biscuit' is a small piece of wood used to strengthen a joint.

Scree the stones, some small, some large, found at the foot of a rocky outcrop, the result of weathering of the main rock face.

Scruple from Latin *Scrupus*, a sharp stone. *Scrupulus* is the diminutive form — a small sharp stone. In modern English the phrase 'He has no scruples' means 'He has no conscience.'

Figure 62: Retaining wall for a bridge showing combinations of split boulders and flat stones to make up level courses. Ardgay, Scotland.

Seconder (southwest Scotland) when copes are not of a uniform size, smaller stones, seconders, are used to bring the shorter copes to an even top line. The term is also used in the 'single' walls of the Mourne Mountains in Ireland for those large boulders built above the foundation, on the second course, as detailed in DARDNI (2013: 27). See Figure 80 for a full description of the Mourne style.

Shims / shim stones small thin flat stones used to adjust the position of larger stones or stabilise them.

Shilly (northern England) angular scree, especially of slate or shale.

Shillet / shellat apparently an English word for the chippings and spalls left after rock is trimmed. This word was collected during a conversation with an English archaeologist. There was no idea as to the proper spelling, like many terms for stonework which have been carried forward to modern times by verbal exchanges only. 'Shellat' is a word from Devon, southern England, describing the weathered slate at the top of a quarry.

Skelb (Scots) a thin flake, slice or splinter of rock.

Sklinter (Scots) a piece of something which has shattered, perhaps bone or stone.

Spallai noted by D'Arcy (2006: 30), referring to the spaces between the large stones in the 'single' walls of the Burren, western Ireland — 'Where it is necessary to rid the fields of them, the small stones *(spallai)* are jammed into the holes…'

Spallies (Mourne, Ireland) smaller wedges of stone used to secure the larger stones in the tall boulder dykes.

Spalls / spells / spales / spawls stone splinters or chippings broken from a larger piece of stone. This makes excellent hearting.

Track ballast a North American term for railway ballast, an ideal size for hearting. It consists of hard angular crushed stone of around 50 mm (2 inches) diameter.

Trig / trigged / trigging (Dartmoor, southwest England) a small stone put under a larger one to stabilise it, hence 'trigging', the act itself. In Cornish hedging (fully described in Section 4.1) a face stone can be 'trigged' or wedged up. A trig is also a stone or brick laid part-way along a builders' line to reduce the sag or the effects of wind.

Void space an essential objective of any stone construction is to reduce internal void space to as near nil as possible. If stones are loosely built, or the hearting is loosely packed, any movement or settlement will be unopposed, and the structure's end is hastened.

Thorson (2005: 74) estimates a void space of 26 percent in a perfectly stacked wall made of melon-shaped boulders. Void space in the range 24% — 26% is generally quoted for British drystone walls. French drystone wallers often quote a figure of 30%. Thorson's estimate matches 'Kepler's Conjecture', a mathematical problem associated with the most efficient way to pack rounded objects such as oranges or cannonballs. It was first set in 1611 by German astronomer Johannes Kepler (1571-1630) and not mathematically proved until 1998. The conjecture estimated 26% airspace in a box of perfectly packed spheres.

Hodgson (1918: 85) estimates stone, dug out and thrown into a loose broken heap, increases its bulk by 50%. This suggests there is 50% airspace in a loose pile of broken stone.

One of the outcomes for a rubble wall building course at The Scottish Lime Centre Trust, Charleston, Fife, Scotland for 13-14 September 2016 states '...the importance of laying stones correctly and the use of pinning stones will be emphasised so that no more than 30% of the volume of the wall is mortar.' Mortar is taking the place of otherwise empty space.

A lot of effort is required to reduce the void space inside a wall to zero. In practice this is impossible; extra minutes spent on hearting is more than repaid by the years added to the life of the wall.

The difference between a well hearted wall and a poorly hearted wall is obvious, the material is either there or it is not. In old dykes the interior often lacks hearting because much of it fell to the lower courses as the wall flexed and the two outer faces moved apart.

This is particularly the case where not enough throughstones have been used, they tend to hold the two faces together for longer.

In some of the old upland march dykes (those used to enclose moorland and define boundaries), the hearting is mainly pieces trimmed from the large face stones. The rough nature of the stone and the way it is used, ('single' walling for example), means proper hearting is not quite as important as it is in the more finessed limestone walls of southern England.

Wedge a small stone placed under or behind a face stone to position it securely.

3.2 (vi) Headers, stretchers, arches and lintels

These components make strong openings, corners and ends possible, and allow construction of more ambitious and complex structures. Some terms apply to specialist masonry components, though the terms have been used in descriptions of drystone work. This is an example of one specialist craft borrowing terminology from another, as craftsmen realise the fundamental similarities between their processes.

Abutment the supporting mass of stone or stonework at either end of a bridge which receives and holds the thrust of the arch or supports the ends of a flat span. The term has also been applied in Kentucky, USA, to a pillar or columns at either end of an opening or gate, as per Murray-Wooley and Raitz (1992: 188).

Arch Leonardo da Vinci defined an arch as '...two weaknesses leaning against each other to make a strength.' The Oxford English Dictionary defines an arch as 'a curved symmetrical structure spanning an opening and typically supporting the weight of a bridge, roof, or wall above it.'

Arches allow the construction of vaulted ceilings with more interior space than can be provided by lintols or corbelling.

Ard doras (SG) a lintel stone, especially in houses.

Advice stane / mairriage lintel (Scots) a door lintel with lettering carved on it, the initials of the couple and the date of their marriage (Figure 64).

Buttress a right-angled extension from a wall, built to provide additional support against a slope or help the wall support a weight. 'Flying buttresses' are elegant stonework arches seen on the exterior of many medieval cathedrals. These were often added years after the original build, to stabilise the structure. Cathedral building stretched the limits of architecture and the skills of the masons. Some cathedrals fell down; others developed faults which tested the masons' ingenuity.

Coonyie / cunyie (Scots) alternative spellings for 'quoin', used as cornerstones or round doors and windows. They are often a decorative feature of a structure, providing a contrast in colour and finish.

Dressings the stone or brick parts of a building, the copings, jambs, sills etc, which are distinct from plain walling. These are often made of sandstone or similar easily carved rock, in contrast to rougher stone used on the face of the wall. Contrasts of colour and texture, in whatever material, give a craftsman opportunity to excel.

Head a masonry term for a lintel, the stone which spans the top of an opening.

Header (North America) a specialised throughstone, a stone laid with its longest edge into and *across* the width of a wall as opposed to a 'stretcher' which runs *along* the length of the wall (Figure 63). The shortest edge of a header stone shows on the face of the wall.

An individual header adds strength and, in combination with a 'stretcher', when they are laid over each other on succeeding courses, can form a solid structure at wall ends, corners and openings Figure 31 shows the use of headers and stretches at a wall end. This arrangement is easily modified to create windows or doors. A lintel across the top completes the opening.

A 'bull header' is a header laid on its edge so the exposed end on the face of the wall is taller than it is wide. The word 'bull' is used in the sense of large, big or aggressive; the stone shows its broadest face.

Inband (Scots) a stone set with its shortest face showing in the face of a wall and running all or partly *across* its width. Also known as an 'inband rybat'. The English form, 'inbond', clearly indicates how the stone bonds into the width of the wall. Inbands are used with 'outbands' on corners, openings and wall ends to create a strong structure (Figure 31).

'Header' is the North American equivalent of 'inband' and 'stretcher' is the equivalent of 'outband'.

Ingo another name for **return**, literally the 'in go' from the front face of a structure into the recess for the door or window.

Lintels / lintols long, often narrow, pieces of stone placed over a gap to create an opening and support the weight above, as in a lunky, door or window. 'Lintol' appears to be the older spelling of the term. Lintols allow the construction of structures with wide

Figure 63: Interlocking quoin stones form a corner. Speyside, Scotland.

openings and high ceilings, with less effort than with corbelling or arches.

In the past it was impractical to import lintols for field walls but worthwhile for houses or farm buildings. Sandstone or granite was often used for finely detailed stonework around doors and windows. It was worthwhile transporting, once fabricated.

A lintol stone must be placed carefully over the opening to ensure a secure 'wall-hold' on each side. Morton (1855: 388) recommends 'Stones for window sills should be one foot wider than the opening for which they are intended: they have usually what is technically termed a wall-hold of six inches on each side…'

Lintel ale a celebratory drink given to masons after fixing the door lintel.

Mairriage lintel / marriage stane / advice stane (Scots) the lintel stone of a door bearing the initials and date of marriage of a couple who set up house there, common in higher-status buildings in the sixteenth through eighteenth centuries (Figure 64).

Odder-stane (Shetland / Orkney) a stone lintel.

Ogeour / oger (Scots) a springer, the starting stone for the arch.

Outband / outban (Scots) a quoin or jamb stone used for bonding, with its long side lying along the main wall face, and its short side in the return or reveal. The English forms of 'outbond' or 'outbound' clearly describes how the stones lie out or away from the end or opening. Outbands are used with **inbands** on corners, openings and wall ends. The North American equivalent is **stretcher.**

Quoin / quinze a cornerstone found at the edge of openings such as wall ends, gateways, doors or windows. They are also found on external corners of a structure (Figure 63). Quoin stones are often specially cut at the quarry for a specific use, with two cut faces to match the angle of the external corner or edge of the wall. As with 'headers' and 'stretchers' or 'inbands' and 'outbands', the way quoins are laid produces a strong bond at what could otherwise be a structurally weak point. These stones are useful and often the first robbed from a ruin. 'Cunyie', 'coin', 'coonyie' and 'quinzie' are alternative Scots spellings.

Figure 64: A cracked, patinaed, mairriage lintel. Munlochy, Black Isle, Scotland.

Return a surface angled away from the face of a main wall, for example the jamb of a door or the face of a window opening, normally at right angles to the surface of the wall. Also called the **ingo**, literally 'inward going' of an opening.

Runner a long face stone used in wall ends or openings, with its longest edge running *along* the length of the wall. Also called a 'stretcher' in North America.

Saving stone (Scots) a stone built over a lintel to ease the load on the lintel and distribute the load of the wall on to the jambs, the vertical sides of the opening.

Safe lintol (Scots) a wooden lintel, for safety, put behind a stone lintel.

Spring (Scots) the arc of an arch. A 'springer' is the stone which forms the beginning of an arch's curve, the support from which the arch springs. Also known as a 'starter'.

Stane couples (Old Scots) stone arches or gables across which rough spars were laid to support the thatch.

Stretcher (North America) a stone with its long edge running *along* the length of a wall. Also termed a 'runner' and equivalent to the Scots term **outband**. A stretcher is laid 'traced' (with its longest edge laid along the wall rather than into the wall), this is not necessarily a weakness; the long stone should securely break several joints under its length and, in a cheekend or corner, where they are used with a 'header', create a strong bookend-like structure.

In North America a 'bull stretcher', or 'shiner', is a stretcher which is also laid on its edge so its broad face shows. The word 'bull' is used in the sense of large, big or aggressive. The stone shows its broadest face. A 'stretcher course' is a course where the stones, bricks or blocks are laid with their longest edge along the length of the wall.

3.3 Dimensions and proportions of a drystone wall

These following epithets are commonly used in North America. Each requirement is a two-word specification; different animals require a different style of fence. Pigs, despite their sensitive snouts, will have no difficulty ripping up a cobbled floor.

A good stone fence should be:

Horse high, bull strong and pig tight.
Horse high, chicken tight and bull strong.
Horse high, bull strong, sheep tight.
Horse high, sheep tight and hog proof.

The seventeenth-century Scots term 'herdwill' or 'herdwall' was the description of a dyke which can keep cattle in or out. A dyke kept 'herdwill' was stockproof and kept crops safe.

Nicholson (1820: 359) advocated removing stones from the land and using them to build a stone wall, and adds dimensions. These instructions are not very precise, perhaps because Nicholson wrote at a time when drystone walls were becoming a common feature in many parts of Britain and widely advocated as part of agricultural improvement. He does make an ambitious estimate for his wall's durability:

> …if they [the stones] be chiefly well shaped, let them be made into walls; for these, if properly made, will last an age, with some trifling repairs. The best method of making these is to dig a trench, where the wall is to be made, to the depth of about eighteen inches; into this throw all the small and bad-shaped stones, until the trench is filled; then on the top of these build the wall, in a mason-like manner, to the height of about five feet, and throw the earth dug out of the trench up against the wall on each side; and in this way it will stand for a length of time beyond the memory of Man.

Nicol (1803: 294) gives dimensions for a wall seven feet high (six feet under the coping plus a twelve inch cope) but warns that a five foot wall might not make a good fence. He wrote from the point of view of a British forester. The walls were required to protect tree plantations:

> The dimensions of what may be termed *a good fence* are, twenty-four inches thick at bottom, twenty inches at top, and six feet high under the coping; which, if properly set on for a fence of this description, will raise it ten or twelve inches higher. The medium height, however, may be reckoned at a foot less than the above; and many are found even *two* feet lower: but just grounds of complaint against their effect, *as fences*, not unfrequently occur.

In North America early-nineteenth century walls were wide and squat — little better than rows of field clearings. Something better was required to keep sheep, cattle and horses in, and out. Fencing was, and still is, vitally important. A farmer's income is tied to the quality of his crops and livestock.

Martin (1887: 20) writes, referring to New England farming, tells us 'A stone wall which affords ample protection against sheep and hogs, may be quite insufficient for horses and cattle.' He recommends increasing the height of a basic wall, using posts and rails. On both sides of the Atlantic other commentators acknowledged higher walls were required for sheep than for cattle. They often favoured tall walls, with loosely built stone on the upper courses, to stop sheep.

D'Arcy (2006: 17) mentions a three metre high (9.8 feet) wall at Leamaneh Castle on the Burren, County Clare, Ireland, which enclosed a herd of fallow deer.

'Tall and wide' makes a good field boundary although this requires more time and material than a low narrow wall. High walls need a wider base, more throughstones and more stone. This makes them more expensive.

The geology of the high sheep pastures of Britain is, generally, basalts, granites and similar hard rock. The lowland arable ground is, generally, softer limestones, shales and sandstones. Geology influences what could be built, and how it was built. The hill dykes of the Scottish Highlands or northern England are massive compared to the finely featured walls in the south. The limestone walls of the English Cotswolds use smaller material in a gentler landscape and enclose docile sheep, compared to the mountain breeds of the north.

According to Seymour (1984: 54-57), 'Using Cotswold stone… [a] five-foot wall would typically be two feet wide at the base and about 14 inches at the topmost horizontal course of stones.'

Raistrick (1973: 17) accounts for the differing dimensions in Pennine walls in northern England, by referring to the stone type. In places where round boulders are most common the walls are built with wide foundations and good batter. Areas with thin bedded stone produce narrower walls with steeper faces.

Smith of Borgue in GVA Galloway (1813: 84,85) acknowledges the need to alter the dimensions of a dyke to suit the available material, and totally depart from the construction style of the normal 'double' dyke to build 'snecks' (short lengths of 'single' dyke) when there are sufficient large stones:

> The thickness of the dyke, and the proportions of it built double, ought to be regulated not only by its height, but also by the size and quality of the stones. For a dyke of sixty inches, where the stones are of a moderate size, twenty-eight inches is a proper width at the grass, leaving a *scarcement* of two inches on each side when the first row of stones is laid. From this the dyke contracts equally till it is levelled for the coping; when (if three-fourths are built double) the width is reduced to sixteen inches. When the stones are large, and very good, the width of the dyke should be a little increased, and the proportion of double diminished. But it is a much better plan, where a considerable proportion of the stones are large, to build snecks of single dyke, at intervals of three yards, or else to increase the number of *through-bands* in every part of the double dyke… When the stones are small, the dyke should be proportionally narrowed, to make the two sides connect more firmly, and afford more *overlaps*.

Sometimes the chosen proportions don't work, as detailed in GVA Caithness (1815: 75) when describing 40 or 50 year old walls on the farm of Thurso East:

> The dyke is two feet and a half broad at bottom, two feet at top, and five feet high, with

a double row of feal or sods on the top. These dykes have been built too broad in the top; and owing to their being top-heavy, (a great error in the building), they burst asunder in many places, which incurs an annual expense for repairs. A four-feet wall, with rough coping, and a regular batter from each side, and thorough bands, would be more durable, and equally serviceable, as a fence.

The stock proofing value of a good cope is beyond dispute. Section 3.2 (iv) of this volume gives full details of various coping systems for drystone walls.

Figure 65: A mortared pillar with good batter and proportions. Aberlour, Speyside, Scotland.

Drystone walling techniques are based on a solid application of physics. Equally important is the instinctive approach of an experienced craftsman working with his or her, local stone. Garner (1984: 7) covers this point well:

If the wall is to be of the standard height of 4 feet 6 inches (1.37m) the foundation will be between 28 and 32 inches (710 and 810 mm) wide. (Dimensions are the result of traditional wisdom rather than science and it is surprising to find that similar optimum measurements were arrived at independently in widely separated areas of Britain.)

The accepted 'formula' (probably an easily remembered generalisation, a conflation of many specifications) says the width at foundation level is half the height of the wall to the top of the 'double' below the cope *and* the width there is half the width of the base. This 'rule', of course, varies and depends on many factors, including local style and available stone.

After analysis of drystone specifications found in various publications an 'average' emerges for a standard 'double' drystone wall with cover and cope:

Scarcement (if any) – 2 inches on each side
Width at ground level for the first course set on the scarcement – 28 inches
Height of double before throughstones are laid -21 inches
Height to the top of the double – 48 inches
Width at the top of the double – 14 inches
Height of the cover and cope – 12 inches
Total height – 60 inches (five feet).
This creates a batter of approximately 1 in 6.

An article from *Sport and Country* magazine for 9 November 1945, entitled 'The Dry Stane Dykes of Galloway', describes a dyking competition at Castle Douglas, Scotland. The author of that piece notes that every craft looks simple, when an old hand is on the job, and quotes part of the dyke's specification, to give an idea of the precision required. Measurements are mainly in inches:

> [H]eight 4 ft. 6 in., 30 in. wide at the grass and 26 in. wide at 'lift,' scarcement of 2 in. on either side from which there must be a gradual taper to 14 in. wide at top of the Double 3 ft. above grass; the Double to have both sides brought up together, with properly blocked stones, laid close together, well-hearted and packed in centre. The outer stones to lie ends-inward so as to stretch into the dyke as far as possible for the better binding of the work. The Double to have one set of through bands 21 in. above the grass at 1-yard centres, projecting slightly on each side' – which, needless to say, is Greek to everyone except an expert dyker!

The size of the cope and cover band can be deduced from the specification. Elsewhere in the article the author tells us the cover band 'must project 2 inches on either side on top of the Double. Above that comes the Single, with no stone less than 10 inches high.'

Incidentally the same article tells us '…a skilled dyker can earn as much as £6 a week, and there is work for at least eight months of the year.' That is the equivalent of about £275 in 2024.

3.4 Batter, camber, line and straightness

3.4 (i) Batter and camber

Batter is not always present is older drystone walls but is one aspect the eighteenth and nineteenth-century commentators enthused about. They thought it essential for a strong structure.

A set of batter frames or A-frames (Figure 19), and the lines stretched between them, define the outer limits of the stonework. For repairs to a fallen section the dyker will normally stretch lines across the gap, accept the dimensions, and build to that.

The width of a drystone wall under the copestones is typically one half of the width of the wall at ground level (see Figure 40). The wall narrows nearer the top as it leans back from the vertical. This is called the 'batter', also known as 'camber or taper' (Figures 65 and 66).

A generally accepted batter for a drystone wall is 1 in 6. For every six units of height the wall leans back off the vertical by one unit. The wall will narrow, on each side of a symmetrical wall, by two inches per foot of rise. The term 'batter' can also be used for widening or expansion with height, for example inside a kiln or a chimney flue.

Batter helps strengthen the wall and, in the case of retaining walls, produces a structure which acts to oppose the pressure behind it. In Leask (1955: 54, 55) we learn how batter developed from practical necessity:

> In the Ireland of today there are many buildings of dry stone and the technique of dry-stone fence or wall building still lives: stability in such structures is still sought and obtained by battering, which gives a measure of security against the slipping or spreading of the stonework.

Leask then refers to batter in church buildings:

> Though adopted for practical reasons in the first instance, the batter seems to have been retained in later works as a grace, an architectural refinement, giving not only a sense of stability but aesthetic satisfaction.

For aesthetic and practical reasons, not all styles of walling have a batter. Myers Johnston points out, in Margaret Brooks (1977: 102), that a garden wall is preferred with vertical sides. It looks better in a garden than a battered style of field wall.

A reverse batter, where the wall slopes out towards the top, perhaps overhanging the foundation, has been used to keep rabbits or sheep out of growing crops.

In brickwork 'hacking' is the term given to laying bricks with the bottom edge of a course of bricks laid slightly in from the top edge of the course below. Standard building bricks are about four inches thick, this transition between courses should give the appearance of a reasonably smooth batter without looking 'stepped'. In masonry, if the stones are thick and the batter angle is too far off the vertical, a 'stepped' appearance, appropriately called 'step walling', is created. It is jarring to the eye. Any effort to create a smooth transition from the top of one course to the top of another, by trimming the stone or using a few thinner pieces over the same height, makes a more attractive wall.

Raistrick (1973: 17) describes how local geology influences the walling style, and batter:

> Where the bulk of the stone is land clearings and rounded boulders, the wall base is made wide and the wall given a good batter, but in the districts... where the thin bedded... flagstones are abundant, the walls are thinner and steeper faced.

Williamson (2002: 107) refers to the Mendips in southern England, where the carboniferous limestone '...provides small and irregular stones... these limestone... walls tend to be less broad and less tapering than in the north of England.' The walls have less batter, the face is steeper.

Seymour (1984: 54-57) describes another area with small flat stone and steeper batter:

Figure 66: Cross-section of low drystone dyke. Note steep batter.
The dyke had been pointed up with mortar. Golspie, Sutherland, Scotland.

> Using Cotswold stone, for example, the batter is traditionally an inch in for every foot of height. A five-foot wall would typically be two feet wide at the base and about 14 inches at the topmost horizontal course of stones.

Todd (1859: 463) contradicts this in a description of the walls in New York State — 'When the stones are mostly small and round, the faces should be laid more battering [sic] than when nearly all flat stones are used.' Todd's opinion is counterintuitive and an interesting topic for debate.

Callander (1986: 48) describes dyking with thicker granite, often in the form of shaped blocks:

> The batter of a dyke is in general 2 inches per foot and that is the extra width that should be added to the base of a dyke for every extra foot in height than the normal four and a half foot dyke.

Raistrick (1973: 20) makes an interesting comment about Pennine walls — 'Where the wall runs along a steep hillside the batter is usually all on the downhill side, and the uphill side is practically vertical.'

This helps control the forces within the structure, an example of good vernacular construction.

MacWeeney and Conniff (1998: 56-60) describe building 'single' walls in The Mourne, an hour south of Belfast, Ireland. The walls are one stone wide and may be 11 feet high. The foundation course, about three feet wide, consists of the largest boulders. The wall is one foot wide at the top. These walls were built using '…sledges, crowbars and the almighty JCB — a hydraulic shovel. A good wall… is a product of oil pressure and blood pressure…' These massive walls are built with comparatively little batter because the stones are huge, they sit solidly. If the normally accepted 'rules' of walling are applied the width of the foundation would be more than five feet.

'On the batter' is a Scots term for a drinking spree. This creates a link, in construction terms at least, with 'getting plastered', another term for a drinking session. JSD makes this connection with his definition of 'batter' in the dictionary's supplement:

> A spree, booze, drinking-bout… This is a humorous application of *batter*, to lay a stone obliquely, or off-the-straight: a term in masonry. Similar terms in masonry are also used to express this state, as *slued,* and *skued…*

The French term for 'batter' is *le fruit* and, for a reverse batter, *contra fruit.*

3.4 (ii) Line and straightness

A wall should have a good 'line' as well as an even 'batter'. It should have no bulges or hollows along its length. A piece in the *Press and Journal* newspaper for 7 February 1935 declares — 'It's an ill-run dyke, ye canna see the scared myauken for jinks!' The scared animal is a brown hare (*Lepus europaeus*). He disappears from view while running down the side of the dyke because it is not straight.

Stonemasons are taught their work should be flat, level, parallel, square and plumb; sometimes it is more important to be 'sweet to the eye'. Drystone wallers work with an irregular material, and can be less precise than the stonemason. The human eye instinctively spots something good looking. 'Sweet to the eye' is often more important than satisfying all the 'rules'.

The old Scots expression 'Be lyne and levell' means 'to build with line and level' — being careful.

Many craftsmen were, and are, proud of their ability to work with no guide line.

'Built bi t' raik o' th eyn', from Yorkshire, England, declares the wall was built by eye only, with no frames or lines. A Scots equivalent is 'Ee sweet', literally 'eye sweet'.

Margaret Brooks (1977: 99) quotes Myers Johnson, a Yorkshire waller who is very proud of a wall he built '...to an eighth of an inch of specification purely by eye. He is teased for his pride but says 'a line and a rule would guide a fool'.'

Figure 19 illustrates the arrangement of equipment to assure line and straightness for a section of drystone wall.

3.5 Building on a slope

The dyker accepts ground is seldom level over any great distance. It is remarkable to see a drystone dyke run, dead straight, across a moor, up a hill, through impossibly rough ground.

Figure 67: This dyke sits securely on the slope, using extra throughstones, large stones in the base, and a stepped foundation.

Building on such a landscape takes skill and some special techniques (Figure 67). This chapter notes some of the building techniques used on slopes. Those specific to 'covers and copes', the top part of the wall, are detailed in Section 3.2 (iv) under the heading 'slopes and inclines'.

On a brief, shallow, slope the courses of stone are typically laid parallel to the ground. There are obvious difficulties if the slope is steeper. The structure could become unstable when stone is laid on sloping stone. If the foundation trench is dug like a staircase, it provides a level base for the foundation stones and reduces the risk of stones slipping. The stones are laid close to true horizontal; they should not follow the slope of the ground.

Williamson (2002: 102) notes a slope of 15 degrees as the maximum incline before the need to 'step' the foundation

Building work usually progresses uphill, from a solid 'cheekend', detailed in Section 4.5. The built stonework braces the uphill work as it proceeds. The stonework is kept tight.

Additional work is often required to strengthen a wall on a slope. Rainsford-Hannay (1957: 76) tells us Pennine wallers in Yorkshire built 'heads' when building up steep slopes – 'In Scotland they would be called cheeks or solid pieces of single walling against which the rest of the work can lean.'

In Yorkshire the strengthening 'heads' are two normal wall-end features, built tight against each other, in the line of the wall. In upland Scotland dykers preferred to build a section of single walling, because the stone in those sheep farming areas is rounded and rough, not flat and slabby like stone in the Yorkshire Dales.

Prevost (1957: 91) relates how, in 1929, two Scottish Border dykers rebuilt a dyke on a *downward* slope of one in two:

> They began their labours from the top and on the steepest slope rebuilt the stone dyke to a height of 3 feet, building up from cheeks spaced at intervals of 29 yards. They heightened the fence with an addition of two strands of wire stretched above the cope stones.

Mundell (1981: 120) gives an example of the precautions used when building on a slope, in the Scottish borders – 'On a steep brae… when rebuilding a dyke a 'raggle' end is built every six yards to keep the dyke from 'creeping down' behind the builder.'

What Mundell means by a 'raggle end' is unclear. It is probably the unfinished end of the dyke, stepped back through the courses, rather than the normal fairly vertical. This provides the maximum resistance to the weight of dyke on the uphill side as the dykers progress.

Another way of dealing with this is to build a short length of 'single' dyke. It has more mass and 'bookends' the new built dyke.

The top line of the dyke, however it is coped, should be even, whatever the lie of the ground; it should not rise and fall with each and every dip. Some sections of a dyke are built higher to accommodate, for example, deep hollows or streams. Throughbands should also sit along a regular line and not follow the ground; in effect their placement is measured from the top, not from the foundation. If the dyke is particularly high at any point, perhaps where it crosses a water course, consider placing additional throughbands.

The angle of repose is the angle, in degrees off the horizontal, at which a heap of material will stabilise without slipping. Gravel has an angle of repose of about 40°, rubble 45°, crushed stone 30°, loose earth 28°, and compacted earth 50°. This is also known as the 'dump angle'. This information has a practical application when considering the stability of a slope.

Gradient is the upward or downward slope from a starting point. This can be measured as a percentage or as a proportion. With the percentage method the increase in height is measured over 100 units of measurement. A rise of ten feet over a distance of 100 feet is therefore a 10% gradient. A rise of 30 feet over 100 feet is a 30% gradient. A negative gradient, for example -20%, is a downward slope. With the proportion method a rise of 10 feet over 100 feet is 10 in 100 which further reduces, by division, to 1 in 10.

Russel (1981: 61) gives a more manageable series of figures for gradients, ones which are more understandable for small projects:

> Any walk with a slope of 5% or more is considered a ramp. When the grade reaches a slope of more than 8.3%, a ramp must be combined with steps for comfortable walking.
>
> 12 inches rise per 10 feet run = 10% slope.
> 10 inches rise per 10 feet run = 8.3% slope.
> 8 inches rise per 10 feet run = 6.6% slope.
> 6 inches rise per 10 feet run = 5% slope.'

4. DRYSTONE STRUCTURES

4.1 Stone structures and wall types

Marcus Vitruvius Pollio wrote *De Architectura Libri Decem* (The Ten Books of Architecture) sometime in the first century BC. According to him a structure should exhibit the three qualities of *firmitas, utilitas, venustas* — strength, practicality and good appearance. These are sometimes called the *Vitruvian Virtues* or the *Vitruvian Triad* and have stood the test of time as a standard for structural excellence.

Words in **bold type** have their own separate entry within this section.

Accretion wall a wall which grows by gradual addition as the surrounding land is cleared of stone. DARDNI (2013: 26) makes an interesting point — 'When deeper metal ploughs became popular, more stones were unearthed and were used to surround the existing wall.'

These are also known as **consumption dykes**.

Auchenhalrig work a style of building with stone, clay and straw, reputed to have been first used at Auchenhalrig in Morayshire, northeast Scotland. It is very similar to some modern 'eco-friendly' building systems. Alexander (1877: 11) quotes details from a contemporary source:

Figure 68: A typical Scottish 'march dyke'. Note the regularity of the covers, copes and throughs.

> This work is built of small stones and mud, or clay, mixed with straw. The proportions of these materials required to make a rood of thirty-six square yards are… about thirty cart loads of stones, ten cart loads of clay or mud, and twenty-four stones weight of good fresh straw. The straw and mud being properly worked together, twenty-two inches are sufficient thickness for a wall of seven feet high – if higher, they should be two feet thick —carried up perpendicularly the same as other walls, and care should be taken never to build more than two or three feet in height in any one part in the same day; if raised more, the wall is apt to swell, for which there is no remedy but to pull it down and rebuild. These walls were equal to the weight of any roof commonly put on mason work, and would, when properly built, and kept well under thatch, last for more than a century.

Alexander goes on to refer to a number of these houses in his district as '…homely yet tolerably comfortable houses, with their walls of rude 'concrete'…'

Avenue an archaeological term for two parallel rows of standing stones, often bordered with ditches and banks of earth leading to a ceremonial centre. One outstanding example is Stonehenge in the county of Wiltshire, England.

Backit dyke (Scots) a nineteenth-century term for a face of stones with earth piled behind. A dyke backed with earth.

Also known as a **ditch dyke**.

Balanced stone walls another term for 'single' walls, a wall built of large boulders, one stone wide. Knott (2004: 92) defines these:

> In Maine many of the remaining rock fences are single-walled, named single-strand or balanced stone walls. These are also found in Massachusetts where they are called lace walls, because of the large spaces in between the rocks, creating a lacy appearance. Held together only by the weight of the rocks, they are nevertheless stable and rarely need repair.

Balla (SG) a wall, hence *balla cloiche* — a stone dyke and *balla dubailte* — a 'double' wall. *Balla* is Scots Gaelic for wall; *Garadh* is Scots Gaelic for a dyke.

Barket (Fife, Scotland) a nineteenth century term for the upright members of a pier, built of timber with a stone filling.

Beaded The occasional widening of a wall along its length, like beads on a string, to use up gathered stone. This is a variation of the **consumption dyke**.

Bedrock walls (New England) Hubbell (2006: 41) uses this descriptive term for walls crossing solid bedrock. There is no danger of damage from frost heave. On sloping bedrock, iron pegs were sometimes punched into the bedrock to add more grip and stabilise the wall.

Beehive hut a corbelled drystone structure shaped like an old-fashioned bee skep – a simple hive made from straw which looks like an upside-down bowl. There are many regional variations in Europe. Similar structures are recorded in northeast USA. Examples include Scottish Hebridean *cleitean* and Irish *clocháns*.

According to NEARA (Summer 2006: 12-13), 'Corbelled stone beehives are a dime a dozen in the Hebrides…' The NEARA article goes on to mention miners' shelters in northern Portugal, structures in Iceland and underground cellars in southwest Sweden.

Similar structures in Bodmin Moor, Cornwall, southwest England, are also mentioned. They were used by hunters for storage and as pigsties and henhouses.

Newman (2000:32) describes circular pig houses in Orkney, typically 1.5m internal diameter, circumference about 7.5m and 1.5m high with corbelled roof. The structure often had an attached enclosed area for a pig run.

Beild / bield / biel (Scotland and northern England) a shelter, a sheep fank, an animal shelter, built to different dimensions depending on the livestock and situation. They rely on the benefits of a stone wall breaking up the wind and providing shelter. There was no roof. The bield may have several entrances, or consist of stone walls built in a cross shape, or short lengths of drystone wall built where shelter is required, if the weather came from a certain direction.

In the English Lake District a 'fox bield' is a corbelled stone enclosure with a hole at top centre to trap foxes. This type of structure is also used by the Inuit in Canada, as described in Adcock (2011: 12).

Bink (Scots) a stone bench by a house door, a stone shelf, a ledge on a cliff.

Blackhouse (Scottish Highlands and Islands) a traditional house built with thick drystone walls with an earth core (Figure 69). A well-built blackhouse had the building stones, and the hearting, tilted to shed the water to the outside. The outside corners were rounded, the inner corners were squared. The roof was thatched with no vent for the smoke from a central peat fire — it seeped through the roof. Early account of these houses tell us they were inhabited by people at one end and, in the winter, their animals at the other. In Walker and McGregor (1996: 28) there is a reference to the high level of ammonia in the air, from animal urine. This reputedly reduced the incidence of tuberculosis among the inhabitants, balanced against the increased respiratory problems caused by the smoky interior.

The 'blackhouse' was gradually replaced by an 'improved' building, the **whitehouse**, in the late 1800's. This was built with thinner stone and lime mortar walls, had windows and chimneys and a modern roof covering, often corrugated iron. These houses were criticised as draughty and damp, compared to the older houses. Some blackhouses were occupied well into the 1950's; many became storage sheds or housing for livestock.

Sinclair (1953: 18) discusses the origin of the name 'blackhouse':

> Whether the term 'black' house (*tigh dubh*) has been applied to distinguish it from the 'white' house (*tigh geal*), for the reason that the former is old, dark and dull in appearance and the latter new and white, or is due to some confusion arising from the phonetic similarity between the adjectives *dubh* (black) and *tugadh* (thatched) is a matter for speculation, but in any event the description 'black' in the derogatory sense is a modern invention.

Fenton (2012: 12) eases any controversy:

> Whatever the origin, 'blackhouse' is a useful technical term for a dwelling in which men and animals lived under the same thatched roof, and which had double-skin walls, and was hip-ended

On a practical point, some of the original 'white' houses were built by improving landlords

Figure 69: Hebridean Blackhouse. Note the stone steps to the top of the wall, for easy access to the thatch. Image courtesy of Inverness Museum and Art Gallery.

who took the chance to make the structure look tidier by applying whitewash. It was a lot easier to whitewash a flat mortared stone wall, compared to a rough drystone wall.

Booley huts (Ireland) ruins of corbelled huts and stone foundations found on upland pastures. The Scottish equivalent was a **shieling**. In the eighteenth century many Irish subsistence farmers practiced transhumance, a seasonal movement of people and their animals to upland summer pastures to take advantage of the grazings. Milk was processed into butter and cheese. This was a valuable trade item, and a food source over the winter.

Transhumance was also a way to keep the cattle out of the arable land, as noted by Evans (1957: 34) — 'In the absence of permanent fences, the majority of the cattle must be moved away during the summer…'

Borra / borradh (Scottish Highlands) JSD defines these as large piles of stone covering a series of small chambers or cells. They were of uncertain history and may have been ancient burial sites or a type of souterrain. Similar structures were noted throughout Scotland and, as often the case, construction was attributed to the Picts, an ancient people from 2,000 years ago. These piles could also be the remains of a shieling, the Scottish equivalent of **booley huts**.

Boucht / boght (Scots) alternative spellings for 'bucht', a sheep pen.

Boulder dyke a 'single' dyke, built one stone thick, largest at the bottom, decreasing in size towards a level top line (Figure 70).

A 'lan-bool', (a term from Caithness, Scotland) is a large rounded boulder taken off the fields. This is like the Scots term 'land-stane' – a stone turned up by digging or ploughing. Loudon (1825: 443) notes:

The wall or dike made with round or land-stones, by labourers, and covered with a coping of turf, is a very indifferent fence… their benefit is not permanent, as the wall is perpetually tumbling down.

Figure 70: Boulder dyke. Lairg, Sutherland, Scotland.

Boundary fence / line fence (Kentucky) a rock wall between two properties.

Boundary wall the perimeter wall marking the extent of an estate or the border between two counties. Also known as a **march dyke** in Scotland. These walls are typically taller, and have more 'presence', than the field walls within the estate. In New England metal pins are often hammered down the middle of these walls to confirm their function as boundary markers.

Boxdrain (Forfarshire, Tayside, Scotland) a drain carefully made with flagstone sides and capstones to provide a channel for water. The stone construction was thought to prevent moles from choking the channel with earth. When dykes were demolished, to increase a field's size, the stone was often buried and formed into drains.

Brae an artificial bank of gravel and stone built across a river as a salmon trap. DSL includes reference to them in the eighteenth century on the River Spey, Highland Scotland, where, at one time, 'braes' were contested through the courts. Salmon were valuable to the Scottish economy.

Such trapping or corralling of fish is recorded in many cultures, from Canada to Australia.

Brairded dykes (west and south of Scotland) hedges or fences stuffed with thorns or other brushwood to prevent cattle getting through to a growing crop. Technically this describes a hedge rather than a stone wall. This is an example of the word 'dyke' used in a general sense for a field boundary.

Breast wall according to Parnell (1833: 200) 'The wall which supports the road is called a breast wall; that which is on the hill side of a road is called a retaining wall.' When a road is built across a slope, the uphill wall is the 'retaining wall' and the wall on the downhill side of the road is the 'breast wall'. **Retaining walls** are discussed in detail, later in this section.

Broch an Iron Age drystone round tower, built around 2000 years ago (Figure 71). There may be 600 brochs around Scotland, mostly along the north and west coasts. The best, almost intact, example is on the island of Mousa, Shetland. It stands approx. 44 feet high (13.3 metres), with solid built outer walls 15 feet (4.58 metres) thick at ground level. The diameter of the broch is approximately 50 feet (15.25 metres) at ground level. Rainsford-Hannay (1957: 56) gives slightly different dimensions and estimates 10,000 tons of stone in Mousa broch. Mousa has thicker walls than the average broch and a smaller diameter. Full details of this structure can be found at https://canmore.org.uk.

Figure 71: Dun Carloway Broch, Isle of Lewis, Scotland. One of Scotland's best preserved brochs. The wall rises to 30 feet (original height unknown). Note the internal passage between the outer walls. © Crown Copyright reproduced courtesy of Historic Environment Scotland.

Brochs were originally thought of as defensive structures; modern research has moved away from that theory. Many ruined brochs were used as quarries. The stone was carted off to build more humble structures.

Bucht / boucht / bought / bught (Scots) an enclosure, a pen or fold for sheep, a cave or hollow among rocks used for enclosing livestock. Originally a small pen where ewes were kept for milking.

Buil (Shetland) a sheepfold or byre. The verb 'to buil' means to drive sheep into a fold, or house cattle in a byre. 'Builling' is the act of enclosing sheep or cattle. More detail can be found in Edmundson (1866: 15). Hibbert (1822: 183) noted disregarded regulations in the sheep pastures of the Shetland Islands:

> [The animals] are allowed to run wild among the hills, herding and housing being almost wholly unknown in Shetland. There is an old law, that was probably introduced by the Scotch settlers, ordering that every scathold have a sufficient herd, and that builling, punding, and herding, be used in a lawful way, before, or a little after, sun-setting and that none scare, hound, or break up their neighbours' punds and buills under the penalty of ten pounds Scots, besides damages but the regulation has not, for a long time, been enforced.

Farmers from Scotland (almost a foreign country to the native Shetlander) had introduced a husbandry system whereby the sheep were herded and enclosed at night. Fines were applied if the sheep enclosures were interfered with. According to Hibbert these old regulations were no longer enforced.

Bur-wall (east Lancashire, England) a wall inclined against a bank, a supporting wall.

Burbenk / borbenk (Shetland) a supporting wall of stone or turf built up against the outside of a weak structure to give it protection and strength.

Burian a mound or tumulus, common throughout Scotland, perhaps the remains of an ancient fortification. Probably from the same root as the English word 'barrow', a burial cairn covered in stone and soil. These structures may be 2000 years old.

Butt a short stone or turf wall, or semi-circular wall facing the flightpath of low flying gamebirds. Lines of 'shooting butts' are common on the grouse moors of Scotland and northern England. Butts act as chest-high hides for guns, waiting for beaters to drive birds in their direction.

Butt and hudd Rainsford-Hannay (1957: 46) describes 'butt and hudd':

Here short pieces of single dyke of the bigger stones are built, with stretches of double dyke between them. The effect is to tighten up the work and divide it into panels. John MacAdam [of Craigengullich] was also the inventor here. An observant fellow, indeed, who really studied how to use the varying shapes of stone to the best advantage for every separate dyke which he built. (Figure 72).

Prevost (1957: 99) includes a definition for 'hud' or 'hudd' from the Chambers Scottish Dictionary:

A part of a stone-dyke built with single stones which go from side to side. Synonymous with sneck. Snecks are sometimes built by dykers to strengthen the dyke if there are sufficient stones available. A butt and hudd, according to the late John Broadfoot of Sanquhar, consists of alternate lengths of stone-dyke built double and single in panels. The big advantage of the 'hudd', also called a 'sneck', is that it strengthens the dyke, limits the extent of any collapse and uses up large boulders without the need to break them. 'Hudd' is a Scots variation of the English word 'hold' — quite apt for this situation.

An older description, sourced from GVA Galloway (1813: 86-87), credits the invention of the 'butt and hudd' to John McAdam of Craigenguillen (note the slightly different spellings from that quoted above) and defines 'hudd' thus:

Hudds or *snecks* are built at the distance of two or three ells. One *hudd* stone will do at the grass; but the more the better. When the double dyke between the *hudds* is built as high as the first hudd stone, a stone sufficiently long is placed so that one-half of it may cover the hudd, and the other half the double dyke. In this manner the hudds are connected with the double dyke on both sides till it is two feet five inches high, when the band stones are laid on.

Figure 72: Butt and hudd. Regularly spaced lengths of 'single' within the 'double'. Dalmellington, Ayrshire.

The 'ell' is an old Scottish measurement which we can consider equivalent to the modern yard or metre.

In DSL, under 'Hud' the term is defined as 'A portion of a wall consisting of stones placed transversely in order to strengthen the structure …Hence *hud(d)-stane*, *-stone*, a stone employed for this purpose…' The definition also refers to Prevost (1957), but transcribes the term as 'a butt and a hudd'. Note the addition of 'a', which leaves the phrase open to interpretation.

MacWeeney and Conniff (1998: 60) perhaps clarify matters when describing Irish stone walls — 'The bottom row of butt-stones, or 'butters', may be a yard thick and two yards long.' These are large stones at the foundation of a single wall, they lie across the total width of the wall. Bearing in mind the close connections between Ireland and southwest Scotland it seems reasonable the word 'butt' was accepted as a synonym for 'hudd'. Perhaps the term 'butt and hudd' developed among southwest Scottish dykers to describe a piece of 'single' work which 'buttressed and held' the 'double' work.

Alan Brooks (1989: 18) gives another term — 'sneaks', '…or panels of single stones within the double dyke to create the 'butt and hudd' style, which is supposed to be tighter than either double or single dyking alone.' Variations are also known as a **hud** or a **sneck**.

Cabhuil (SG) a stone dyke across a stream with an opening to allow fish through. A fish trap.

Cabane (France) a small drystone hut with a corbelled roof. Various styles exist in rural France, and all around the Mediterranean.

Cairn / carn / carne / cyarn / kairn a pile of stones, shaped like a cone or pyramid, or with vertical sides. See Section 4.6 for more on cairns, boundary stones, heaps and markers.

Capstone wall / hard cap wall Hubbell (2006: 38) illustrates this style with an example from Connecticut, New England. The wall is capped with flat slabs reaching right across the top, with a small overhang. A well built example of these walls can also act as walkways

Figure 73: A section of two-tiered Carlow fence. County Carlow, Ireland.

between houses; ladies avoid the mud.

Carlow fence (County Carlow, Ireland) a local style of fence constructed with long granite lintels resting on wide 'v' notches cut in low granite uprights known as 'cliffs' (Figure 73). The lintels are typically 2 to 3 metres long and 24 x 26 cm thick. Fence height is around a metre, often backed up with a hedge. Variations can be two-tiered; about a metre high with the lower tier nestled inside the higher. Conry (2000: 57) suggests most of the Carlow fences were built in the second half of the nineteenth century.

Carraig (SG) a wall built below the high tide mark, to trap fish as the tide recedes.

Cashel the circular wall enclosing a Celtic monastery, also the name for an Irish 'rath' or fortified farmstead. A general search through Irish dictionaries produces the similar-sounding word *caiseal*. It is defined as an ancient stone fort or an unmortared stone wall.

Catch dam / drain dam (Kentucky, USA) a face of stone against a slope, to control erosion, a retaining wall.

Cathair an Irish stone fort, circular in shape. Also known as a **cashel** or a **rath**.

Catwa / catwall (Shetland and Orkney) the stone wall dividing a cottage into two apartments. It extends up to wallhead height, not to the roof ridge. Perhaps named for the cat who uses this as a walkway.

Chambered cairn an archaeological term for a chambered tomb covered with stones.

Chevaux de frise literally translated as 'Frisian horses'. This method of obstruction and defence was developed in the Frisian Islands (off the North Sea coast of the Netherlands, Germany and Denmark) to compensate for their lack of cavalry. The ancient fort of Dun Aonghasa on Inishmore, one of the Aran Islands lying off Galway, Ireland, has an excellent example. A 30 feet wide field of upright sharp-edged stones guards the entrance.

Löbbecke (2012: 326) calls these upright stones at Dun Aonghasa 'Spanish riders' (Figure 103).

Modern chevaux de frise are more portable, made of wood, metal and wire.

Chinked wall (New England, USA) Hubbell (2006: 67) and Thorson (2002: 125) describe this wall as a well-laid smooth-faced wall, using large flat-faced stones with smaller stones on the face, to fill in the gaps and level up the large stones. Hubbell writes – 'Wedge-shaped stones are best locked into place by pushing them into the gaps from the inside.' Chinking from the outside also improves the appearance of the wall but the small stones will most likely fall out.

Chip and block a style of walling found in Devon, England, a local variation of the **Cornish hedge**, a stone-faced earth bank. Smaller stone chips, behind the face of the wall, keep the bigger blocks secure and ensure they bond better with the earth packed into the centre of the structure. The chips are never used on the face of the wall; they would inevitably fall out and loosen the stonework.

Cist a stone coffin, made with flat slabs of rock, dating from the Bronze Age, 2500 BC — 800 BC.

Clachan / clachain, (Scots from Scots Gaelic) Dwelly and JSD have long definitions for this term — a village or hamlet with a parish church, inn and smithy. The term was also used for a druidical stone circle. Christianity recognised the need to capture the allegiance of the locals and set up churches on the site of old stone circles. 'Clachan' developed as a term for a church. The Scottish Gaelic phrase 'going to church' literally means 'going to the stones'.

Another meaning of the term is a set of stepping stones.

'Clachan' also applies to a cluster of houses in Ireland but, from personal correspondence with Patrick McAfee, Dublin mason and writer, unlike the Scots Gaelic definition, the Irish 'clachan' is a collection of houses without church, school or shops. MacWeeney & Conniff (1998: 89) describe a clachan as '…like a child's box of bricks spilled on the floor.'

Clapper / clapper bridge / clapper walls a simple bridge with large horizontal stone slabs supported on vertical stones to lie just a few feet above the normal water level. They were common on old trade routes in Exmoor, Dartmoor and Lancashire before the modern road system was built for heavier loads and wheeled traffic. Most clapper bridges will be over 500 years old. Some examples exist in the Scottish Highlands.

In south Cumbria, northwest England, walls made of vertical sheets of slate are known as 'clapper walls'.

Clawdd (pronounced 'clow-th', plural *cloddiau*) a general Welsh term for a hedge, embankment or stone wall, but more particularly a style of stone-faced earth bank found in northwest Wales and Pembrokeshire in southwest Wales (Figure 74). The face stones are laid on edge, long axis into the wall, protecting the earth core. The *clawdd* can be topped off with turf. A hedge is often allowed to grow on top. If stones are slightly tilted towards the centre of the *clawdd* this encourages rainwater to flow there and nourish the plant growth.

Gertrude Jekyll (1843-1932) the Victorian garden designer recommended the construction of *clawdd*-like structures, this time with horizontal stones which, as detailed in Jekyll (1901: 12), '…should all tip a little downwards at the back, and the whole face of the wall should incline slightly backward, so that no drop of rain is lost…'

Figure 74: Welsh Clawdd. Note the large stone securing the end. Image courtesy of Sean Adcock.

In other localities up the west coast of Britain similar constructions are called 'Pembrokeshire hedge banks', **Cornish hedges** and **Cumberland banks**.

Adcock (2012) has produced a fine monograph on Clawdd construction, full details in the bibliography.

Clawdd cerrig (W) a stone wall. The Welsh proverb *Yn ddigywilydd fel clawdd cerrig* has been translated to 'as impudent as a stone wall'. Why should a wall be rude? Enquiry to the National Library of Wales confirmed *ddigywilydd* can also mean 'shameless or without embarrassment'. This alternate meaning is more generous, closer to 'proud'. The wall is stoically performing its duty; it is not defiant or insolent.

Clawdd cysgod (W) a wall built to shelter animals from the weather.

Clawdd pillari (W) literally 'pillar hedge'. These fences are made of long narrow slabs of slate placed close together in the ground like fenceposts. These uprights are strengthened and held in line with wire intertwined between the stones, at the top and bottom (Figure 75). Good examples exist near the disused Penrhyn slate quarry near Llanberis, North Wales.

Clay and bool (northeast Scotland) a wall made of 'bools' (rounded boulders) with clay as the mortar. According to Murray (2008: 41) the technique '…was typical in the area of Morayshire around Garmouth and south into Banffshire in the nineteenth century.'

A coat of harl or good pointing and a whitewash kept the weather away from interior of the wall.

Clearance dyke / clearance wall a wall constructed out of stone cleared from the land around it, called a **consumption dyke** in Scotland. **Clearance cairns** are described in Section 4.6.

Figure 75: Clawdd Pillari. Cwm Croesor, Gwynedd, Wales. Photo courtesy of Sean Adcock.

Figure 76: Cleit. St Kilda,Outer Hebrides, Scotland.

Cleit / clait (SG) (plural *cleitean* or *cleytan*) small corbelled or lintoled drystone structures with a turf roof covering. There are over 1000 of them on St Kilda in the Scottish Hebridean Islands. They are assumed to have been used for storage, mostly seabirds, turf or hay. Cleits are a construction type often defined as **beehive huts**.

According to Stell and Harman (1988: 46) the cleit in Figure 76 '…measures internally 4.67 m in length by 2.01 m in maximum width near the W[est] end, where it achieves a height of about 2.44 m; it is narrower and lower towards the doorway at the E[ast] end.'

A few cleits on St Kilda have smaller structures attached, accessible through an internal opening. Some larger cleits may have been used as housing.

DSL defines a cleit as a 'penthouse'. This is not in the modern sense of a luxury top floor apartment. It stems from old French and English terms for a shed or a building attached to, or 'pending upon', another.

Clochan / clochain / clochaun round or square corbelled **beehive huts** with domed roofs, found throughout south-western Ireland. There are fine examples on the Dingle Peninsula in County Kerry and several stunning examples on Scelig Mhichil (Skellig Michael) off the coast of Kerry, southwest Ireland. These structures were part of a monastic community which existed until the late twelfth century.

Cobblestone construction a style of masonry from upstate New York, USA, dating from 1825—1860. Cobblestones, water-worn egg-shaped stones up to ten inches long, were set in mortar to form walls 20 inches thick. The cobblestones were often veneered up against a structural fieldstone wall.

Freeman (2005: 12) describes how it took a mason 2 or 3 years to complete a house because only 4 courses, totalling 18 to 24 inches, were built in a day. Several houses would be worked on at once. For this labour a mason, in 1850, received $1.00 to $1.50 plus board, for a 10 to 12 hour working day. In a recent conversation, an experienced lime mortar practitioner suggested the work might have been faster if the builders had used hot lime mortars, which set quicker.

Slocombe (2012: 20) includes an illustration from England for a similar looking wall, captioned 'Pebble walling used alongside refined classical detailing in a Regency house in Brighton. The pointing between the pebbles is a modern cement-based material.' Slocombe says pebbles '…have been used for construction in coastal towns such as Brighton and Happisburgh in Norfolk.' The beaches there provide tons of this attractive material.

Compound walls (New England) Hubbell (2006: 36) defines these as:

> …those built of more than one material. Generally, the base is made of stone, three to three-and-a-half feet high — the maximum height a heavy stone can be lifted comfortably. To bring the wall to the required legal colonial height — usually four feet or more — other material was added which generally consisted of wooden rails, although almost anything would do.

Examples of this can be seen in Scotland where a low dyke, originally built to enclose sheep, is heightened with posts and wire netting, to keep out deer.

Consumption dyke / consumpting dyke / consuming dyke / clearance wall / consumption wall / accretion wall a wall which 'consumes' the stones cleared from the surrounding fields (Figure 77). Thorson (2002: 125) gives detail of a similar 'disposal wall' where — '…two single walls are built several feet apart and the gap between them in-filled with stone.'

Consumption walls are not always finely built structures. The prime aim was to remove stones from the land without the effort of moving them too far. Monymusk in Aberdeenshire, Scotland, has two spectacular consumption dykes. The biggest, the West Dyke, or Broad Dyke, is 27 feet wide at the top, 6 ft high, 500 yards long, running east to west with a flagged pathway along the top. The East dyke is 7 feet wide at the top and 334 feet long, also with a flagged walkway along the top. Rainsford-Hannay (1957: 71) estimates these walls were probably built in the early to mid-nineteenth century.

Garner (1984: 18) includes an image of a rough Welsh consumption wall 8 feet (2.4m) high and 4 feet (1.2m) thick. The 2005 edition of the same booklet, page 22, shows a tidily built consumption wall near Harlech, Gwynedd, Wales, measuring 6 feet (1.8metres) thick. No height is given, it looks about five feet high.

Conry (2000: 91-95) details various consumption walls or clearance walls found in County Carlow, Ireland. A typical measurement is given as 1.67 metres to 1.9 metres high and 3.10 metres to 3.30 metres wide, with little or no batter. One series of walls are '…reservoirs for stones unearthed during the reclamation of the land.'

Some consumption walls in Cumbria, English Lake District, have a 'beaded' effect, where the wall widens to accommodate the stone gathered near that spot.

Callander (1986: 13) writes:

> Numerous stories exist of places where the ground level dropped by as much as 12-18 inches when the stones were cleared out and it was reckoned a field of only three to four acres would supply sufficient stones to enclose it with a four foot dyke, 36 inches wide at the base and 20 inches wide at the top.

Stilgoe (1982: 175) describes how stone was cleared from the fields in the most efficient way:

> The stone walls of New England and the Middle Colonies were built by men interested far more in land-clearing than in fencing. They piled the rocks not in heaps but in rows equidistant from the center of their rectangular fields, along each edge; rock clearing explains in part the small size of colonial fields. Men understood the great effort required to clear fields larger than one or two acres, and slowly – especially in New England – they learned how much longer they spent clearing one eight-acre field than eight one-acre ones.

The rows of stones described by Stilgoe looked like untidy walls; they were primarily rough consumption walls. According to Thorson (2002: 154):

> ...many of New England's so-called fences had little to do with agricultural production ...more than anything else they were linear landfills, packed full of agricultural refuse that would not decompose.

Thorson (2002: 159) explains how fields became smaller:

> If the walls round a field filled up before the supply of fieldstone was exhausted, a good alternative for holding the excess stone was to create another wall, usually at right angles to the first, in order to maintain square fields.

These small fields were difficult to use with larger farm machines, different livestock or new agricultural processes.

Dutton (1808: 152) notes the effort applied to moving stone in County Clare, Ireland:

> Great exertions are made... in clearing their land from stones; for this purpose walls ten feet thick are not uncommon, and pyramids of stones of a large size remain as monuments of Irish industry, and not, like the pyramids of Egypt, everlasting proofs of human folly.

Consumption pile large heaps of stones cleared from the land, piled rather than laid in rows.

Copestone wall Thorson (2005: 106) uses this term to for New England walls with:

> ...the final tier of stones on edge across the double wall to produce a fencelike top (related to spikes, hedges, and broken glass) that prevents climbing or sitting on the wall.

Figure 77: Stones, gathered to clear an area for better grazing, were formed into a wide dyke. Highland Scotland.

Hubbell (2006: 68) notes this as:

> …a very British wall building technique where flat, book-sized stones are set on edge along the top of the wall. The additional weight helps anchor the wall and discourages adventuresome sheep.

This is the classic **locked top** cope commonly seen in Britain. Both authors contrast this to **capstone** or **hard cap** walls where the cope consists of flat slabs laid flat across the top of the wall. Section 3.2 (*iv*) has full descriptions of these coping terms.

Coping walls Conry (2000: 87-91) gives details of these substantial walls in County Carlow, Ireland:

> Basically they consist of long straight dry stone walls built of granite stone with battered sides and a wide coping of irregularly shaped stones, standing on their edges on the top of the wall to deter livestock, particularly sheep, from crossing them.
> A typical example is 1.6 metres (5 feet 3 inches) high. The width at grass level is 90 cm, this tapers to a width of 45 cm under the copestones, at a height of 1.20 metres. The cope on top of this is vertical, 40 cm high and 55 to 70 cm wide, overhanging approximately 12 cm on each side. One wall featured a box drain incorporated into the foundation, along the length of the wall. Some of the walls use big hewn stone, like the granite walls of

Aberdeenshire in Scotland.

Corbelled structures an individual corbel, corbal or corball is a stone which projects intentionally from a wall, often to support a weight. The word is said to derive from a Latin word for raven, *corbellus*. The projecting stone is thought to resemble the beak of the raven.

When built from two sides, or from all sides of a circular or oval structure, the process is known as 'corbelling'. Each succeeding course overhangs the one below it until the walls meet to create a space with a domed roof. Corbelled structures can support their own weight during construction, quite different from an arch which depends on temporary framing until the setting of the keystone. Corbelling is an important feature of 'beehive' structures, found in many rural environments where there is loose stone and a need for shelter. See **clochan** and **cleit** above.

Corbelled structures come in all sizes. Löbbecke (2012: 162) describes the small isolated corbelled structures in Aveyron, France — 'They were built by shepherds or road workers and provide a place to sit for one person. They look like a petrified cloak, like a shepherd's coat.'

Corbelled structures include Italian *trulli* (singular *trullo*), Sardinian *nuraghi*, French *cabanes*, Minorcan *barraques* (singular b*arraca*), Scottish Hebridean *cleitean* (singular *cleit*) and Irish *clocháns*. Those are collectively described as 'beehive structures'. The interiors, the upper part at least, is basically the same, rounded and domed. The exterior varies with the local building style.

Trulli, from the Apulia region of southeast Italy, have conical roofs, are whitewashed inside and out and often built in tight fitting groups. Barraques in Minorca, Spanish Balearic Islands, have an exterior tiered like a wedding cake.

One of the finest rectangular corbelled structures is The Gallarus Oratory in County Kerry, Ireland. It has survived remarkably well since early Christian times. Round corbelled structures are stronger than rectangular ones because the circular shape leans in and supports itself.

Borut Juvanec (2010: 96) theorises that round corbelled structures throughout Europe conform to a common formula. Put in its simplest form, the theory proposes a corbelled structure is basically an equilateral triangle, the base being the diameter of the structure, measured from the centre of the wall to the centre of the wall on the other side. A line from each of these points produces a triangle with three equal sides, with the peak of the roof as the apex. Juvanec tells us the interior of these structures is rounded, the exterior can be square or stepped, depending on the local style.

Evans (1957: 115-117) notes the existence of ancient and modern examples of corbelled construction, from the Scottish Hebridean islands to the western Balkans, and makes an interesting point:

> Whether or not the origin of this style of building in western Europe should be placed, as I am inclined to believe, in the megalithic period, there can be no doubt that it represents, in various parts of the Mediterranean-Atlantic route, the survival of an ancient method of building. It has had a long life in exposed hilly districts where suitable materials lay to hand, especially where the interests of simple shepherds and cattlemen were dominant.

Villages of houses with domed or conical stone or mud-brick roofs are a feature of south-east Italy and Syria, and of the dry belt stretching eastwards to India.

Löbbecke (2012: 10, 35) considers corbelled structures date back to 5000 BCE. Stone was used because timber was in short supply. She makes an interesting point about the age of these stone structures:

> It has been consistently shown in the various countries of southern Europe that the corbelled dome huts still visible today were not built until the eighteenth century; only in very isolated cases can one be dated a little earlier. The height of their dissemination was in the nineteenth century and they were seldom built in the 20th century.

With few, but significant, exceptions this corresponds with the construction of agricultural drystone walls in the British Isles.

Cornish hedge / Devon hedge a style of wall found in Cornwall and Devon, southwest England, built like the Welsh clawdd or **Cumberland banks**, stone-faced with packed subsoil interior. The fill is rough subsoil because loam-rich topsoil produces a weaker structure. The structure is typically 5 feet (1.5 metres) high with a base of the same width. The width at the top is half the height. The hedge has a curved batter, steepest nearest the top. When these hedges reach maturity, they can have bushes or trees growing on top.

Garner (1984: 25) calls them '**stone hedges**' and writes:

> Stone hedging is the only branch of the craft where it is considered correct to lay stones tilting towards the centre. Where vegetation is relied on to bind the work rain must be encouraged into the centre.

Water finds its way to the middle of the structure and nourishes the roots of the plants and bushes growing on top of the bank.

Brooks (1989: 16,91) details a 'slate hedge' from Veryan in Cornwall, England, a type of Cornish hedge. The face stones (slate) are small and fragmented and laid in a herringbone pattern (Figure 32).

The term **stone hedge** has been used to describe Cornish hedges with stone, instead of subsoil, in the middle. This style developed in upland

Figure 78: Dyke built with shaped granite stones. Royal Deeside, Scotland.

areas of Cornwall where stone was in plentiful supply. The website www.cornishhedges.co.uk gives a detailed description of these hedges and their history.

Cotswolds (Victoria, Australia) a local term for a drystone wall, most likely derived from the English region, famous for its limestone walls.

Cow path walls / cow run (Vermont, New England) parallel stone walls forming a lane or passageway between barnyard and pastures. They control the passage of cows.

'Rekster' is a Shetland Islands equivalent from Scotland.

Crannog / crannoge / crannogue (Scotland and Ireland) an archaeological term for a timber dwelling, built off the shore of a loch, on stilts, or on an artificial island of stone and earth. Crannog building is associated with the Celtic Bronze and Iron Ages, roughly 1000 BC—1000 AD.

Cray / crae / cree (Scots) a pen or enclosure for hens, sheep or pigs.

Creel-barn (Scots) a barn with the lower wall built of stone and the upper part wicker work. A creel is a basket or fish trap woven out of willow or hazel twigs. The building will have been named because of the wicker work, not necessarily because creels were stored there.

Cribstone Bridge a unique structure connecting Orrs Island and Bailey Island, Maine, USA. Also known as the Bailey Island Bridge. It was built between 1927 and 1928 from 12 feet long granite slabs, often called 'kerb stones', also known as 'crib' or 'cribbing'. 10,000 tons of granite are laid in a latticework, to allow the tide through. The bridge is 1150 feet long with an 18 feet wide concrete road along the top. Apart from a 52 foot concrete span over the shipping channel it is built 'dry' with no mortar or steel pinnings. The structure was repaired in 2009–2010.

Local legend says it was based on a Scottish design.

Cromlech an old archaeological term for a simple megalithic burial chamber with several large stone uprights and one or more capstones. This was then covered in stone and soil. Since construction, up to 6,000 years ago, the covering has eroded away leaving a structure looking like a huge multi-legged stool. The term has been largely replaced by the word **dolmen.**

Cross-dyke (Scots) a dyke running downhill at right angles from the **head dyke**, the overall boundary for the farm or estate. Cross-dykes divide the arable land into fields and are built smaller than the more substantial head dyke,

Crub / cru / croo / kru / krue / kro (Scots) an enclosure or fold for sheep, hens, pigs or cows, built with stone or stone and turf. Most likely a shortened version of **planticrue.** In the north of Scotland, and in the islands, smaller versions were used as nurseries for growing the likes of cabbage plants. In Shetland a 'cru' or 'crue' is a sheep pen. In Cornwall, southwest England, these small enclosures have been called a 'crow' — another variation of the spelling.

Cruive / criv / craive / crave / crove / croove / cruve / crive (Scots) an enclosure for animals, especially poultry or pigs. The term also applies to a wicker salmon trap. A cruive-dyke was a rubble dyke extending across the river to hold the wicker traps. In the eighteenth and nineteenth centuries these traps were obliged to stay open for periods over the weekend to preserve future stocks for net fishing at the mouth of the river.

Cumberland banks (Cumbria, northwest England) one of a type of wide-based stone-

Figure 79: Downstream side of a groyne, built to protect the riverbank and improve fly fishing for salmon. The bank of stone is held solidly in place by a timber bulwark. Strathnaver, Scotland.

faced earth banks found mainly in the western edges of the British Isles. For that reason, they are sometimes referred to as 'Celtic hedges'. **Cornish hedges, Pembrokeshire hedge banks** and Welsh clawdd are other varieties. In Cumbria these structures are also known as '...Kes's or Kessies', according to Brooks and Adcock (1999: 76).

Cuthai (SG) a cattle fold.

Cyarn an old Scots spelling for 'cairn'.

Cyclopean a description of mortared and drystone walls built from large close-fitting stones. They are named for Cyclopes, the mythical race of one-eyed giants who built massive walls out of large stones, too heavy for mortals.

Daek (Shetland) a local dialect variation of 'dyke' which is also used to describe peats built like a wall, for drying. 'Innadaeks' is the ground within the dyke. 'Ootadaeks' is the ground on the hill, outside the dyke.

Daek-end (Shetland) describing where the dyke is built down into the water, to prevent animals from walking round and trespassing.

Daich a term for a watercourse or ditch in Lancashire, England.

Dale / deall (Aberdeenshire, Scotland) a term used for any division of land or the dyke marking it. This has resonances with the Yorkshire Dales, an area famous for its wall-lined fields.

Dam-dyke / dam-dike (Galloway, southwest Scotland) the barrier confining water in a dam. Probably a mound of earth rather than stone.

Danish fences (County Derry, Northern Ireland) the local name for pre-Christian walls which separated the very early enclosed farms from the rougher land outside, as detailed in DARDNI (2013: 6). The walls were formed from heaped stone slabs set diagonally or vertically. In time the bog covered the stonework leaving only the vertical stones visible when the peat was removed for fuel.

Dead fence a field boundary made from wood or stone, unlike a living hedge. Dickson (1805: 110) believed a dead fence could not be built high enough to provide shelter. He advocated hedges for shelter.

Deck / dekk / dek / deak Shetland variants of 'dyke'. 'Deike' is another variation, from Roxburgh in the Scottish Borders. Incidentally 'deik' is an Orkney verb meaning to run away and hide where you can't be seen.

Devall (Scots) a 'sunk fence' or **ha-ha**, from the old Scots word 'deval' meaning 'a sloping surface'.

Dick / dik (Caithness, northern Scotland) a dyke. These spellings reflect the local accent.

Die wall (Scots) from Pride (1996: 30) — 'a low wall, generally for supporting purposes, *eg* a dwarf wall at the edges of steps or a sleeper wall to support and keep floor joists above ground level.'

Dike a wall, an alternative spelling of 'dyke'. Both 'dyke' and 'dike' can be used as a verb to describe the building process or the enclosing of an area with drystone walls. 'Dike' can also mean a causeway or track. That is closer to the meaning of the word as understood in the Netherlands or the Fens of southeast England, where it is a watercourse.

Dike (Lancashire) a drystone wall. The term is also loosely applied to any fence, or a drainage course or ditch. The spelling 'dike' is found in older Scots publications.

Disposal wall / rubble wall (New England) Hubbell (2006: 30) illustrates this as two low outer walls, appearing to be only a couple of feet high, with a roughly tossed interior, estimated six or seven feet wide. Hubbell eloquently describes how '...the disposal wall is built to present an attractive way of containing nature's rocky bounty.' This is a variety of **consumption dyke** or **accretion wall**.

Ditch in most areas this is a trench dug into the ground to carry off surplus water. MacWeeney and Conniff (1998: 28) define this term from the Irish point of view:

> In Ireland, for unknown reasons, the word 'ditch' came to apply to the raised bank, rather than the trenches from which it was formed. The word is now used to denote almost any form of raised boundary, including an unmortared – but not a mortared – stone wall; a mortared stone wall is known as a 'wall'.

Stephens (1891: 230) adds further information:

> The low dry-stone wall is a common field-fence in Scotland, and is there named a *dyke*,

Figure 80: Irish single stone ditch. The 'scraw' is the top layer of soil and grass. It is removed to form the foundation trench. 'Butt' stones are the large foundation stones. 'Seconders' are the stones above the butt stones. They are secured with smaller 'cog' stones. The 'toppers' are the top course, in effect the cope. The stones are large, up to 175 kg (385 lb). They are placed carefully – with even batter from side to side, and lying across the joint between the two beneath. Diagram extracted from Field Boundaries, a publication of the Department of Agriculture and Rural Development, Northern Ireland.

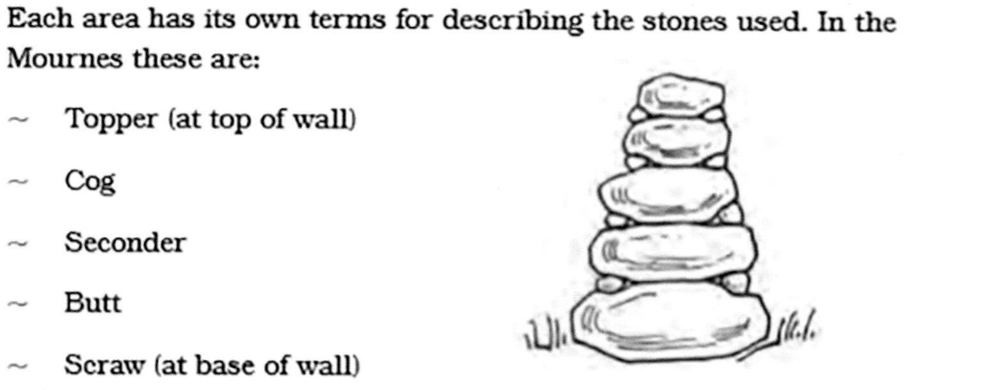

to distinguish it from wall, which implies a structure of stone and lime. A dyke in Ireland means a ditch, and a ditch a turf-wall.

In accounts of Irish walling, references to 'ditch' may confuse the casual reader. Substitute 'wall' or 'dyke' for 'ditch' and the descriptions are more like drystone walls in Scotland or England.

DARDNI (2013: 26) describing dry stone walls in the Mourne Mountains, County Down, Ireland, defines 'single ditch' as '…the most common type of wall. It is normally 5-6 stones high. A variation is the 'lacey' or compact single ditch.'

DARDNI (2013: 27), details how a typical 'single stone ditch' (known elsewhere as a 'single wall') is built in the Mournes. See Figure 80.

DARDNI goes on to define the 'double ditch' as a style with many variations:

> It may be filled with small stones, or earth and small stones. It may have interlinking cross stones where flat stones from derelict buildings or quarries are inserted at intervals in the wall. In some areas, where there are insufficient stones to make walls stockproof, a fence is erected on top of the ditch.' Elsewhere this is known as a 'double wall' or 'double dyke.

A combination of 'single ditch' and 'double ditch', one on top of the other, is also describes what sounds like a variation on the Galloway dyke or Aran style of *feidín* wall:

> …in certain areas of the Mournes a filled double ditch with a single ditch on top can be found. The single wall was constructed with boulders, left over after the double ditch had been built.

There is another combination, a type of consumption wall:

> single ditch can be sandwiched by a double ditch. This type was originally built as a single ditch after land reclamation with a wooden plough. When deeper metal ploughs became popular, more stones were unearthed and were used to surround the original wall.

Patterson (1880: 30) defines 'ditch' as 'a fence, generally of earth.'

Ditch dyke a basic barrier of ditch and stone. Callander (1986: 14) defines them as:

> …where two feet of stone were placed along the top of one of the ditch's edges and the earth thrown behind them. Two developments from this were the sunk dyke or ha-ha and the half-dyke or backit dyke.' These are all covered individually in this section.

Dolmen an archaeological term for a simple megalithic burial chamber with three or more uprights and one or more capstones. They have the appearance of a flat slab sitting on a set of legs after the stone and earth covering has eroded away. Dolmen date from 5,000 to 6,000 years ago. 'Cromlech' is an older archaeological term.

Löbbecke (2012: 38) tells us 'Dolmen (Breton for 'stone table'), which were erected of monumental stone slabs, were very widespread internationally.'

Double dyke / double stone dyke the 'standard' style of dyke built with two outer skins and hearting and throughstones between, as opposed to the 'single dyke' which is one stone wide, larger stones at the bottom, smaller towards the top.

Henderson (1898: 196) distinguishes between the 'single' and 'double' dyke, based on available stone:

> Where boulders are used in the construction of the dyke, as usually happens on moor and rough pasture-land, it is said to be a single dyke. A double dyke is one built of quarried stones — stones not quarried proper, but raised from the nearest exposed faces or outcrops. Where stones have to be quarried in the ordinary way and then conveyed some distance to the site, the fence becomes too costly for ordinary purposes. It is only at those places where suitable rocks crop out every here and there in easy touch with the proposed line of fence, or where the ground to be dealt with is pretty freely studded with boulders, that the dyke comes in as a profitable fence.

The principles of the double dyke pattern were open to interpretation, depending on the site and available material, and flexible enough to produce the Galloway dykes and the 'gentler' Cotswold and Mendip walls.

Double-faced dyke a double dyke, a two-faced freestanding wall, the faces connected with tie stones and hearting in between. This is known as a 'double ditch' in the Mourne mountains, County Down, Ireland. DARDNI (2013: 26) confirms dry stone walls are known as 'ditches' in that part of Ireland.

Figure 81: Caithness drystone dyke, using the local flat stone. The thorny yellow bush is variously known as 'whin', 'gorse' or 'furze', once a common hedging plant throughout Scotland.

Double stack / double faced wall Hubbell (2006: 23) presents this as an eighteenth century New England wall which was '…considerably wider [than a **single stack** wall], made up of two parallel rows of large stones with a filling of a variety of material in between.' Hubbell suggests many single stack walls became double walls as more and more rock came to the surface and was added to the sides of the wall.

Dry-dyke / dry-dike (Scots) a low wall built of stones without mortar. The DSL has a 1572 reference from Peebles in the Scottish Borders — 'the haill communite to help to big it with dry stanis.' This translates to, 'the whole community to help to build it with dry stones'.

Section 2.7 discusses the evolution and etymology of 'dry' versus mortared construction.

Dry fence / Dry laid fence (Kentucky) a rock fence built with no mortar. Evans (1957: 110) uses the term 'dry fence' for Irish walls — 'These 'dry fences' are common in County Kerry and also in north Antrim…' Irish masons emigrated to work on the Kentucky walls and will have taken the terminology with them.

Dry masonry (North America) masonry work laid without mortar.

Dry-stack (USA) a term for stonework where mortar is used, and is not visible. The 'dry-stack look' is achieved by using mortar and holding it back from the leading edges of the stone.

Dry stack as seen in a Seattle, Washington, USA, on a notice in a boatyard advertising the overwinter storage of small boats in racks, out of the water.

Drystone dyke / drystane dyke / dyke the most common Scots terms for a drystone wall, shortened to 'dyke', the name for a stone wall built without mortar. The DSL records the term 'dry' — referring to stones built without mortar, dating from around 1540. Section 2.7 discusses the evolution and etymology of 'dry' versus mortared construction.

Drystone / dry stone / dry-stone three written variations of the same term.

Dry stone wall / drystone wall / dry wall a wall built without mortar. Also known as a 'dyke' in Scotland, a 'stone hedge' in Cornwall, and as a 'ditch' in Ireland.

Dry wall in North America this term is used for walls built without mortar, often qualified with the statement 'generally used for retaining walls', because they are porous. The concept of a mortar-free wall is challenged because strength is equated with cement or concrete.

'Dry wall' also describes a wall which is built with mortar but the mortar is kept back from the front face to make it look as though the wall has been built 'dry', without mortar.

'Dry wall' should not be confused with the North American term 'drywall', known elsewhere as

Figure 82: A limestone wall, showing wallend, throughstones, coverstones and cope. Kentucky, USA.

'sheetrock' or 'plasterboard'.

Dud's dyke (Scots) a fence used as a clothes line, around a piece of ground used as a bleaching green. 'Duds' is a Scots term for clothes.

Dump and hole Callander (1986: 35,84), defines this as a 'style of dyking that does not aim at courses' with a description of this as a basic style of construction undertaken by tenants who enjoyed a cheaper lease while they were improving the land. The tenant's benefit was short lived:

> …usually the only reward was to be faced with a rent beyond their means when the improved holding came up for re-lease.' This encouraged a system whereby the tenants leased, improved and, forced by economic pressure, moved to unimproved ground to repeat the cycle.

Dumped wall / tossed wall / thrown wall (New England) Hubbell (2006: 29) and Thorson (2002: 125) present this as a disorderly wall with no deliberate arrangement. Pioneers carried the stones to the edge of the field and dumped them in a row, using no skill and producing no formal structure. This led to wide structures, not particularly well built. Over time they were rebuilt into reliable walls.

Dun a generic Scots term for any ancient fortification or defensive structure. Their history is lost; many were **brochs**. Most survive as piles of stone.

Dún Aonghasa / Dún Aengus (Inis Mór, Aran Islands, Ireland) this outstanding example of a'promontory fort' deserves its own entry. It consists of three outer rings of thick drystone walls backed on to the edge of a cliff, some 100 metres high. It was probably originally 'D' shaped.Construction of the site began about 3000 years ago. One outstanding feature of the site is a 30 feetwide barrier, a chevaux de frise, guarding the entrance (Figure 103).

Dyke / dike the Scots term for a drystone wall, an evolution from the original sense of the word as a ditch. Winchester (2016: 17) refers to the lack of documentary evidence for medieval field boundaries in northern England:

> Where they are mentioned, field boundaries are often described using the Latin term *fossatum*, meaning something dug — its root is *fossa*, 'a ditch or trench', probably translating to the vernacular 'dyke', which was used both in the sense of a ditch and of an embankment.

He goes on to mention a 1543 reference to a site in Ribblesdale, in the Yorkshire Pennines where the use of '…*murus sive fossatum* ('wall or dyke') suggests that 'dyke' was being used to refer to a field wall…'

Winchester (2016: 18) provides older evidence for the term:

> …on the hill pastures of Newby, near Clapham, [north Yorkshire] the name 'le Standik' ('the stone dyke') was recorded in around 1170 and confirms the existence of at least some stone boundaries at that early date.

DSL defines dyke or dike as 'A wall (of earth, turf, or stone) made for enclosure or defence or as a dividing line. Very common after 1530.'

Stephens (1891: 230) has a more specific usage:

> The low dry-stone wall is a common field-fence in Scotland, and is there named a *dyke*, to distinguish it from wall, which implies a structure of stone and lime. A dyke in Ireland means a ditch and a ditch a turf-wall.

The poem, 'Davie the Dyker', from *The Scottish Gallovidian Encyclopedia* by John Mactaggart (1824), is preceded by a definition:

> Dyke — A fence. Anciently dykes were built for confining mankind to certain portions of the earth; as those the Romans built in Scotland, to keep at bay the daring Scots; and that one by the Chinese to hold out Tartars, considered one of the wonders of the world; now they are only used to inclose brute cattle; some people make a trade of building dykes, one of these must have been *Davie the Dyker*...

A short version of the poem can be consulted in Rainsford-Hannay (1957: 95-97) and it is included in this book as an appendix..

A 'dyke' in Lincolnshire, England is a wide ditch used for a fence, and for drainage. Dyke is a Scottish term for drystone wall, a shorter form of 'drystone dyke', also spelt as 'dike' and appearing in a variety of spellings, mostly now obsolete. 'Deak', 'dick' and 'dic' are notable as the varieties from Caithness and the northern island groups of Orkney and Shetland.

In Nova Scotia, Canada, a dyke is a seawall made from a bed of spruce boughs covered with soil and rock, topped with sod to form a waterproof wall, sometimes miles long. They had flapper valves set where brooks flowed into the ocean to allow fresh water to flow out at low tide. This structure was introduced by the French in the 1600s and developed until their expulsion in 1755. The French word *dique* is a sea wall or dyke, as found in Holland.

In Cumberland, northwest England, a 'stick dyke' was a fence made of dead wood and brushwood and a 'watter dyke' was a water filled ditch — wide and deep enough to form a barrier, in effect a fence.

The verb 'to dyke' also applies to the building of a drystone wall or the enclosure of land with a dyke. An area which is 'dyket about' is enclosed with dykes.

'Dyking' can also have the simplest of meanings. Steed (2011: 105) quotes the price for thirteen loads of Dundry limestone for a new spire in Bristol in 1491. This converts to £1.28 in modern decimal currency. The transaction includes the cost of 20 pennies (approx. 8½p) for stacking the stone:

> For xiii lode of frestonys 25s 7d.
> For dykng of stonys 20d.

Dyke-end / daek-end / dike-end (northern Scotland) a dyke built down into the sea, past the low-water mark, to stop cattle from walking round the end of the wall and trespassing on to neighbouring land.

Dyke share (Cumbria, northwest England) a share in the responsibility for repairing a stone wall.

Dyke side (Scots) the land alongside a dyke.

Dyke-slough / dyke-sheuch / dykesheugh (Scots) a ditch or open drain alongside a dyke at the 'end riggs' — the outer edges of the field, the margin which is seldom cultivated.

If ploughs delve too near the dykes, they can loosen the foundation.

In Northern Ireland a 'sheuch' or 'sheugh' is a ditch or open drain which is often the line of a property boundary. Patterson (1880: 89) defines a sheugh as a ditch and includes the saying, from Ulster, Ireland:

> I always let the *sheugh* build the dyke,' ie, I always let what was dug out of the ditch make the raised fence, a saying, my spending never exceeded my earning....

Figure 83: A landscape of walls and small fields, with ivy growing up the stones. Inis Oírr, Aran Islands, Ireland.

Dykie / dikie (Aberdeenshire, Scotland) a low, little wall, or a small ditch. Many Scots words add '-ie' or '-y' to form the affectionate diminutive. Therefore 'dog' becomes 'doggie', an affectionate term for a small dog.

Edge fence (Kentucky, USA) a fence made up of diagonally coursed stones laid onto the ground with smaller stones, on top of these, also on edge, to bring the wall up to height (Figure 84). These stones are laid with their edges *across* the line of the wall, unlike flagstone

Figure 84: Edge fence. Kentucky, USA.

fences where the edge is *along* the line of the wall. Murray-Wooley and Raitz (1992: 191) describe this type of fence as one with diagonal coursing '…laid with rock edges or most narrow faces towards the ground and coursed at an upright angle between 25 and 40 degrees from the vertical.'

Murray-Wooley and Raitz give Kentucky synonyms of 'rick-rack', 'rip rap', 'rough and ready', 'stacked fence' and 'stand up fence'. Murray-Wooley and Raitz also quote 'up-and-down fence' as an example of diagonal coursing from Cornwall, England and 'vertical coursework' as an example from the Isle of Man.

Rainsford-Hannay (1957: 87) describes a 'stone faced bank' on the Isle of Man:

> Stone slabs were put edgeways into the bank; they had been chosen of a slight wedging shape. Slabs, some 2 to 3 feet long, 3 inches thick at one end tapering to 1 inch at the other, were placed alternately thick side to thin side along the wall edges. They fit so tightly that not a weed nor a blade of grass could show on the outside. The longer they stand, the tighter they fit, with a full weight of bank above them.

Edge fences are ideal for land which floods. The water flows through and pressure does not lift any horizontal stone.

Dutton (1824: 164) gives details of 'those kind of keyed walls…' in Galway, Ireland. They are built of long thin stones of differing sizes wedged together, on their edges, to a height of four feet. The wall is strong and '...they keep out hunters, who though very allowable in their proper place and season, are very unwelcome visiters [sic] to a well improved place, or to a breeding sheep farm.'

D'Arcy (2006: 14) identifies the 'slab wall' as one specifically associated with the Burren in western Ireland. An image shows slabs laid on edge across the line of the wall with smaller slabs set between the taller uprights to bring the structure up to an even height. It has every appearance of an edge fence.

The term **wedge wall** (Figure 106) is sometimes used. It is a good description of the way stones are 'wedged' into each other. It may also be a corruption of 'edge wall'.

Estate walls good quality walls built around farms or mansion houses. They differed from agricultural walls because they were built to a design rather than developing from rows of rock as the fields were cleared. Ornate estate walls gave the landowner a chance to show off their wealth and provide privacy. Thorson (2005: 73) describes these in northeast USA:

> Well-built double walls, complete with a rubble fill and capstones, are often referred to as estate walls ...an expensive capital improvement requiring more financing than even well-established farmers could afford. Hence, the wealth manifest in estate walls came generally from industry, rather than agriculture.

Hubbell (2006: 68) also has a definition:

> The estate wall, or wall of affluence of the turn-of-the-century period is a carefully laid-up structure, most often constructed of quarried stone.

Face-wall a wall which 'faces up' a solid bank or slope. They are substantial, not just a veneer. **Retaining walls** and **breast walls** are built stronger than face-walls because they resist the force of unstable material behind them.

Faced dyke / face dyke / sunk dyke a basic form of boundary with one stone-covered face. It is built by digging a trench two feet deep and raising a mound with the excavated material. The soil is allowed to settle for some time before a stone face is built up from the base of the trench. A hedge was often planted on the earth bank behind the stone face. It is, in effect, a type of **ha-ha** and like a **Galloway hedge**.

GVA Dumfries (1812: 153) advises the face dyke should be 'well banded' with long stones (it must have sufficient throughstones) and its face should be a little concave to reduce the chance of bursting out in the middle.

Famine walls / penny walls (Ireland) these walls were reputedly built during times of famine or economic depression, commissioned by a landowner to provide a basic income for destitute labourers and, at the same time build enclosures, for developing agriculture.

Fank / flank (Scots) a general term for a stone enclosure, a sheep shelter or part of a series of pens used to manage sheep at clipping time and other occasions when the flock is gathered together. See Figure 85 for a Welsh version. Modern fanks are built in a more rectangular style.

The word shares common roots with the German verb *fangen*, 'to catch'; the Irish *fang*, 'to entangle' and the Scots Gaelic *fangaich*, 'to enclose'.

Figure 85: Corlan Anafon sheepfold, Carneddau, Wales.
Image courtesy of National Trust.

Sheep fanks have special features which allowed them to enclose and handle sheep safely. There are no throughstones sticking out on the inside, to save the sheep from injury. The inside walls are often near vertical, to discourage climbing. The wall often stands on a wider scarcement, so the constant movement of sheep does not undermine the foundations. A fank built as a shelter is circular or had rounded corners; sheep were free to move around if the wind changed direction.

Farmer's wall (New Hampshire, USA) a rough wall, often just one or two stones thick. Hubbell (2006: 26) described them:

> These typical single-stack farmers' walls were built as the seemingly endless procession of rocks rose to the surface each spring.' The walls he is referring to were on ground first cleared in 1650.

Fauld / fold / faud (Scots) a sheep pen, hence 'fa'-dyke' a shortened form of 'fauld-dyke'. JSD has 'fald-dyke' as a 1437 Aberdeenshire term for a '…wall of turf, surrounding the place appropriated for a fold.' Turf was an accepted material for building an agricultural boundary before reliable drystone dykes were developed in the eighteenth century.

Figure 85 shows an eighteenth century Welsh version, built of stone. It is in an area where sheep are grazed on common ground (shared pasture). Sheep were gathered and separated, by ownership, into the smaller pens. The 'V' opening at the top, middle, helped funnel sheep

Figure 86: Feidín wall from east County Galway, front elevation and cross-section. Note the batter on the bottom half, which is built very similar to a conventional 'double' wall. Image courtesy of Pat McAfee.

into the system. Similar structures exist in Iceland and Africa.

Fauldin (Scots) a fold for cattle or sheep.

Feal dyke a dyke of turf and stone. Section 4.4 describes stone structures incorporating turf and earth.

Feidín wall / fidín wall / figín wall / balla feidín / clai fidín (western Ireland) this name first appears in east Galway in the 1800's as a term for a 'double' wall built with small stones – *feidíns* in the Irish. This presumably contrasts with earlier walls built 'single' with large boulders, or slabs set on edge.

The cope, the top half, of these *feidín* walls consists of large stones, laid across the width of the wall. The cope can be one or more stones high, depending on available stone and height required. D'Arcy (2006: 27) gives more detail:

> In the classic *balla feidín* the base courses of fist-sized stones are covered with flat horizontal stones and stacked above with more upright-orientated stones. Variations on the theme can be seen in parts of the north Burren. Some eye-catching patterns result from the angular stacking of stones near the top of the wall.

Evans (1957: 108) has a similar illustration to Figure 86, but does not use the term *feidín*. He describes the wall as an 'East Galway ditch of Galloway type', using the word 'ditch' for the Irish meaning, a drystone wall. 'Galloway type' refers to the Scottish **Galloway dyke**. One striking similarity between *feidín* walls and Galloway dykes is the heavy open-built top; this was reckoned to deter livestock because of its apparently unstable appearance. Galloway dykes and *feidín* walls are a combination of large stones and small stones, used to best advantage.

According to Evans (1957) and McAfee (2011), drystone walls may have been introduced to Ireland from southwest Scotland by improving landlords in the nineteenth century. This is highly likely. Irish farming practices had been moving forward since the eighteenth century, with liming of fields, drainage and new hedges and walls, as evidenced by Hutton (1892). Contemporary accounts record the spread of drystone walls through western Ireland in the early nineteenth century as part of agricultural improvements. Irishmen who worked as seasonal labourers in Scotland would have been familiar with the latest developments in farm fencing.

McParlan (1802: 59) commented on the poor nature of fences in County Donegal. In the mountain areas they were not stockproof:

> …either a person stands out to watch the cattle …or the cattle are tied by ropes… to prevent their reaching the potatoe (sic) or corn sowings…'

The fences in the lowland areas were no better. McParlan noted there were no hedges '…except about gentlemens' houses…

Dutton (1808: 152) described stone walls in County Clare as poorly built and requiring '…constant mending.' Within two decades the situation, reported from neighbouring County Galway, was much improved, according to Dutton (1824: 163):

> Formerly the usual wall was composed of stones piled up without order, which may be called filligree (sic) work, easily built up, but as easily thrown down… within the last twenty years, however, a very material change for the better has taken place…

Good prices for produce and bank loans had created '…a powerful impulse to every kind of improvement… [including] …good double stone walls…'

The style of these walls evolved to suit the local stone as agricultural improvement moved west to the Burren and the Aran Islands. Here there was abundant limestone, often in slabs

Figure 87: Feidín wall, Aran Islands, front elevation and cross-section. Note vertical 'mother' stones and no batter on the lower half. Image courtesy of Pat McAfee.

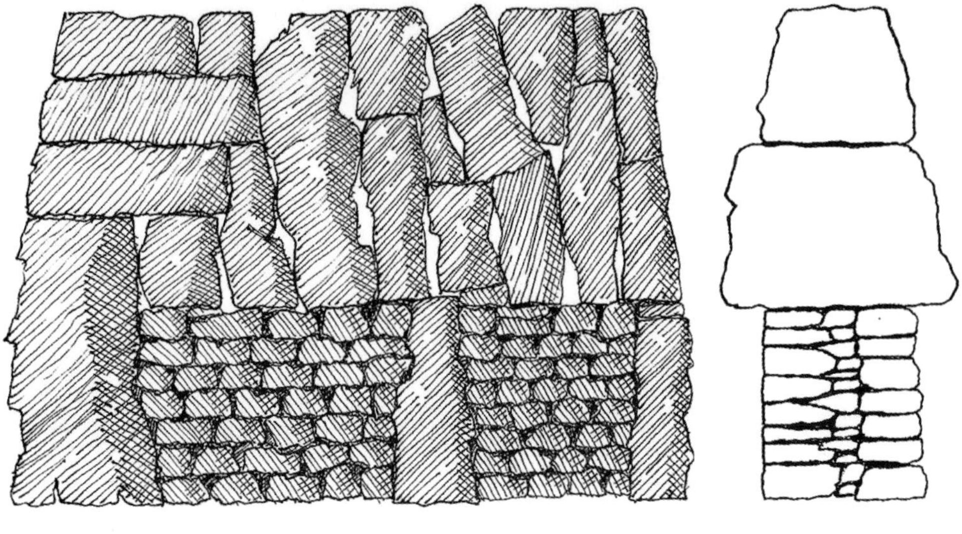

big enough to act as wall ends or occasional uprights which supported smaller stones. The limestone slabs produced a more open style of build, with a more angular appearance.

D'Arcy (2006: 27-28) describes *claí fidín* on Inis Oírr, the smallest of the Aran Islands – 'Where available, an upright or standard (*Cloch seasaimh*) is inserted at fairly regular intervals to provide stability.' These uprights have become known as 'mother stones' (in Irish, *clocha mháthar* or *clocha máthar*).

MacWeeney and Conniff (1998: 136) give more detail:

> ...o the islanders, the *claí fidín,* a wall of little bits... was the hardest to build and the most prized, composed of densely packed bits of rock at the bottom between upright 'mother stones'. Such a wall with the larger stones set improbably on top, spoke of the almost magical skill of the builder.

Laheen (2010: 105) describes *fidín* walls on the Aran Islands as a type of 'single' wall with smaller stones at the bottom, larger ones in the upper section:

> The single drystone wall is one stone deep and generally about 4 feet high. It is constructed by clearing away an area of soil about 2 feet wide down to the rock... then placing the large vertical upright stones about 4 feet apart. These are known as *clocha máthar* (mother stones). In between the mother stones the wall is filled with small stones; on top of this foundation, stones are placed randomly. The lower level of stones is tightly packed, to strengthen the foundation, to stop rabbits getting through, and because close to the ground wind pressure is not an issue. The lower part is usually about 2 feet high. The upper part has gaps between the stones to create the wind filter. This is the basic format for single drystone wall building, and it is the many variations of this basic layout that gives the various drystone walls their aesthetic appeal. This is also known in Aran as a *fidín* wall. However not all walls display these characteristics.

Many *feidín* walls are built 'double' in the bottom half, it is tighter than when built 'single'. This, of course, is only practical if there are enough small stones. *Feidín* walls continued to be a highly regarded agricultural improvement well after the nineteenth century, as confirmed in Laheen (2010: 105) — 'This type of wall was insisted upon by the Department of Agriculture when it was grant-aiding wall building in the twentieth century.'

This ability to vary the style of build according to the available material is elegantly noted by D'Arcy (2006: 28) when describing the varieties of stiles (crossing points through or over a wall) in western Ireland — 'All demonstrate the inherent 'stone knowledge' of the builder, a patience born of hard physical labour, and a discernible affection for the material at hand.'

This style of wall had an added advantage – the smaller stone, tightly built, in the bottom half, was an efficient rabbit barrier.

A 'family' of terms has emerged over recent years for the Aran style *feidín* wall. Figure 87 illustrates these. Mother stones (*clocha mháthar* or *clocha máthar)* are those set upright at regular intervals, with their long edge across the line of the wall. Smaller stones, children (*na paisti*) are laid horizontally between the uprights to a height of about two feet. Father stones (*na hAitheracha*) act as the heavy cope which lies across the top of the small stones.

According to personal correspondence with Patrick McAfee the first two terms, for 'mother' and 'children', have historic validity and 'father' may have been created more recently to complete the 'family'. This is a natural development, an adaption of existing terminology, not unlike some of the descriptive drystone terms created in the eighteenth and nineteenth century.

The term 'Irish family wall' has been used in North American stone festivals.

The walls of the Aran Islands, including the *feidín* walls, are indeed art, as described in Scully (2019: 82):

> I could see that they were pure culture, made with an exquisite sense of order and design, their durability based on a few large stones at regular intervals standing upwards, everything resting subtly on them, all the smaller stones, carefully selected, depending on them.

Fell wall a wall between the arable and the hill grazings. Also called a **head dyke**. Fells are the rough pasture land or moorland in the north of England and southern Scotland. 'Fell' appears in many place names.

Fence historically this was a description of a boundary whether it was a wall, hedge, bank, ditch or dyke. Nowadays it commonly defines a barrier made of wooden posts and wire.

One contortion of the word 'fence' is found in Muilenburg and Swineford (1975: 31). They refer to the unsuccessful attempts to grow 'hedge fences' in Kansas. GVA Derbyshire

Figure 88: A low flagstone dyke, with metal supports. Brough, Caithness, Scotland.

(1813: 83,84) refers to 'stone wall fences' and 'wall fences'.

Macdonald (1887) distinguishes between 'dyke-fencing' and 'wire fencing'. The first is built of local stone; the second is made from post and wire.

Fend (Scots) a sea-dyke, or breakwater, like the English word 'fender' and no doubt linked to the word 'defender'.

Flagstone dyke / flag dyke / flag fence / flag wall / slab fence a wall constructed of large thin sheets of stone, quarried in areas where the rock can be split into thin flagstones (Figures 88 and 89). The rock type is typically siltstone, fine sandstones or limestones – all sedimentary rocks with a small particle size.

Typically, individual flags are embedded, one third of their height, into the ground, with their long edge along the line of the wall, perhaps with occasional fenceposts to keep the stones stable. If required, a series of posts holding a single strand of wire gives the flags extra height.

Haslingden Flags from quarries to the north of Rochdale in Lancashire, England, were used to build fences from the 1700s to early 1900s. The flags were held together with diamond shaped plates bolted in pairs between the flags. The fences were typically 1.5m high, using flags one metre wide.

Caithness flags are common in that county in the north of Scotland. Individual flags are set with a few inches overlap, sometimes held together with metal clamps. Caithness flagstone is a versatile material. Large pieces have been used for byre partitions, room dividers and to make lean-to additions for cottages.

Flag walls were also built in the Cotswolds, southwest England. In Cumbria in the English Lake District the local style of flag fences uses Brathay flags, sedimentary blue slate

Figure 89: Limestone flags in a field wall. Inis Oírr, Aran Islands, Ireland.

flagstones. The village of Hawkshead in Cumbria has fine examples.

Conry (2000: 59) describes a flagstone fence in a corner of west Clare, along by the Cliffs of Moher and a boundary fence between some fields on the estate. These were made from the local Liscannor flagstones, the only walls of this type in Ireland.

Flag fences are typically only a few inches thick and are the key to the riddle 'Where can you find a wall a mile long, a yard high and an inch thick?'

Flag walls are also known as 'tombstone walls' because the individual flagstones look like gravestones.

A slab fence, using thicker pieces of stone, is described by Conry (2000: 60-66). See Figure 98. This 'Scotch Fence' is specific to part of County Carlow, Ireland. The fence is made of granite slabs set on edge, typically 1.25 metres high. One of these fences had an iron peg '…set in every 15th to 17th slab, carrying a line of plain wire to give extra height to the field fence.'

The granite slabs are cut to reasonably similar sizes and were more expensive to produce than limestone or siltstone flags from, say, Caithness or Liscannor, which can be prised off the bed of the quarry with basic tools. These slabs required shaping but sedimentary stone is easier to shape than granite.

Conry (2000: 59) describes a use for second grade slate near the villages of Mynydd Llandegai and Penisa r Waun in north Wales. Long narrow slabs of slate were planted upright into the ground, like fenceposts, with wire twisted along the tops and bottoms of the slabs to strengthen the structure. Gaps were left between the individual stones. These fences were called 'Slate Pillar Fences' or 'clawdd' pillari in Welsh. This translates literally to 'pillar hedge' (see figure 75).

In south Cumbria, northwest England, walls made of vertical sheets of slate are known as 'clapper walls'.

Muilenburg and Swineford (1975: 33) include a type of flag wall in Russell County, Kansas:

> Shell rock taken from the nearby hills in large sheets was stood on end forming three sides of the corral. It stood as a solid fence six to eight feet high enclosing a sloping hillside area of about 200 square feet…

The age of the corral is not given, it was recorded in 1872 as '…falling down and showed no signs of recent use.'

The difference between an **edge fence** and a flag fence or **flagstone dyke** is the placement of the stone. The edge fence has the stones set with their long edge *across* the line of the fence while the flag fence has the edge of the stone *along* the fence line.

Slate, granite and limestone have been used as fenceposts, with wires strung through holes or wire loops to create a stockproof barrier.

Flat fence (Kentucky, USA) a description of walling with stone laid in careful horizontal courses, in areas where diagonal coursing is the norm. Diagonal coursing has stones on edge or leaning.

Fleet dyke (south of Scotland) an old term for an embankment or wall built to prevent erosion or flooding, made of turf, 2-3 feet high set a few yards back from the bank of the river. 'Fleet' appears to be a variation of 'flood'.

Flow dyke according to JSD, a term from Banffshire, northeast Scotland, for a small drain carrying off water.

Fold an enclosure for sheep, a fank or pinfold. **Fauld** is the Scots equivalent.

Frame walls Loudon (1825: 444) gives details of a building method, apparently British, which used a frame of boards, the height and width of the wall. This was filled with stones gathered from the adjoining fields. Liquid mortar was then poured in to fill the spaces. The structure was left a few days until strong enough, after which the frame was dismantled and moved along the line of the wall for repetition of the process. This method sounds similar to modern concrete slipform construction.

Freestanding wall a wall which stands free with both sides visible; it is not built against a bank or supported by another structure.

French drain in its simplest form, this is a trench filled with small stones. It was popularised by Henry French (1813-85), a lawyer from Concord, Massachusetts, USA, who advocated this system in his book, *Farm Drainage*, first published in 1859. The original field systems in New England consisted of many small fields. Ambitious farmers recognised the need to remove some of the stone walls, enlarge the fields and improve drainage. These requirements were simultaneously satisfied by digging deep trenches and filling them with dismantled walls.

French walls a Canadian term, from Nova Scotia, for stone walls believed to have been built by the Arcadians in the seventeenth and eighteenth centuries. The Arcadians were the original French immigrants who settled the eastern regions of Canada.

Gabions / gabion baskets from the Italian word *gabbione* meaning 'big cage'. Modern gabions are rectangular wire baskets filled with stone. They are used to form retaining walls and the principle has been incorporated into buildings. One outstanding example is Hertzog & de Meuron's design at the Dominus Winery, Napa Valley, California.

Gabions are free draining and quick to erect. They can collapse if not carefully filled. Stone must be built carefully to maintain the basket's shape and strength.

Galloway dyke / Galloway dike in the eighteenth century this term was a generic name for drystone dykes. They were developed in Galloway, in southwest Scotland, and quickly spread throughout Scotland and the north of England. More recently the term is applicable to that regional style, prominent in the southwest of Scotland, where the bottom half of the dyke is built 'double' and the top half is built 'single' (Figures 58 and 59). This is the most efficient way of dealing with large stones and produces a strong fence on wild moorland and sheep pastures.

Rainsford-Hannay (1944: 24) details the origin:

> Dry stone dyking, in a comprehensive way, began in about 1712, when the brothers M'Kie of Palgown, in the extreme west of the Stewartry of Kirkcudbright, set the example by starting to enclose many square miles of otherwise useless moors. They allowed free-rented pieces of land to people who would work for them in the summer. Early in summer these people with their women and children took to the hills with poles, tents, and blankets. In two or three days they had built themselves turf huts, and heather formed their beds. The brothers themselves superintended, and in a few years the work, which at first looked like the labours of Hercules, was done.

This is within the area known as Galloway. The roadside sign for the farm now reads 'Palgowan'.

GVA Galloway (1813: 91) gives some background:

> The McKies of Palgown had for ages farmed the patrimonial estate, and were distinguished for hospitality, and superior activity and intelligence as storemasters. Alexander and Anthony, who had received a very liberal education, were endowed with good sense, and

Figure 90: Dykes at Palgowan, Galloway, Scotland.

> great activity of mind and of body, preferred the occupation of their ancestors to every other. Enlarged in their ideas far beyond the common peasantry, in those rude times, they formed plans of improvement, which to others would appear wild and romantic.

The distinguishing feature of the Galloway dykes, the one most remarked upon by early writers, are the top courses of rough stones set on edge. This was an improvement on earlier copes consisting of turf or a piling up of small stones. This cope, known as the 'Galloway cape', often makes up the top quarter or top half of the dyke.

Loudon (1825: 443) describes 'The Galloway dike or wall' as one half 'double' and one half 'single' and refers to '…rugged round…' stones in the upper half. He goes on to account for the main feature of the dyke:

> Its tottering appearance is so well calculated to prevent sheep, cattle, or other animals,

from approaching it, that it is seldom indeed that any attempt is made to leap over it. This circumstance, together with the ease with which the stones are procured, in most of the situations where the Galloway dike is used, renders it a valuable fence.

This style of dyke incorporates two important elements. Smaller stone, tightly built on the bottom half, makes it rabbit proof; Larger stones, in the top half, make cows and sheep think twice before jumping.

GVA Perth (1794: 115) describes two variations of drystone dyke from Galloway, one built 'single' for its full height and a more common style where the wall is built half 'double', half 'single':

> ...where the ground is high and the stones fit for the purpose, the fences are Galloway-dykes. When the stones are very large, they are in some cases built single from top to bottom; but more commonly about half of the wall is made of such small stones as have a bed, and is built double. The upper part of the wall is made of single stones ... and every row decreases in size, towards the top of the wall... These walls may be made of the coarsest stones; and when they are properly laid... they stand better than double walls, unless these are cemented with lime and have plenty of band stones.

The second option, half 'single', half 'double', corresponds to what we nowadays consider as a 'Galloway dyke' – the iconic regional style, a drystone appellation.

GVA Mid Lothian (1793: 38-39) describes the Galloway dike in detail:

> A very particular and formidable looking fence is used in the moor-land parts of the country, where the pastures being generally fed off with sheep, a thorn hedge is unsuitable. This is what is called the *Galloway dike,* built of whin stone, usually 5 feet high. It is the manner in which it is constructed that renders it singular. To the height of 2½ feet, it is built neatly, but without mortar; in breadth from 20 inches to 2 feet. Upon this is laid a thin course of flat stones, which project three inches over the under part. Above these are placed, in seemingly careless manner, a set of rugged round stones, which in building are brought gradually to a taper, ending at last in stones of three or four inches diameter at the top; but so thin laid, that the light is seen in numerous large vacancies, and the wind blows through them as though a hedge. From its tottering appearance, cattle of all kinds are shy of approaching this kind of barrier; even hares avoid it. Being well founded, and having little top-weight, it stands well, and is easily repaired when it falls into disorder; which, as no living creature likes to attempt it, is very seldom the case.

Early writings about dykes refer to them repulsing cattle. By the 1800s sheep are more frequently mentioned. This reflects the changes in agriculture around that time.

Nicol (1803: 299) makes a good point regarding dyke construction, the importance of using rough stone with the minimum of shaping:

> The chief art in building the *Galloway-dyke* consists in *matching the stones at sight*, so as they may *bed* well, and hold together; for, the prices generally allowed on this species of building will not admit of *the tool* being used. In some parts I have observed a very handsome and durable method of erecting these walls; *viz.* building one half, or at least

two feet, with flat slaty stones, and finishing with large blocks at top.

The reputation of Galloway dykes and their builders spread throughout the country. In the appendix to GVA Berwick the author advises employing Galloway men for building Galloway dykes:

> ...if these are attempted, a set of dykers from that country ought to be employed, otherwise it is only laying out money to little purpose, and in the end you are disappointed. The perfection to which that branch of mason craft is now arrived in the west of Scotland, is astonishing, which in other places is only badly imitated.

GVA Hebrides (1811: 164) speaks highly of the Galloway dyke as '...a species of inclosure commenced in 1720 in that southern district of Scotland, and now well known and much esteemed over this kingdom...'

Further praise is found in GVA Galloway (1813: 83):

> But by far greatest part of the fences... are the drystone walls, or dykes, for the building of which the workmen of this country are so remarkable, that walls of that kind are known through the greatest part of Scotland by the name of *Galloway dykes*.

These rugged drystone walls were ideal for enclosing hill pastures, often over remote hillsides, far from the view of casual visitors, as described in Rainsford-Hannay (1957: 53):

> The tourist travelling along Highland roads cannot expect to see these Galloway-dykes at near hand. Nearly always they run away up in the hills and are usually march dykes.

In Corangamite (1995: 82) Bill Harlock uses the term 'Galloway' for a wall at Pomborneit North near Lake Corangamite, west of Melbourne, Australia. It seems to lack the finesse of the Scottish version:

> It's what they call a Galloway. It's a Scottish wall and it's built with one stone on top of the other but it's built like a snake because if it was built in a straight line it would fall over, so they bend it.

This theory has a precedent in the English 'crinkle crankle' wall, also known as 'crimkum crankum', 'serpentine' or 'wavy wall'. They date from the eighteenth and nineteenth century and are one brick thick, with a regular lateral wave along its length. Advocates thought them stronger, cheaper to build and, perhaps, suitable for defensive structure.

Corangamite (1995: 25) describes a Galloway Dyke in Victoria, Australia, with an interesting explanation for the origin:

> The name Galloway Dyke originates from Scotland. According to Scottish walling history the original Galloway style featured single walling from bottom to top... but this proved too difficult for most later wallers who started to build only the top half with a single thickness.

This piece of walling history has not been traced. The 'difficulty' was presumably a reference to the suitability of the stone for 'single' dyking, rather than the skill of the dykers.

One final point on construction, from Munday (2012: 72), when referring to an Australian wall built similar to Galloway dykes — half 'double', half 'single', he writes — 'As a rule, boys would build the lower section of the wall then the craftsman would follow with the larger stones on top.'

Comparison is drawn between the Irish *feidín* wall and the Scottish Galloway dyke. It is reasonable to assume dykers from southwest Scotland were employed by landlords in the west of Ireland who were keen to improve their estates. These dykers, no doubt, tried to replicate dykes from their home area using the Irish stone and adjusted the style to suit. Well into the twentieth century many Irish came over to Scotland to work for the summer season on farms. This work must have included dyking.

Galloway hedge / sunk fence reputedly invented by Hamilton of Baldoon in 1730 (Figure 91). The dyke's lower part acts as a retaining wall. When it reaches the top of the embankment thorns are planted at the back of the dyke and laid through the dyke to stick out the front. The dyke then continues to full height. With time the thorns grow vertically to produce a formidable barrier of stone and hedge.

Rainsford-Hannay (1957: 46-48) describes a Galloway hedge with the thorns growing through the stonework just below the level of the throughbands. He thinks this stock-proof obstacle is an ideal combination of hedge and wall because the stones protect the vulnerable base of the hedge and the hedge protects the vulnerable cope of the wall.

Evans (1957: 109) includes a drawing of an Irish variation, the 'hedged ditch'. It consists of a bank of earth, faced on one side with stone and quicks (hawthorn) planted at the base of the wall. The hawthorn grows up in front of the stonework to create a daunting barrier.

Garadh the Scots Gaelic term for a wall or dyke, which develops into:

Garadh cinn — the dyke separating the arable land from the moor or common grazing.
Garadh dubh — a turf wall separating the moorland from valuable autumn pasture.
Garadh-sleibhe — a moor boundary wall.
Garadh an arbh — a dyke surrounding arable land.
Garadh cloiche — a stone wall.
Garadh-criche — a boundary wall.
Garadh dorma — a dyke of stone or turf between crofts.
Garadh gorm — a turf dyke.
Garadh gleiddth — a defending wall.
Garadh na h-eaglaise — a churchyard.
Garradh Iaruinn — which literally translates as 'iron dyke', a local name for walls found near South Erradale, Gairloch, Wester Ross, Scotland. The walls were built with pieces of bog iron, the iron-rich hardpan found in the peat bogs, a source of raw material for eighteenth century iron smelting.

Gavel dyke / gavel dike (Cumberland, northwest England) a section of headwall allocated to a farmer for him to maintain, even though he may not have land adjoining the headwall. The headwall separates the arable ground from the common hill pasture. The farmer benefits from the common grazing; it is therefore reasonable for him to share in the maintenance of the wall. This definition is based on Adams (1976: 129).

Gerbick / gerback / gorbak / garbak. (Orkney and Shetland) a ridge of earth thrown up to mark the boundary between two patches of arable ground.

German fences (Texas, USA) Knott (2004: 1) identifies these as stone fences built by German immigrants in southwest Texas in the nineteenth century, apparently influenced by British styles of walling from further east in Tennessee, Kentucky, Missouri and Arkansas.

Giant's grave (Scots) a general term for prehistoric sites whose origins are lost in the mists of time, for example cromlechs or chambered tombs. The ruins of many structures are attributed to giants or mythical ancient tribes.

Girth JSD defines this as '…denoting the circle of stones which environed the ancient places of justice.' These were natural amphitheatres, ideal for gathering and ceremony.

Gravity wall a type of **retaining wall** which relies on its mass and its batter to counteract the pressures from behind. Many retaining walls are face walls; they provide a securing skin to a reasonably solid embankment. A gravity wall has a real job of work to do; it opposes the live pressure of the background.

Groyne stonework at the side of a river, carefully placed boulders, calculated to save the bank from erosion or in some way improve the fishing by influencing the water's flow. Examples exist in many Scottish salmon rivers (Figure 79).

Ha-ha a structure which is well known in England, less so in Scotland. Often found in the ground of the landed gentry where the addition of a fence or wall interrupted the view.

It consists of a deep trench, faced up with stone on one side and sloping away on the other side. When looking out from inside the property the wall is invisible. The wall is more obvious when looking towards the mansion. This was also termed a **sunk fence** because it 'sinks' out of view when seen from a distance.

Hadrian's Wall a defensive wall built by Romans, from the Solway Firth to mouth of the River Tyne, 73 miles (117.5 kilometres), an approximation to the Scottish border, though all the wall is in England. Work started in AD122 and lasted six years. Most of the wall was dressed stone and lime mortar with a large part built of turf on a stone foundation. Accounts vary, a maximum height of 15 feet (4.6 metres), and width of 10 feet (3 metres) is quoted. A lot of the stone has been recycled into local roads, field walls and houses.

Haifreis-dyk (Shetland) a shared dyke, a common responsibility, for example the head dyke separating arable fields from the grazing ground on the hill.

DSL records a judgement from a 1602 Shetland Sheriff Court judgement (1602 Shetland Sheriff Court (ed.) 27), a case against several individuals:

> …for breking the foldis dume in noche bigging up the haifreis dyk with the gudis thairof as thay wes dempt, thairfoir ilk and of them are decernit to pay thairfoir ane dumra and to big up the dyk foirsaid within the space of acht dayis, under the paine foirsaid.

The individuals did not heed a previous judgement to rebuild their portion of the shared dyke. They were fined and had to rebuild the dyke within eight days. This shows the importance attached, at least by the authorities, to the proper maintenance of the head dyke to avoid damage to growing crops.

Figure 91: The hedged ditch, a cross-section. Extracted from Evans (1979: 109).

Half dyke / half-and-half dyke (Upper Annandale, Scottish Borders) Prevost (1957: 99) describes this as 'A stone-dyke built so that the *double* and the *single* both measure half the total height. Thus a half-and-half dyke, half double and half single.' This description fits what we nowadays think of as a **Galloway dyke**.

Half-dyke / halfdyke (Scots) a stone face against a bank of turf and soil which has previously been allowed to settle and consolidate. GVA Aberdeenshire (1811: 221-222) describes the construction thus:

> Sometimes a wall of three or four feet at the bottom – the same in height, and eighteen inches on the top, of turf (provincially [called] feal) is used as an inclosure, but is seldom durable. At other times, after a backing, as it is called, of turf and earth has stood 6 or 8 months, a stone breast or front, of about eighteen inches at the base, fourteen at the top, and four feet high, is built close to the turf, but made to rest on its own base. This is commonly called a half-dyke, and when well executed, is a warm inclosure, and very durable. It costs from fivepence to tenpence per ell, of thirty-seven and one-fifth inches.

Half wall Todd (1859: 468-469) details three ways of building a 'half wall fence' in localities where stone is not abundant, and some fencing timber is available. The combinations of a lower section of stone and an upper part of timber produced a stock proof barrier. Todd goes on to describe a wall built on top of a ridge of earth which had been left to settle for a year. This is like the description of a Scots 'halfdyke'. The creation of the earth ridge reduced the amount of stone required to attain height.

Figure 92: View down on Village Bay, St Kilda, Scottish Hebrides. The head dyke (in the middle) separates the arable ground from hill grazing. The scree in the foreground provided ample building material.

Wiggins (1840: 284) quoting from a New York

source notes that round stones '...will not stay long in their place...' and they '...should only be used in what are denominated half walls, which are to be crowned with wood.'

Hard cap wall / capstone wall Hubbell (2006: 38) illustrates this style with an example from Connecticut, New England. The wall is capped with flat slabs reaching right across the top with a small overhang. Well built examples of such walls doubled up as walkways between houses; ladies avoided the mud.

This contrasts with a **copestone wall** where the copestones are set vertically, on edge.

Hasty wall (Kentucky / Pennsylvania, USA) a basic stone fence where the stones are thrown together to form a rough wall; a wooden rail fence brings the fence up to height.

Hawg walls / hogg walls (North Carolina, USA) long walls running into the forests, used to herd pigs back to a holding area.

Head dyke / hede-dyk / heid-dyke / heiddyck (Scots) the boundary wall which separates the arable ground from the rough hill ground (Figure 92). In southern Scotland, and northern England, the head dyke is known as a **fell wall**, named for the 'fell' or moorland it is built on.

GVA Perth (1794: 106) includes a description:

> Because it runs across the head of every farm, it was called very properly the head-dyke. Above the head-dyke the young cattle, the hill-horses (and frequently all the horses), the sheep and the goats were kept in summer; the milch cows were fed below. Little attention was paid to boundaries beyond the head-dyke.

Prior to agricultural improvements in the eighteenth century a typical head dyke was built of turf or roughly placed stone and expected to be stockproof for the summer only. After harvest the livestock on the hill ground were allowed through the head dyke to graze the aftermath on harvested fields.

The tenants within the head dyke understood the importance of maintaining the dyke. Tenants were often allocated specific stretches of dyke, theirs to keep in good repair.

Head wall an English term equivalent to **head dyke**.

How (Orkney) a mound, a tumulus, a knoll as exampled by Maeshowe, a Neolithic chambered cairn and passage grave in Orkney, Scotland.

Hud / hudd a Scots term for a section of stone-dyke built 'single' with large stones, hudd-stanes, which reach across the width of the dyke. **Butt and hudd** is illustrated in Figure 72.

Prevost (1957: 99) described these as '...sometimes built by dykers to strengthen the dyke, if there are sufficient large stones available.'

GVA Galloway (1813: 86) credits John McAdam of Craigenguillen, (who also invented the **locked top** cope style) with the development of hudds, also known as 'snecks'. These were '...paces built single at short intervals: a very useful contrivance: for if any accident happen to a part of the dyke, these snecks prevent the evil from spreading far.'

The hudd, as a unit, performs a similar function to the throughstone. GVA Galloway (1813: 87) goes on to detail construction. (The ell can be regards as equivalent to the modern yard of 36 inches (91.44cm):

Hudds or *snecks* are built at the distance of two or three ells. One *hudd* stone will do at the grass; but the more the better. When the double dyke between the *hudds* is built as high as the first hudd stone, a stone sufficiently long is placed so that one half of it may cover the hudd, and the other half the double dyke. In this manner the hudds are connected with the double dyke on both sides till it is two feet five inches high, when the band stones are laid on.

Intake walls (Lake District, England) late eighteenth and nineteenth-century walls which enclosed hill land and allowed its improvement into grazing ground.

Jammy walls (Cumbria, England) a **single wall** which looks flimsy and liable to fall. 'Jammy' is a British slang term meaning 'lucky'. Winchester (2016: 55) explains the term:

In northern England little single walling is found and ambivalence towards it is perhaps suggested by the term 'jammy' walls, which is encountered in south-west Cumbria and implies that it is only luck that keeps them standing!

Kail yard (Scots) a kitchen garden for vegetables, such as kale, a cultivar of the species *Brassica oleracea*. The walls of a house whose roof had fallen in often became a sheltered garden. Scottish art and culture are sometimes criticised as 'kailyard', suggesting a provincial, insular attitude.

Keft a bank of earth and stone topped with a continuous line of hawthorn or gorse, found on the coastal plain of Cumbria, northwest England and on the Isle of Man. These are another variety of the stone-faced earth banks found up the west coast of Britain.

Kes's / kessies a variation of the **Cumberland bank**.

Krubb (Shetland) a small drystone enclosure.

Kru / krue / kro a small drystone enclosure, the word has similar origins to **cro, planticrue** and **crub**.

Lace wall a term used, particularly in USA, for a 'single' wall, built with large irregular boulders. There are many gaps between the ill-fitting stones, hence the fancied appearance of lace. The walls may look fragile; they are heavy and solid. Walls of this style are also called **single stack**, 'single strand' or **balanced stone** walls in North America.

This wall name is particularly associated with Martha's Vineyard in Massachusetts. One alternative name is 'singing walls' because the wind, blowing through the loose construction, can, by chance, produce a musical note. It is very difficult to build a wall which intentionally produces musical notes, nevertheless it was once believed 'singing walls' were built by ancient cultures for the music.

North American sources mention how the fragile appearance of lace walls scare sheep; they are less inclined to attempt a jump over. This is very similar to the sheep-proof claims for 'single' walls and Galloway dykes in Britain.

Dutton (1808: 150) describes walls on the Burren, Count Clare, Ireland, an area famous for its limestone:

In the rocky regions stone walls are necessarily the only fence; they are made very differently, according to the fancy or abilities of the proprietor; the usual way is by stones

> piled on each other without any order like filigree work; they are called Burrin walls, and form a very unstable barrier…

D'Arcy (2006: 33) quotes from Lord Chief Barron Edward Willes' letters to the Earl of Warwick in 1761. The walls in County Clare are described as:

> …walls of a very singular construction. They are piled one upon another, but not flat and like the stone walls in Gloucestershire, solid; but the stones are placed in such a manner, that there are interstices thro'out the whole length of the wall of about 5 or 6 inches in diameter, so that … it very nearly resembles a net in the sea to catch fish…

Laid drain (Scots) a passage for water, six to ten inches wide, made by a row of stones down the side with flat slabs across the top. Another style had stones leaning towards each other at the bottom of the trench, in a tent-like shape.

Laid wall Thorson (2005: 39) and Hubbell (2006: 60) use this term to describe a deliberately coursed New England stone wall where the intention is to produce strength and good looks. This is an improvement on the rough 'tossed' wall where stones were thrown into a rough linear pile.

Lair (Scots) a fold or enclosure for sheep or cattle.

Laird's side (Scots) the lord or landowner's side of the wall, the side visible to whoever is paying. Imagine the side of a wall built near the mansion house, in constant view of the landowner. The best dykers in a team build that side. The side facing away from 'the big house' is left to apprentices.

Lane way (Rhode Island, USA) two parallel stone walls leading cattle to pasture. Also known as 'cow run', 'cattle way', 'cattle guide', 'cattle run' across New England.

Lea-laik (Ayrshire, Scotland) not really a man-made stone structure but JSD's description deserves a mention — 'A natural shelter for cattle, such as is produced by glens or overhanging rocks.'

Leas (SG) (plural *leasan*) a drystone enclosure created for growing willow, a useful material for making all kinds of baskets or screens. **Salley garden** is the Irish equivalent, from the Irish word *saileach*, willow.

Limestone plank fence Winchester (2016: 58) identifies these as stone slab fences in the Cotswolds, England. A type of flagstone fence.

Line fence (Kentucky, USA) a rock wall between two properties.

Lipped-dyke an old Scots term for a drystone wall where the stonework was pointed up with mortar. In the early nineteenth century this was thought to strengthen the wall and add to its longevity, at a time when drystone walling was still not fully accepted. **Teething**, and uses of mortar, are discussed in Section 2.7.

List wall (Somerset, England) a wall topped with turf.

Dickson (1805: 112) gives another definition, for no specific locality — 'In some places walls are made partly with a cement, and partly dry, and are termed *list walls*…'

DSL defines 'list' as a hem or edging, the bounds or limits of a town or the palisades round a jousting site. These prominent sites needed well built walls for strength and prestige.

Loaning dyke (Scots) a wall which bounded the 'loan' or cattle path en route from the common grazings to the township, where it widened into a small green, used for milking. An English equivalent of this term is 'lane', which may have morphed into 'lawn' in more modern times.

Long and short work Munday (2012: 59) includes an illustration from South Australia (Figure 93). Every ten metres or so a 'bookend' of large stones reaches across the width, with smaller stones filling the spaces between. The wall is capped off with horizontal slabs.

Retaining walls in France and Spain and the **feidín** walls of the Aran Islands, Ireland, incorporate uprights to stabilise a wall of smaller stones. British walls have their equivalent in the use of occasional wallheads to strengthen the wall and limit the extent of collapses.

Harris (1983: 333) has an alternative definition and includes an illustration of 'long-and-short work' with a description — 'In rubble masonry, quoins which are placed alternately

Figure 93: Long-and-short-work. South Australia.
Photo courtesy of Kristin Munday.

horizontally and vertically.'

March dyke / mairch-dyke / march dike (Scots) a boundary dyke running between estates. These can be substantial structures, maybe six feet high with two rows of throughstones. The individual field walls within the estate could be eighteen inches lower than the march dyke (Figures 14 and 68).

Maceria Harris (1983: 343) defines these as 'In ancient Roman construction, a rough wall having no facing; constructed in a wide variety of materials.' Anyone who has seen field boundaries in parts of Ireland or western Scotland, made from available stone, supplemented with timber, a bed end and the odd fishing float or oil barrel will understand this.

Mosaic wall according to Hubbell (2006: 67), this is a New England wall with irregular shaped stone tightly fitted together to produce a more formal appearance '...through selective use of size, shape, and color.'

Mallorca and the Mediterranean area have many examples of this type of wall. Laying stone 'flat and level, length into the wall' seems alien to these skilled wall builders but their structures are strong enough to endure seismic shock and floodwater pouring through them.

Mountain wall a drystone wall, especially in Wales, marking the boundary of the fields around the houses and the mountain, grazing. **Head dyke** or **fell wall** are other terms for this wall.

The Mourne Wall built between 1904 and 1922 in the Mourne Mountains of County Down, Ireland, to enclose a catchment area for Belfast's water supply. The wall is 22 miles long (35km) and, because it is in Northern Ireland, reputedly the longest in the UK. It is a popular circular walk over fifteen mountains. Old images of the wall show hikers walking along the top. On average the wall is 1.5 metres (4 ft 11 in) high and 0.8 to 0.9 metre (2 ft 7 in to 2 ft 10 in) thick. The wall is built in a variety of styles, depending on the available stone and the terrain.

McAfee (2011: 100) gives more detail. 'It is …laid dry to a height of up to 2.4m (8ft). At the base the wall is about 900 mm (3ft) and battered both sides with a large projecting coping on top.'

Offa's Dyke (*Clawdd Offa* in Welsh) named after the Offa, King of Mercia 757-796. A series of earthworks marking the boundary between Mercia and Wales, built from near the mouth of the river Wye to near the mouth of the river Dee. In places it is 65 feet (19.8 m) wide and 8 feet (2.4 m) high. The dyke was up to 150 miles (240 km) long, depending on archaeological opinion as to whether the whole length was built at one time or was part of a series of barriers constructed in different eras.

One-face wall (Scots) a retaining wall, facing up a bank of earth or other material.

Orthostat an archaeological term for a large stone or slab set vertically.

Wildgoose (1991: 214) identifies 'double orthostat walls'. These consist of two parallel lines of vertical slabs, in-filled with rubble and boulders. Wildgoose's study of the walls in Royston Grange, Derbyshire, details a progression of walling styles from medieval to modern times.

Outgangs (Lake District, England) long walls leading down to sheep pens. They funnel sheep down to the pens for clipping etc.

Pack (Lancashire) a rough wall made of stone, dirt or blocks of coal to support the roof in a coalmine.

Park dyke / perk dyke (Scots) a park is a field, therefore a wall round a field.

Parpen wall / parpall-wall (Scots) a substantial masonry partition, built one stone thick. Each stone has a good face at both ends because both ends are visible in the final structure.

In mortared masonry the term 'Yorkshire parpoints' is a style of ashlar construction, also called 'parpen ashlar'.

'Parpane' or 'parpane-wa' was a term from Aberdeenshire, Scotland, for the parapet of a bridge. 'Porple-wall' is a Scots term for a partition wall.

The term 'parpen' is also used for a throughstone.

Parrick / parrock (Scottish Borders) a small timber enclosure or pen which DSL defines as especially '…one used for familiarising a sheep with a strange or neglected lamb…'

Patina water, frost, biology and pollution will alter the appearance of a stone structure. Millions of individual finger rubs will, over time, wear down the faces or toes of the hardest

Figure 94: The patina on the damp left-hand side contrasts with the dry right-hand side. Water, from the roof, has washed out the mortar and now supports plant growth in the cracks.

statue.

Various dilutions of buttermilk or manure are recommended to hasten the patination of cut stone. Nature does the job efficiently, but slowly. Stone 'greens up' quickly under a tree because it is shaded and more humid. Stonework facing south is cleaner; the sun dries the stone and inhibits growth of algae, mosses and lichens.

When water flows across a zinc coated corrugated iron roof, and drips over stone, the tiny zinc content inhibits moss and lichen growth. Copper roofing often stains stonework green.

Peatclaig (Scots) a stone walled enclosure for storing peats.

Pech's cairn a Scottish term for a burial cairn or barrow. Local legends attribute construction of many old cairns and structures to the Picts, a tribal people dating from the third to ninth century CE. They left no written evidence. We know something of their culture from many inscribed stones and monoliths.

Pembrokeshire hedge bank (western Wales) a freestanding stone-faced earth bank, one of a series of styles found up and down the west coast of England and Wales. Often the bank is topped off with a thorn hedge as an additional deterrent to livestock. **Cumberland banks** and Welsh **clawdd** are other examples of this type.

Penny walls / famine walls those built at the time of the Great Famine in Ireland (1845-52). Anecdotal evidence says men were paid one penny a day to build walls. There are many cases when, in times of hardship, the landed gentry recognised their responsibilities and set unemployed men to work for a small wage, on projects which improved the local infrastructure.

Figure 95: A planticrue. St Kilda, Scottish Hebrides.

Pickie dyke / picki dike Orcadian terms for stonework attributed to the Picts. Orkney abounds with ancient stonework, for example Skara Brae, Maeshowe, brochs, souterrains and standing stones.

Pinfold / poind fold / poyndfalt (Scots) an enclosure for animals. It was often the 'pound' for stray animals. The animal was released after paying 'pin money'. 'Pin' is used in the sense of a small incidental expense. This system of fines also operated in New England.

In modern times we still find use for such an enclosure, for stray dogs or badly parked cars.

Pirca Harris (1983: 422) includes this term from South America, 'A type of crude wall construction using dry-laid unshaped stones; found in the Andes.' Other sources mention the use of fieldstones with mud as mortar.

Plantation fence (Kentucky, USA) Murray-Wooley and Raitz (1992: 23-37) describe this early type of drystone wall, no doubt influenced by British styles. These fences surrounded pastures, barnyards, graveyards and gardens and date from '…as early as 1777… but these fences did not come into wide use until the early nineteenth century.'

The wall had a foundation, were built 'double' with throughstones, coverbands and vertical copes. The wall had a batter.

'Plantation' in this sense means a farm. The **turnpike fence** was a later style of Kentucky fence, built along the roadways.

Planticrue / plantiecrue / planticrub / plantiecrub / plant-acruive / plantiecruive / plantiecrew / planta-crew / plantikrob (Scots) a small drystone enclosure for growing plants, for example cabbages or seedlings before they were strong enough for planting out in the field. **Crub** is a short form of the term.

These small enclosures can be square or circular with no entrance, to keep animals out. Some individual structures have walls with a negative batter; they curve outwards nearer the top, an added deterrent to livestock (Figure 95).

Löbbecke (2012: 107) describes low curved semicircle drystone walls in Lanzarote, Canary Islands, Spain. They protect low-lying grapevines and vegetable plants from the wind and create a microclimate. A layer of black lava gravel helps retain heat and humidity.

Post and panel walls (County Wicklow, Ireland) described in Ó Maitiú and O'Reilly (1997: 73,74) as:

…very characteristic of the district. They are composed of rough irregular stones built to a height of about 1–1.5m (3'–5') and have 'scantlings' – narrow stone posts about 1.5m (5') high at intervals of about 4m (13') and perforated to take barbed wire. The panel of the name refers to the loose stones between the scantlings.

The local stone broke easily into long narrow pieces which were useable as fenceposts. Figure 120 shows stone fenceposts as part of a wall in County Wicklow.

Stone fenceposts are also found in western Ireland, with slabs of limestone; in north central Kansas, with limestone; and Wales, with slate.

Puidge / pudge / pooge (Scots) a small enclosure or building, often associated with pigs and hence a description of somewhere messy or in a state of disorder.

Pund (Shetland) a small fold for sheep, probably derived from the more familiar 'pound' as are other Scots synonyms, 'poyndfalt', 'poind' and 'poynd'. In Shetland there are four large nineteenth century stone and mortar structures described as 'pony punds' (Figure 96). The interiors had smaller pens and roofed lean-to shelters along the inside of the walls. There are high peaks in the corner, supposedly to break the strength of the wind. Anecdotal evidence suggested these high peaks were also there to stop ponies from climbing up the corners and escaping. This imitates overhanging covers and copes on drystone walls, used to exclude rabbits or wolves, or vertical walls on the inner face of sheep fanks to stop sheep

Figure 96: Pony pund, Swinster, Shetland. Note the old field markings and the fence reaching down below the tide mark, to restrict livestock.

from climbing out.

Evidence from Tait (2002: 23-25) clarifies matters. Only one pund held ponies – the one on Noss. The Marquis of Londonderry bred Shetland ponies there for his coal mines in Durham. The other punds were used for sheep. The peaked corners appear to have been a design 'feature', they were unnecessary for security because of the internal roofing.

Putt / put (Scots) a buttress to support a wall. A jetty or stone buttress projecting into the water with the intention of protecting the riverbank, or for attaching salmon nets to.

Raik dyke (Scots) a stone dam built across a 'raik' – a stretch of river used for salmon fishing.

Random dyke / random rubble dyke a dyke built with no attempt at any formal coursing, using irregular stones of various sizes. Building with rough stone takes skill and the final structure can be just as tight as one built from flat stone. 'Random' is not meant to suggest a slapdash construction method, the term applies to the stones.

Ráth (Ireland) a type of ring fort or fortified farmstead, an earth bank with a wooden palisade and a ditch outside. According to Jenner (1992: 53-55) there are an estimated 30,000 in Ireland dating from around 1000 CE. Traditionally they are regarded as 'fairy forts'; it is considered bad luck to disturb one.

A stone structure was called a **cashel**; when these became larger and more defensive it was termed a 'caher' or a 'dun'. Collectively the cashel, caher or dun are often referred to as 'stone ring-forts'.

Readyke (Scots) a ditch round a hill, for defensive purposes.

Reaves low stone banks running parallel, in groups, in Dartmoor, possibly property boundaries from late Bronze Age or early Iron Age, perhaps 3,000–4,000 years old.

Ree / reed / reeve Mactaggart (1824) gives this definition for 'ree':

> A round sheep-fold, where sheep are put into on snowy nights, to hinder the snow to *ree* or wreath them up; as the wind, by whirling round this circle, lets the snow not wreath in it.

The shape of the enclosure kept the snow from building up and burying the sheep.
'Sheep ree' is unique to the south west of Scotland as a name for a sheep enclosure.
'Reed' nowadays is a term for sheep or cattle yards throughout Scotland.

Rekster (Shetland) an enclosure for sheep, or a walled track. A 'rekster-dyke' helps control animals along a route as do 'cow path walls' in New England.

Retaining wall / retainment wall / sustaining wall a wall built across the face of a slope to stabilise it and reduce erosion or slippage (Figure 97).

Trautwine (1937: 603-613) notes the difference between a **face-wall** and a retaining wall. A retaining wall holds ground back. A face-wall 'faces up' a solid slope:

> A retaining-wall is one for sustaining the pressure of earth, sand, or other *filling* or *backing*, deposited behind it after it is built; in distinction to a **face-wall,** which is a similar construction for preventing the fall of earth which is in its undisturbed natural position, but in which a vertical or inclined face has been excavated. The earth is then in so consolidated a condition as to exert little or no lateral pres[sure], and therefore the wall

may generally be thinner than a retaining one.

According to Trautwine 'Experience, rather than theory, must be our guide in the building of both kinds of wall.' Richard Fish in O'Reilly and Perry (2009: iii), in a volume about retaining walls in civil engineering situations, some going back to Victorian times, realise the same point:

> Whether empirical rules were ever devised, or whether it was due to an innate understanding of the materials, is unknown. Conventional analysis would often suggest that such walls should not be standing…

Trautwine recommends foundation widths for retaining walls should be based on '… railroad practice, when the foundations are not more than about three feet deep.' The advice is paraphrased thus:

> For a mortared wall of cut stone the foundation width should be 35 percent of the height. For a mortared rubble or brick wall the foundation width should be 40 percent of the height. For a well-scabbled dry rubble wall the foundation width should be 50 percent of the height.

Trautwine is worth reading, though the theoretical geometry and mechanics are beyond the needs of the drystone waller or those building small drystone structures. The waller should be guided by the rock and local traditions. Whatever the style, it is important to have a lot of throughstones, reaching well into the ground behind the retaining wall, and have solidly packed stone and hearting behind the face stones. Soil or gravel will not provide sufficient support.

The use of fabric behind low retaining walls is debateable. It will get in the way of any planting and become an inconvenience if the wall needs repair. Fabric is not usually necessary for garden retaining walls up to four or five feet high. The need for fabric in a design may have originated from a desire to control water behind the stonework, especially for engineered retaining walls, several metres high. Drystone retaining walls are freedraining and should be wide enough to allow water to fall to the bottom of the wall, well before it emerges through the face. A good retaining wall is, in effect, a vertical French drain. Good drainage, on the slope behind the wall, also helps.

Mundell (2009) is another useful source combining the theoretical and practical aspects of drystone retaining structures. His PhD thesis describes a series of experiments conducted at the University of Bath.

A domestic retaining wall (as opposed to a tall wall requiring an engineer's input) can be built as a single-faced wall built up against the banking or a two-faced wall built in front of the banking. The ground is then pulled down to fill up the back. The coping could be turf, vertical copes or flat slabs — whatever suits the landscape's design. The reader is recommended to the Dry Stone Walling Association's leaflet 'Technical Specifications for Simple Retaining Walls', freely available from their website – www.dswa.org.uk, as at August 2024.

Figure 97: Retaining walls. Deia, Mallorca, Spain.

The batter on a retaining wall is often steeper than on a freestanding wall. This shifts the centre of gravity back from the face, increasing the resistance to the forces behind the wall.

Todd (1859: 467) recommends the stones should be laid at right angles to the battered face, lying perpendicular with their long edge tilted down towards the centre of the wall. This resists the thrust of frost and allows the stones to settle back down again after the thaw.

In the south of France retaining walls normally have all the stones, including those in the foundation, lying perpendicular to the battered face.

A **gravity wall** is a type of retaining wall, relying on its mass and slight batter, to hold back the material behind. A **breast wall** is another type of retaining wall.

Rett / Retta-dyke (Shetland) a wall built to herd animals and funnel them into a rett, an enclosure. 'Rekster' is another term for a Shetland sheep enclosure.

Löbbecke (2012: 344) includes an image of an Icelandic *rett*, a central yard surrounded by smaller pens which, 'In the fall, the farmers from one community… herd the sheep into the large enclosed yard, where they can then sort them in the adjacent pens according to their identifying marks.'

See Figure 85 for a Welsh version of this sheep management system.

Reuch-stane dyke / rough-stane dyke / ruchstane dyke (southwest Scotland) a drystone wall, built out of 'reuch' stone, rough stone or boulders used in their natural state. 'Roch dyke' is another Scots variation. A 'roch dyker' is a drystone waller. 'Roch' is pronounced with the same ending as the Scots word 'loch'. 'Roch' in this case derives from a Scots variation of 'rock' and a suggestion of rough, undressed rock. There is no connection with the word 'rough,' meaning sloppy or untidy work.

Revetment a covering of stone or similar material to protect an embankment, as at a riverbank, seashore or side of a road.

Rexter dyke (Shetland) a wall used to guide livestock towards a holding pen.

Rick-rack (Kentucky) a wall style with diagonal coursing, an **edge fence**.

Rickle dike (Scots) a rough drystone wall, almost a pile of stones, a loosely built drystone wall.

DSL has the word 'rickle' derived from a Norwegian dialect word *roykla* meaning 'to throw (stones or peats) together to form a heap; to build (up) quickly or unsystematically.'

'Ricklie' or 'rickly' means shaky, loose, disjointed, dilapidated or piled up in a loose manner.

A badly built dyke might be judged no better than 'a rickle of stanes' or 'gey rickly'.

JSD defines a rickle-dike as a north of Scotland style which could fit the description of a **single dyke**, or a **Galloway dyke** — 'A wall built firmly at the bottom, but having the top only the thickness of single stones, loosely piled one above the other.'

JSD then goes on to quote GVA Inverness (1813: 114):

> The larger stones are laid in the foundation; and in every opening between the top of these, the next stones in respect of size are laid longitudinally across the wall, and so carefully, that they neither lean to one side nor another; and so on, until the fence be of that height which is required.

A 'rickler' is an unskilled stoneworker who would 'rickle up' a drystone dyke. Rainsford-Hannay (1957: 95) includes John Mactaggart's nineteenth-century poem 'Davie the Dyker' wherein the dyker's work is praised:

> A rickled rood ne'er left his hand,
> His dykes for centuries will stan'.

'Rickle' also describes a rough style of coping made of up of small stones, detailed in Section 3.2 (iv).

Ring another old generic name used by Scots for the remains of ancient stone structures.

Ring-dyke (Cumbria, England) Adams (1976: 129) defines this as a dyke which surrounded the cultivated area of a township, like a head dyke separates the arable ground from the common grazings.

Rip rap (Kentucky) a wall style with diagonal coursing, an **edge fence**.

Riv (Banffshire, northeast Scotland) an enclosure for pigs or poultry.

Rock wall Hubbell (2006: 100) explains the difference between stone and rock and tells us he has never heard a wall builder in New England refer to his product as a 'rock wall'.

Thorson (2005: 17) offers contrary evidence from New England walls:

> The term *rock wall*... is fairly common, because there are many areas where the 'stones' are large, blocky, and irregular. Hence, the term *rock wall* connotes material chaos just barely constrained by the order imposed by a line of stone. The term *stone wall* connotes more order, principally in the shape and size of the stone.

Knott (2004: 21) refers to 'rock fences' in Texas.

It is important to understand of the difference between a 'rock' and a 'stone'. Rocks are large pieces, often still attached to the earth's crust. In a building context stones are smaller pieces, worked by man or nature to become components for something bigger.

Roog (Shetland) a mound or knoll, a prehistoric stone cairn or barrow.

Rough and ready (Kentucky) a wall style with diagonal coursing, a type of **edge fence**.

Rough-stane dyke an English version of **reuch-stane dyke**.

Round / rounde / roundel / roonall (Roxburgh, Scottish Borders) a circular or semi-circular dyke or wall, made of stone or stone and turf, used as a shelter for sheep. Similar structures are found throughout the north of Scotland, no doubt taken there by shepherds from the Scottish Borders when sheep farming was expanding throughout the north in the eighteenth and nineteenth Centuries. Examples are also found in Ireland.

Rubble wall one built, dry or mortared, with rock of irregular shape and size, just as they come out of the field or from the quarry. Rubble walls can be built random, with no real coursing, or brought up to occasional level courses with shims or packing to accommodate throughstones, copes or lintols. 'Snecked rubble' is a style of building common in Ireland and Scotland, described in the DSL as:

> ...uncoursed rubble, in which the stones are used as they occur, the interstices between the larger stones being filled with smaller pieces. When this is done with great nicety, and so as to preserve perfectly the horizontal and vertical bond by the complete interlacing of the amorphous stones, the operation is termed snecking, and the work is called snecked rubble.

The term applies to mortared work but those drystone wallers acquainted with rough stone will find they build to similar principles. Rankine (1862: 389) provides a link between mortared and dry rubble work — 'Dry stone walls should be built according to the principles ...already laid down for rubble masonry... with the single exception that the mortar is to be omitted.' (See Figure 29).

Ruck foun (northeast Scotland) a northeast Scottish term for a circular stone foundation for a hayrick or grain stack. It keeps the hay or sheaves off the wet ground.

Rumbly dik / rumbly dyke (Shetland) a loosely built wall. In Shetland 'rumble down' means 'to demolish'.

Rumbling sivers / rumbling syres / rumbling suiver / rummlinsires / rumbling cundie a drainage ditch filled with loose stones instead of pipes or drainage tiles. The stones were carefully arranged to allow passage of water.

DSL gives dimensions with their definition — '...being about three feet deep, three or four feet wide at top, and two in the bottom.'

'Sivers' will be related to the word 'sewer' and 'rumbling' for the sound of stone rolling down the channel in a flood.

Rummel / rummle / rumle (Shetland and throughout Scotland) a wall, dyke or building which has been hurriedly erected or is in a rickety state.

Rustic a masonry style where the joints and beds are dressed square and the face is roughly textured. A specification may ask for a 'rustic' drystone wall. Clarification should be sought before construction starts to confirm what is meant. Is it a traditional wall or a rough folly, one which looks like a ruin?

Salley garden (Ireland) a walled area of waterlogged ground for growing 'sally rods' — willow rods for weaving baskets, creels and lobster pots. Willows are trees and shrubs from the genus *Salix*.

The Irish Gaelic word for willow is *saileach*. The poem 'Down by the Salley Gardens' by WB Yeats (1865-1939) tells of a rendezvous in such a place.

The Scots Gaelic word *seilach* is a willow copse, where willow grows.

Saxeane an old Scots term meaning 'made of stone', from the Latin *saxeus*, 'made of stone'.

Scone house Rendall (1997: 38) describes this structure from the farm of Garso in North Ronaldsay, Orkney, Scotland:

> ...a scone house or store for the pats of dried cow dung which were collected for fuel ...The house is a detached drystone building, approximately 18 feet by 8 feet, with a flagstone roof, and is built just inside the sheep dyke which separates the farm from the shore. There is no proof that the house was built for this specific purpose but I was told by my informants that formerly every house on the island had its own scone house.

Scotch fence a granite stone slab fence recorded in County Carlow, Ireland, detailed in Conry (2000: 60-68). The fence may have been designed and built by Scottish stonecutters though it '... might be more appropriate to call them Indian Fences because there is so much of this type of solid granite fence in many parts of India.'

Figure 98: Scotch Fence. County Carlow, Ireland.

The fence (Figure 98) consists of granite lintels, or slabs, standing upright, side by side, to give one long continuous fence. Typically, the fence is 1 to 1.25 metre high. Individual slabs are 8 to 35 cm thick and 12 to 65 cm wide. One of these fences had an iron peg '...set in every fifteenth to seventeenth slab, carrying a line of plain wire to give extra height to the field fence.'

Hubbell (2006: 92) has an illustration of a similar fence in Westport, Massachusetts, built from alternate high and low granite slabs.

Ireland, New England and Scotland all had thriving granite industries in the eighteenth and nineteenth centuries. Undoubtedly there was a lively exchange of ideas, investment and personnel between the quarry companies.

This construction is similar to a **flagstone dyke**.

Scots dyke an earthwork, three and a half miles long (5.25km) between the rivers Esk and Sark, built after the report of the Commission on the Marches in 1552 to confirm part of the Scottish-English border.

Sheep / sheep dyke the need to improve and protect livestock bloodlines was undoubtedly an important factor influencing the spread of drystone walling across marginal land in the eighteenth and nineteenth century. Fortunes were made by careful breeding and management of sheep, primarily for wool. The Scottish Blackface (Figure 99), was the first sheep breed to spread through Scotland. The Cheviot (with better wool) became the

Figure 99: A Blackface ewe, typical of the hardy breed which spread throughout Scotland.

predominant breed in the low ground of the Scottish Borders and Caithness.

The **Galloway dyke** is described in agricultural writings of the period as a 'sheep dyke'. Rainsford-Hannay (1957: 51) confirms:

> ...there is a most decided consensus of opinion about the way *galloway-dykes* 'deter and terrify sheep'. One occasionally sees a dyke on a skyline, and the stones at the top making an almost lacelike pattern, especially with a westering sun behind them. Perhaps they do exhibit a tottering appearance.

Sheep-crue / sheep-cree (Shetland and Caithness) a sheep pen. A variation of **crub** or **planticrue**.

Sheep-fank (Scots) an enclosure or set of enclosures for sheep dipping, clipping and other management tasks.

Sheep-fauld a sheep fold. 'Fauld' is a Scots form of the English 'fold'.

Sheep faw (Scots) a shelter for sheep on lower ground below the moors.

Sheep hadin / sheep haddin / sheep hadden (Shetland) describing a field wall as high enough to keep sheep on the grazings and out of arable ground. Perhaps a corruption of 'sheep-hold-in'.

Shieling (Scottish Highlands and Islands) small huts made of stone or turf on summer pastures. Turf walls were built on a stone foundation to protect the walls from water and damage. The term applies to the pastures and the buildings on them. Shielings are an example of the world-wide practice of 'transhumance', an agricultural system using seasonal upland pastures to produce storable dairy products from cows and sheep. The shieling huts often had stone shelves built into the structure for standing milk, butter and cheese.

The Irish equivalent is a **booley hut**.

Show walling work with emphasis on a well finished face rather than strength, associated with walling competitions.

GVA Galloway (1813: 87) warns against work which looks good while hiding a multitude of faults:

> There are few operations connected with husbandry where more depends on the skill, attention and fidelity of the workman than building of dykes; but few, unfortunately, where the want of these will more easily escape without being detected. It occurs too often that dykes executed with great neatness, and apparently in a very substantial manner, exhibit, in a few years, all the symptoms of decay and ruin; when, however, it is too late to obtain redress from the faithless contractor.

Side dyke (old Scots) a dyke which divides land of different ownership, hence the expressions 'to rin side-dykes with', 'to neighbour with' or 'to march with'.

Singing wall a New England term for a type of **single wall**, so called because they produce a musical note, when the wind blows through them at a certain strength or angle. 'Single' walls have a very open build, in contrast to the more tightly built 'double' walls. 'Singing wall' is likely a straightforward corruption of 'single wall'. It is unlikely these walls were built specifically for their musical potential. They are also known as **lace walls** in New England. Figure 100 is an excellent Irish example.

Loudon (1825: 443) details a British equivalent, describing the rough stones in the upper part of the Galloway dyke '…[built] so open as to leave a considerable vacuity between each; by which means a free passage is afforded to the light and wind, which blows through them with a violent whistling noise.'

Single wall / Single dyke a drystone wall which is, for its total height, one stone wide, largest stones at the base, smaller with each succeeding course until the level top line (Figure 100). It contrasts with a **double dyke** which has two outer skins of stone, linked by throughstones, with smaller hearting stones in the middle.

'Single-work' or 'singling' describes that part of a wall which consists of single stones taking up the total width of the wall. This might be in the upper half of a **Galloway dyke** or a stretch of single walling built to use up large stones in a wall, which is otherwise built 'double'.

Figure 100: Single wall. The Burren, Ireland.

Single dykes are an ideal solution for enclosing moorland full of large course-grained boulders which are easier to move than break. They can be flimsy, like these walls on the Burren, County Clare, Ireland, described by Dutton (1808: 150):

> In the rocky regions stone walls are necessarily the only fence; they are made very differently according to the fancy or abilities of the proprietor; the usual way is by stones piled on each other without any order like filigree work; they are called Burrin walls, and form a very unstable barrier; a beast scratching against one of them often brings down many perches, but this, from custom, is little regarded, for in a few minutes all is built up again…

Winchester (2016: 55) also writes about the precarious nature of the build and refers to 'jammy walls'. 'Jammy' is a British slang term meaning 'lucky'.

In northern England little single walling is found and ambivalence towards it is perhaps suggested by the term 'jammy' walls, which is encountered in south-west Cumbria and implies that it is only luck that keeps them standing! **Boulder dyke, lace wall** and **balanced stone wall** are other examples of this style.

GVA Inverness (1813: 114) describes a 'single' dyke, its advantages and ease of construction:

> The double stone walls without lime, are not near so effectual a fence against sheep, as the single stone walls, provincially called *Galloway* or *snap* or *rickle* dykes. The larger stones are laid in the foundation; and in every opening between the top of these, the next stones in respect of size are laid longitudinally across the wall…The walls have such a tottering and alarming appearance, that all kinds of cattle are terrified to attempt them: and as an additional recommendation, they require fewer stones, are more expeditiously built, and last as long as the double stone walls without lime. In fact, no kind of stones are too shapeless for being used, and no person so unskilful as not to succeed in the building, since the principal art required, is no more but to lay the centre of gravity of every stone on a firm place of the stones below; which any man may understand, who had ever made one stone to lie above another.

Constructing these walls sounds like a true trial of strength and organisation.

Oram and Stelfox (2004: 86) give some insight in their description of the building of 'single' walls in the Mourne Mountains, County Down, Ireland, where these walls are called 'ditches':

> The first or bottom stone to be placed is known as the butt stone and this is rolled into place by hand using a crowbar on a plank known as a prise. When the bottom stones were in place they used a piece of wood approximately seven foot long by ten inches wide to start adding on the other stones. Some of these stones had to be shaped and dressed using a large hammer. The hammer was called a scabbel hammer… from the old word

Figure 101: Stone storage shed. Nybster, Caithness.

'to dress'. The use of the plank (prise) was commonplace until the late 1960s. After that a fork was attached to a tractor, followed by the modern day hydraulic digger.

Single stack / single strand / farmer's wall (New England, USA) low, loosely built walls of piled stone, largest stones at the bottom, progressively smaller with height. An example of a **single wall**. Hubbell (2006: 23, 26) tells us, 'If the supply were sufficient and a pioneer had the time, he might put some order into the stone piles surrounding his fields…'

They were '…built as the seemingly endless procession of rocks rose to the surface each spring.'

Skeo / skee / skio (Shetland and Orkney) a little drystone hut in which fish or meat are wind dried. 'Skeo' is also a Shetland term for a badly built house.

Löbbecke (2012: 38) refers to Icelandic *fiskbyrgi,* a hut built 'single', one stone thick, to allow wind through, for drying fish.

Slab fence a stone wall using flat slabs of stones, set into the ground. Like a **flagstone dyke** or **Scotch fence**.

Slab wall a type of **edge fence**. D'Arcy (2006: 14) identifies this wall as one specifically associated with the Burren in western Ireland. An image shows slabs laid on edge across the line of the wall with smaller slabs set between the taller uprights to bring the wall up to a uniform height. It has every appearance of an edge fence.

Slave walls / slave fences (USA – particularly Kentucky) a novice looking at miles of intricately built stone walls may assume they were built by forced labour, not by wage-earning craftsman.

Murray-Wooley and Raitz (1992: 84) describe how many of the Kentucky walls were built by the Irish in the nineteenth century and:

> By the twentieth century, the infusion of new money into the Bluegrass provided the motives and the means to employ crews of black stonemasons, who had learned the trade assisting Irish turnpike fencers... While the work of black stonemasons remains in living memory and may thus be the source of the 'slave fence' myth, these builders were not slaves.

Smyoo-hole (Orkney) a drainage channel with a stone cover.

Snap dyke / snap fence / snap-topped dyke (southwest Scotland) this is an early description of what we today know as a 'double dyke' with vertical copestones. DSL connects 'snap' with the old English word 'snape', meaning to 'taper' or 'bevel'.

The distinguishing feature of this dyke was the cope, twelve inches, or more, of roughly triangular, flattish stones set on edge with an upward taper. This was a type of dyke developed for sheep farming, a means of producing a stockproof fence with a strong permanent top course. It was a definite improvement on a topping of turf, thorns or similar ephemeral material.

In GVA Galloway (1813: 83) Smith of Borgue describes a 'double dyke' from Galloway with the, then, innovative vertical cope:

> ...the most common in the lower part of the country, are what are called, *snap, or snap-topped dykes*. These are, for the most part built two-thirds, or three-fourths of their height double; ie, the two sides are formed of two different sets of stones, resting against each other, and connected together by stones, which, from time to time, are laid across the dyke. On the top of this double dyke a set of long stones are laid horizontally across the dyke, projecting a little on each side; and above these the remainder is built single, which, from the irregular figure of the stones, leaves apertures through which the light appears.

Smith is describing a 'double' dyke with coverstones and copestones.

Sneck / snick / sneg / sneak (Galloway, southwest Scotland) a short length of wall built from large stones which reach across the total width of the wall. This strengthened the wall and restricted the extent of any collapse. This feature was also a handy way to use up a surplus of large boulders.

Alan Brooks (1989: 18) gives the additional spelling of 'sneaks' for these panels of **single** dyking.

GVA Galloway (1813: 86) declares the inventor as John McAdam of Craigenguillen (who also invented the locked top) and gives more detail:

> ...snecks or hudds, ie, spaces built single at short intervals: a very useful contrivance: for if any accident happen to a part of the dyke, these snecks prevent the evil from spreading far.

In the same document, GVA Galloway (1813: 85), The Reverend Smith advocates the use of snecks:

> When the stones are large, and very good, the width of the dyke should be a little increased, and the proportion of double diminished. But it is a much better plan, where a considerable proportion of the stones are large, to build snecks of single dyke, at intervals of three yards, or else to increase the number of through-bands in every part of the double dyke…'

Butt and hudd and **hudd** are other descriptive terms for sections of 'single' within a long dyke.

Soothouse (Achill Island, County Mayo, Ireland)

Aalen et al (2011: 247), give details of this drystone building:

> Soot-houses were simple stone huts, the timber and sod roofs of which were slowly burnt each winter until they collapsed: the ash was then used as fertilizer. Examples are known only from Achill Island, where they remained in use until the early twentieth century. Almost certainly, soot-houses once had a wider distribution.

An alternative method made use of the soot collected in the thatch of houses. In the spring the thatch was stripped off (sometimes along with turf walls) for use as fertilizer. The practice is recorded throughout Ireland and Scotland. This 'spring clean' cleared the accumulation of muck within a house, especially when part of the building was shared with livestock.

Perry (2012: 17) describes 'The Pixie Houses of Devonshire', small corbelled structures whose purpose was '…to provide a dry and safe place for storing ashes raked out daily from the open hearths of the farmstead, ready for use on the land.'

Souterrain a stone built semi underground structure from the early Iron Age, 2,000 years ago. They typically consist of a narrow passage with narrow chambers off to each side.

Spanish riders also called **chevaux de frise**. See Fiigure 103 for an Irish example.

Spite wall / spite fence Hubbell (2006: 55) includes an example of an eleven foot high wall built by Edmund Proctor, a farmer in Westminster, Massachusetts. He liked to work on Sunday and preferred his neighbours not see him.

Figure 102: A 'sneck' of 'single' within a length of 'double' in a Galloway dyke. This is an opportunity to use large stones; best result for least effort.

In Ireland, MacWeeney and Conniff (1998: 108) relate how a farmer built his boundary wall a few feet inside his own land, allowing him to sue his neighbour when animals trespassed. This forced the neighbour to build another parallel wall, inside their land, to avoid litigation.

Muilenburg and Swineford (1975: 49) explain parallel fences in Kansas, USA. These fences had posts, cut from the local limestone, with barbed wire strung between. They were built a few feet apart, as a precaution, by neighbouring farmers who were unsure of the exact boundary line. These were called 'spite fences'. The space between was called 'the devil's lane'. The parallel fencing also had the practical purpose of ensuring the boundary was stockproof and gave the individual farmers peace of mind if they did not trust the neighbour to keep their fence in good order.

MacWeeney and Conniff (1998: 80-83) give the historic background to the 'jealous wall' at Belvedere House in County Westmeath, Ireland. The mock Gothic ruin was built in the eighteenth century as a folly by the first Earl to block out the view of a larger house his brother had built a half mile away.

Stachle / stathel / staddle / staidel / staffle the foundation for a stack of grain or hay, to keep it off the damp ground and away from vermin. The most modern design is a circular cast iron or stone frame set on downward pointing stone piers. 'Staddle stones' are a familiar sight in old farmyards, they look like stone mushrooms. The terms derive from the Old English word *stathol*, a foundation.

Stacked fence (Kentucky) a wall style with diagonal coursing, an **edge fence**.

Stand up fence (Kentucky) a wall style with diagonal coursing, an **edge fence**.

Stane dyke (Scots) a drystone wall.

Stane krib (Scots) a jail.

Steinreich a medieval German term, literally translated as 'stone rich'. It applies to people who were so rich they built with stone; the lower social classes built with wood.

Stell / stell dyke a drystone enclosure built to provide shelter for sheep on open ground. Sometimes called a 'shelter stell'. (Figure 104). It will typically have no gate and be circular, or, less often, square. The size depends on the need. Small stells may be built up against a large rockface. The walls can be 5 or 6 feet high, A stell built to the proper proportions of wall height to stell width is not supposed to collect snow. A flock of sheep tend to keep to one area of ground and quickly learn where the shelter is.

Borchgrevink (1982: 89) quotes Charlie Jardine's description of a stell:

> ...out in the hills for shelter for sheep in stormy weather... It can either be round or square or a dyke with four wings running in north, south, east and west. A cross dyke, it's like a cross on the hill. It doesnae matter which way the storm comes, the stock has always got shelter somewhere on one side.

Hogg (1807: 124) recommends 'The best form... is that of a complete circle or octagon, (the latter not being so apt to rush,) with a door in it, at which the sheep may go in or out.'

Hogg thinks any collapse of the stonework, a rush, will most likely be contained within the straight edges of the octagon.

According to Adair (1998: 43) stells came into the Scottish Borders around 1760. He also mentions an Irish version of stell, built like a cross; sheep got shelter whatever the wind

Figure 103: A defensive field of 'Spanish riders' or 'chevaux de frise'.
Dun Aonghasa, Inis Mór, Aran Islands, Ireland.

direction. Cross shaped drystone structures are also found in the north of Scotland, no doubt built when sheep farming expanded from the south.

The simplest version of a stell is a straight dyke built on exposed slopes, across the direction of the prevailing wind.

Callander (1986: 86) defines a 'stell-dyke' as:

> a circular dyke forming a sheep enclosure normally 20-40yds. in diameter with a 3' entrance on the lee side. Mentioned in the Napier Report as being advocated after the disastrous winter of 1620. A flank or beild.

The Napier Commission (appointed 1883) researched into the living conditions of crofters in the Highland and Islands of Scotland and led to greater security of tenure for these small farming units.

Blackwood's Edinburgh Magazine (1819: 75) refers to the big storm of 1620 — 'The most dismal of all those [storms] on the record is *the thirteen drifty days*. This extraordinary storm, as near as I have been able to trace, must have occurred in the year 1620.'

Thousands of sheep were smothered in the snow. One farmer, Sir Patrick Scott of Thirlestane, lost all his sheep '…save one black ewe, from which the farmer had high hopes of preserving a breed; but some unlucky dogs, that were all laid idle for want of sheep to

Figure 104: A circular stell approximately 17 metres (58 feet) wide, enclosing about 240 square metres (2,600 square feet). The railway line, which passes behind, was completed in 1874. The sheepfold may have been built around the same time. Both structures contributed to the lively sheep trade in the area. Kinbrace, Sutherland, Scotland.

run at, fell upon this poor solitary remnant of a good stock, and chased her into the lake, where she was drowned.'

A 'sorting stell' is a similar structure with internal divisions for managing sheep. Figure 85 shows a fairly complex Welsh example.

'Stell' has also been recorded in the North riding of Yorkshire, by Pease (1928) as a term for a 'large open ditch, wide and deep… made through low-lying country by enlarging natural streams or watercourses.'

Stem (Scots) a dam of stones, used to form a watering place for cattle or to act as a fish trap.

The Great Bernera Lobster Pond build by Murdo Morrison in the 1860s is a good example. It cuts off a small bay of Loch Risay (*Loch Riosaigh* in Gaelic) on the island of Great Bernera off the west coast of Lewis in the Scottish Outer Hebrides. The Gaelic name is *Tob Blar Meadha*, Ordnance Survey map reference 173373. The enclosure was originally built to keep lobsters alive in the summer until the market for them had improved and they could be transported with less risk of perishing in the heat. The wall is 75 yards long from end to end, up to 20 feet deep and 3 feet across the top. Other ponds exist; this is the largest.

Steugie / stuggi / stuki (Shetland) the extensions of a sheepfold wall which aids the penning of sheep by hemming them in as the approach the fold. These were also known as **soadin dykes** or **rexter dykes.** In modern Scots a 'stookie' is a plaster cast on a broken limb. There is a similarity between restricting the movement of livestock and holding a damaged bone in place.

Stob-dyke (Scots) a fence made of wooden posts and wooden rails, another use of the word 'dyke' for a barrier.

Stone-faced earth bank a style of wall with two outer skins of stone and a compacted earth centre. Examples include **Cumberland banks**, **Cornish hedges**, Welsh **clawdd** and **Pembrokeshire hedge banks.** This type of structure is common in the wetter, west coast regions of the British Isles.

Stone fence the North American term for a stone wall. Thorson (2005: 59) differentiates between a 'wall' and 'fence' in New England:

> Normally, the term *wall* is used when the typical structures are massive, low, and made of slabby stone. The term *fence* is used when the structure is a single stack that meets or exceeds chest height. To prevent further confusion, I recommend that we consider stone fences to be those that are both freestanding and roughly chest height to an average person (3.5 to 4.5 feet high, which was the minimum legal height of a fence during colonial times).

In Britain the term 'fence' in the stone sense is applied more particularly to flag fences, slate fences or stone slab fences – tall thin stones set vertically on edge, tight against each other, with their long edges in line with the line of the fence.

Winchester (2016: 58) says stone slab fences are known as **limestone plank fences** in the Cotswolds.

Stone hedge a variation of the **Cornish hedge** with a stone centre, rather than subsoil. This hedge has no throughstones; it relies on the batter and proper placement of stone for its strength. The solid stone interior is more practical in upland areas of Cornwall where stone is more easily found than soil.

Garner (1984: 25) gives a detailed description, though his description is for a hedge with an earth interior:

> ...the characteristic feature of the south-western peninsula [of England] is the stone hedge in its various forms, basically an earth bank faced with stone and topped off with turf, brushwood, stone or a permutation of these. Typically the base and the height measurements are the same and the hedge tapers to rather less than half the width of the foundation. The faces of the hedge are slightly concave.

Stone house (Scots) the clubhouse near an ice rink where the curling stones were kept.

Stone land (Scots) a large mortar and stone-built tenement with a slate roof. A multi storey dwelling unit, perhaps dating back to the eighteenth century. The Royal Mile in Edinburgh has several good examples. Sometimes there are two or three stories under the current basement level.

Stone thrust (Scots) a small pier or quay.

Strothie / strodie (Shetland) a narrow lane between two parallel walls.

Stump fence (New England, USA) the first form of fence set up by pioneers. After clearing trees, the pulled stumps were arranged in a line to form a barrier which eventually rotted down.

Styag (Shetland) a pen for geese.

Sunk dyke / sunk fence / sonk dyke (Scots) a wall of stone or turf at one side and earth at the other (Figure 105). **Face dyke, Galloway hedge** and **ha-ha** are variations on this theme.

Surrey-fence defined in GVA County of Perth (1794: 114):

> In a few places there is what is called a Surrey-fence, and by some persons a snap-dyke, which consists of about thirty inches of a single stone wall, raised on the top of the ditch, behind the thorns, instead of stakes and rafters, to defend the hedge and ditch.

A ditch is dug out, a thorn hedge planted on one bank and a stone wall built behind that, to protect the hedge and ditch. The structure may have originated in Surrey, southern England.

GVA Surrey (1809: 452) refers to the planting of ozier (willow), '…in many instances they are made to serve as a fence, planted on a mound of earth in moist situations…'

Sustaining wall another descriptive name for a **retaining wall.**

Sweat house a common structure in parts of Ireland. Also found in northern European and North American cultures.

Joyce (1913: 625) includes an illustration of a drystone structure with interior measurements 5½ feet long, 4 feet wide and 5 feet high with a low door.

He then details treatment of rheumatism:

> A great fire of turf was kindled inside till the house became heated like an oven; after which the embers were swept out; and vapour was produced by throwing water on the hot stones. Then the person, wrapping himself in a blanket, crept in and sat down on a bench of sods, after which the door was closed up. He remained there an hour or so till he was in a profuse perspiration; and then creeping out, plunged right into the cold water, after emerging from which he was well rubbed till he became warm. After several baths at intervals of some days he usually got cured.

Joyce quotes from Prof Henry Hennessy: 'What are called Turkish baths in Ireland and Great Britain have been designated Roman-Irish baths in Germany and Bohemia.' This suggests the use of sweat houses was common in Ireland and famed throughout the rest of Europe.

Syver (Scots) a ditch, drain, water channel, and specifically, a field drain lined with stones and covered to form a small culvert. Related to the word 'sewer.'

A **rumbling siver** is a similar system of drains, with narrow ditches filled with loose stones, carefully arranged,

Taith-field / tath-fauld / tathfaud / tothe-fold / toathin-fauld (Scots) an enclosure

Figure 105: Probable remains of a 'sunk dyke', replaced by a post and wire fence. Badenoch, Scotland.

where animals are kept for the purpose of manuring. 'Taith' is animal dung. A landlord in the eighteenth and nineteenth century often included, as a condition of the lease, a requirement that tenants' cattle be enclosed in the landlord's fields overnight; this fertilized the ground.

During a personal conversation, John Aitken, retired shepherd, recalled herding yearling sheep through Aberdeenshire in the late 1940s. A local farmer was upset at the sheep eating all the grass, until he saw the benefit of the dung. He asked if the sheep could return the next year.

Terracing areas as diverse as European vineries, Asian rice plantations and in the South American Andes have developed terraced landscapes. This system allows best use of a gradient by building a barrier of stone, wood or mud brick across the slope to hold back soil, in a 'stepped' landscape. The flow of water is slowed, leading to better irrigation and reducing erosion.

Delius et al (2001: 86) illustrates the development of southern African terracing from a single row of stones to a substantial wall. Over time more stones were added to avoid undermining, strengthen the barrier, and hold back the increasing height of soil behind the wall.

Thrown wall (New England, USA) a roughly built stone wall, often 'single', relying on mass rather than architectural merit. A **dumped wall**.

Tombstone wall a wall built of vertical slabs of stone, tight together. Named for their similarity to gravestones.

Toon dik / toun-dyke / toon dyke (Shetland) a community's boundary wall enclosing

Figure 106: Wedge wall, held secure at each end. The Burren, Ireland.

the arable ground, separating it from the hill grazing. A variation of **head dyke**.

Tossed walls (New England, USA) roughly built walls, the result of field clearing when the farmers carried the stone to the edge of the cultivated area and tossed them into a pile. 'Laid walls', where the stone was more carefully placed, were an improvement on 'tossed walls' which were primarily meant to clear the land not create property boundaries. A **dumped wall**.

Town pound a stone-built enclosure in eighteenth and nineteenth-century New England for holding stray animals until their owners paid the fine. Etymologically related to **pund**, originally from the Norse.

Triple wall according to Thorson (2005: 73) this wall develops when another face of stones is built up against an existing 'double' wall, perhaps when additional stones are cleared from the land. It is a kind of **consumption wall**.

Trombe wall a wall built as a heat sink, to absorb heat from the sun and, with clever ventilation, release it inside the building during the cooler night time. The wall is built behind a glass exterior wall for maximum absorption of heat. This heating system was developed by Frenchman Felix Trombe in the 1960s, using ideas patented in the 1880s.

Victorian gardeners used stone as a heat sink. Fruit trees were grown against high stone walls. Sometimes the walls had heated cavities. Retaining walls in vineyards help produce a micro climate which protects against frosts and promotes vine growth.

Trullo (plural *trulli*) a corbelled, conical structure from south east Italy. Löbbecke (2012: 258) describes how trulli were built without mortar so they were easily demolished when the inspectors called, and the owners could avoid paying property taxes to the king.

Tumble walls (Burren, Ireland) the remains of what were probably ancient 'single' walls enclosing small irregular fields, as per D'Arcy (2006: 14-15) referring to research by Plunkett-Dillon (1985-86). These old walls likely provided material for modern walls on the same sites.

Tumps a Cotswold dialect term for tumuli, Neolithic age drystone structures and tombs.

Turnpike fences (Kentucky, USA) drystone walls built along the roadways from the mid 1800s.

Murray-Wooley and Raitz (1992: 37-41) explain the difference between this type of wall and the **plantation fence**. The stone in a turnpike fence is laid 'traced', with the longest edge along the wall, and the walls often have little or no batter. Throughstones strengthened the structure. The cope is '…formed of upright rocks on the fence's outside edges. The valley between the two uprights is filled with spalls.'

Contractors could build a turnpike fence quickly; the construction required less skill than the plantation fence.

Vaccary walls those walls, made with large flat slabs set upright with their edge along the line of the wall, were reputedly built in medieval times to form a fence enclosing pastures for cattle. They are most common in the Pennines especially in Yorkshire and Lancashire. The term derives from the Latin *vaccaria*, from *vacca*, a cow.

Vauch / wauch (Scots) a bank or mound of stones set up in a river as a kind of salmon trap.

Wa / waa / waw Scots variations of 'wall'.

Walking wall described by Thorson (2002: 125) as a New England wall '…whose broad capstones were laid like a sidewalk.' This was ideal for getting from house to house in the 'mud season' of early spring.

Wall-fence Loudon (1825: 443) uses this term for walls built from stone and earth, to differentiate them from earth-based structures like 'ditch fences', 'drain fences' or 'hedge fences'. He dismisses '…those of the earthy kinds… [because they] …are not by any means durable…'

Wart (Orkney and Shetland) a small stone-built shelter from an Old Norse word *vartha* meaning a pile of stones raised as a mark or beacon. 'Wart' is also an Orkney term for a small stone shelter built by boys when watching over the cattle on the common grazings.

The word appears in many place names in Caithness, Orkney and Shetland.

Water dyke / watter dyke (Scots) a floodwall or embankment. In Cumberland, northwest England, this is a term for a water filled ditch which acts as a boundary.

Wedge wall possibly a corruption of **edge wall** or 'edge fence', a wall with stone slabs laid across the line of the fence. The stones wedge down into each other, vertically or at an angle (Figure 106).

Weem (northeast Scotland) a local term for an Iron Age underground structure.

Wet wall (Kentucky) a stone fence built with mortar between the stones.

Whinny dyke this term comes from a conversation with a South African who had watched two young men build a 'whinny dyke' at the Black Isle Agricultural Show, just north of Inverness, Scotland. The wall was described as an 'ordinary double dyke with cope'.

Unfortunately, the spelling of the term or the origin of the dykers was not confirmed. The term has associations with 'whinstone', a generic name for any hard grey rock in Scotland.

Whitehouse / white house (Highlands and Islands of Scotland) a house built of stone and lime mortar which replaced the older drystone and thatched **blackhouse**. These houses had thinner walls, a chimney and fireplace – an improvement on the blackhouse which had an open fire and vented smoke through the thatch. The thinner walls and roof meant they were often not so 'cosy' as the older houses they replaced.

The Historic Environment Scotland description of the blackhouse at Arnol, Isle of Lewis, (https://www.historicenvironment.scot/visit-a-place/places/the-blackhouse-arnol/history/) gives some detail about the progression from blackhouse to white house in that area:

> All homes built in Arnol up to 1900 were blackhouses – double-walled dwellings simply known as taighean ('houses').
>
> A different type of house appeared when new health regulations required byre and dwelling to be separated by a wall. The single-walled dwellings cemented with lime mortar were such a contrast they were called taigh-geal ('white house'). The term taigh-dubh ('blackhouse') was used to describe the old houses from then on.

The White House in Washington DC, USA, is so called because the original building was whitewashed. Collen Williamson (1727-1802) from the village of Dyke, Morayshire, Scotland, was one of the master masons in charge of the early stages of construction of the White House.

Winter-dykes (Scots) a clothes horse, a wooden frame used for drying clothes in front of the fire in the winter. JSD defines the term further by saying it is an outdoor frame for drying clothes, sometimes incorrectly used as a term for a frame to dry clothes indoors.

Patterson (1880: 117), relating to Ulster, Ireland, includes the above description and also defines 'winter dyke' as 'two strong fences of stones or earth crossing each other at right angles. These are erected on exposed pastures to shelter cattle left out in the winter.'

Yair / yaire / yare (Scots) a semi-circular stone or stake and wattle structure, perhaps four feet high, built into the estuary or seashore to act as a fish trap when the tide receded.

Yaird / yeard (Scots) a churchyard, a stackyard in a farm, a garden – especially a cottage or kitchen garden.

Yaird dyke / yard dyke (Orkney, Shetland) a garden wall, the wall round a yard. Note here the use of 'yard', similar to the North American term for the ground surrounding a house.

Yird / yirth (Galloway, Scotland) a heap of large boulders forming a kind of den or small cave which became a trap for sheep if they fell in.

Yird-house an early Iron Age souterrain, named by people in subsequent ages who were not aware of its age or origin.

Zocos (Lanzarote, Canary Islands, Spain) low semi-circular stone walls built in large groups, to protect vines and trees from the wind.

4.2 The advantages and disadvantages of a drystone wall

The text in italics in this section is introduced by the author for emphasis. It is not part of any original quotation.

4.2 (i) Advantages

A drystone wall is satisfied with a *simple foundation,* dug no more than a few inches below grass level. This reduces costs and creates a flexible base in areas with seismic activity or frost heave.

The wall is *fireproof* and, in high moors, can act as a firebreak when muirburn gets out of hand. Stone walls have controlled wild fires in Australia but Munday (2012: 97) gives a case where a stone wall in South Australia 'caught fire'. It had been built over an old brushwood fence; the dried material in the centre of the wall ignited when stubble was burnt. There is evidence that New England walls, and similar structures in Australia, were built by pre-colonial populations, to control fire as part of their land stewardship process.

A drystone wall can be built on rocky ground. The ground is cleared of loose stone as construction progresses. It is difficult to drive fenceposts into rock.

A drystone wall can be built on any type of ground. In the late eighteenth century and early nineteenth century thorn hedges had their advocates, the drystone dyke was the better option especially where, as pointed out in GVA Fife (1800: 137) – '…the ground is high and cold, the soil not proper for rearing thorn-hedges, or where sheep are kept, and consequently thorns cannot thrive.'

A drystone wall harbours less vermin than a hedge. Some Australian walls were built specifically to control rabbits; built tight enough so that rabbits could not get inside, or through, the stonework. Corangamite (1995) often refers to rabbit management. Dykes are hunting ground for the stoat (*Mustela erminea*) and the weasel (*Mustela nivalis*). Their slim shape permits access through small openings, to seek mice and the like.

A wall takes up less room than a hedge. Raistrick (1973: 15) comments that walls take up less space than hedges and do not steal nutrition from the crop. GVA Argyll (1798: 64) favours dykes where '…stones abound almost everywhere…' It is *easy it is to repair* a fallen dyke and 'It also takes up less room than any other [fence]; by which a considerable quantity of ground is saved.'

A drystone wall requires minimum maintenance compared to a fence or hedge. A post and wire fence might need replaced every twenty years and a hedge re-laid every fifteen to twenty years, depending on how quickly it grows and how it is trimmed.

A stone wall is permanent if built properly. Baker (1913: 2) includes an anonymous quote which is applicable to any stonework:

> The first cost of masonry should be its only cost. Though superstructures decay and drift away, though embankments should crumble and wash out, masonry should stand as one great mass of solid rock, firm and enduring.

A drystone wall can stand for four years or forty years. *If it collapses, without loss of life, it is not necessarily a tragedy*, assuming there is sufficient skilled labour to rebuild it. The material is already on site and one craftsman has all the skills to complete a repair.

A drystone wall uses local materials with the minimum of tools.

A stone wall is cheaper in the long term, all things considered. Deane (1797: 105) writes:

> When ground is wholly subdued, and the stumps of its original growth of trees quite rotted out, if stone can be had without carrying too far, stone walls are the fences that ought to be made. Though the cost may be greater at first than that of some other fences, they will prove to be the cheapest in the end.

A qualified acceptance of stone walls is given by Stephens (1853: 593):

> Dry-stone dykes are not nearly so picturesque objects in a landscape as thorn hedges, nor do they afford so much or so comfortable shelter to stock. They are *easily and quickly erected*, however, in a country where the materials suited to their construction are abundant; and after being completed, they satisfy the mind of the farmer that *little trouble will be experienced from them for a series of years to come.*

Lambert (1829: 152) thought:

> Stone walls are not so handsome or so agreeable a fence to the eye as quicks or thorns, but they are a more profitable one to the farmer; as a stone wall improves the land for some distance on either side, whereas a thorn fence not only occupies more ground, but by extracting its nutriment from the soil about, rather injures the surface on either side. *A wall is also a drier and better shelter for cattle*, as thorns, unless kept clipt, will drip upon them.

A drystone wall is flexible and able to respond to freeze and thaw cycles and reasonable earth movement without showing distress or collapsing. Does this suggest a drystone wall is earthquake proof? Aitken and Rippingale (2001) record a personal experience of a drystone wall which moved with the ground as a severe earth tremor, measuring 6.8 on the Richter scale, rippled across the landscape in Seattle, Washington, USA on 1 March 2001. The wall survived intact; nearby concrete structures cracked and showed distress.

Lassure (2006) notes this, and other evidence, and concludes that drystone walls '…have no propensity, by virtue of their being mortarless, to be more earthquake-resistant than mortared walls.' He acknowledges that the quality of build affects the strength of the wall:

> To all appearances, certain features have to be taken into account to explain the varying degree of resistance to earthquakes offered by dry stone walls. It is obvious that a wall built according to the book (with throughstones in sufficient numbers to bond both faces, carefully inserted pinnings within the wall, staggered joints) will provide a higher resistance to tremors than a poorly bonded contraption (no through stones, no staggering joints, voids in the core).
>
> It also stands to reason that a wall made of large stones will be less likely to fall to pieces than one of smaller stones, or again that a wall built of flat stones will be less susceptible to crumbling than one of rounded fieldstones with minimum frictional resistance. Only

a systematic investigation of the damage inflicted on dry stone walls in relation to the quality of their craftsmanship can provide better information on the subject.

No mortar is used in a classic drystone wall. This means a lower build cost than mortared walls and no problems with mortar in damp, frosty or hot weather.

Good dykes add value to a farm.
GVA Galloway (1813: 83) thought good thorn hedges or stone dykes were superior to a turf wall topped with gorse:

> It would not be too much to say, that most farms in Galloway, where proper inclosures have been made, are thereby increased one-fourth, one-third, or, perhaps, in some instances, one-half in their value.

Schrank (1995: 24) quotes from an 1821 report proposing improvements to an estate in Orkney —

> He advocated a five-foot-high drystone dyke which he estimated would cost from £100 – £150 per mile. *The tenants would defray the costs by paying increased rents.*

Drystone walls are self-draining and, when built as retaining walls, can take a lot of water pressure before they fail. On the island of Mallorca, in the Spanish Balearic Islands, retaining walls have been seen with water pouring through them; they did not collapse. A freestanding wall can survive the surge of floodwaters until debris builds up against it.

A drystone wall is an attractive hand-crafted structure with variations in texture and colour, this makes it unique and aesthetically pleasing. This is a modern virtue rather than a historical one. Function was more important than form in the eighteenth and nineteenth centuries.

The advantages of a completed wall are immediate. Hedges are an alternative but they take time to grow. According to Sinclair (1813: 45):

> Stone walls... have one advantage... according to a common saying, 'they are major from their birth;' or, in other words, are in their greatest perfection as soon as made. They have, however, this disadvantage, that they are getting worse every day... stone walls though not so ornamental as hedges, yet are preferable in point of utility; for the benefit of the inclosure is immediate.

Lambert (1829: 153) gives another, unanticipated, advantage of walls over hedges – the temporary fencing protecting the growing hedge from cattle would '*...be burned for fuel by the poor during the winter.*'

Another advantage for a dyke, compared to a post and wire fence, is mentioned in the *Dumfries and Galloway Standard and Advertiser* for 20 August 1939, in an article for a drystone dyking competition — '*A dyke takes longer to build but the money goes in wages spent locally*, while much of the cost of a wire fence goes on imported material.'

Macdonald (1887) advocated fencing of hill pasture because *it restricts the movement of animals and helped prevent the introduction of disease.* He adds:

> Where there is an abundance of stones at hand, dyke-fencing is, if anything, preferable to wire-fencing… *it is less liable to suffer from heavy snowstorms or frost than wire…* [and] …affords more shelter to sheep. Dyke-fencing is specially (sic) commendable on exposed ground, and if it cannot be raised to a sufficient height conveniently, a wire or two… [can be] … run along the top of it in order to prevent the sheep from leaping over.

A properly build drystone retaining wall prevents erosion, slows the flow of rainwater to improve irrigation and increases the viability of marginal farmland. The development of terracing, in southeast Africa, from an initial line of stones to a higher, stronger, barrier is detailed by Delius et al (2001: 86).

Drystone walls are an important part of the cultural landscape. Destruction of dykes, as part of agricultural improvement, can be a political statement which comes to nothing, as noted by Hazard (2008: 76) when he refers to the rebuilding of stone walls in Estonia after independence from the Soviet Union — 'The Soviet regime dispersed thousands of Estonians and bulldozed hundreds of miles of dykes for its grand and ultimately futile plans of collectivisation.'

4.2 (ii) Disadvantages

Drystone can be expensive in the short term compared to a post and wire fence or a wooden rail fence. Because of this William Mackintosh (1729), in his postscript, saw little future in stone walls. He thought a hawthorn hedge and ditch were a more profitable and longer lasting benefit:

> Stone Inclosure, I think, looks very bleak, and a little ornamental, if Inside or Outside of it is not planted with Trees… Tho' Stone Inclosing is much more commendable than no Inclosing, and indeed an Improvement, still it is a very dear one, and no Hope nor Prospect of its ever becoming a national or universal Improvement.

The structure is difficult to move without total destruction and reconstruction. Thorson (2002: 126-127) quotes the landscape historian John Stilgoe who 'disliked the stone walls because they seemed – and often were – poorly planned and poorly layered, liable to topple at the first frost and certainly *wasteful of space.*' Thorson mentions other writers who thought the walls were ugly or obstacles to cultivation and considered how to remove them. Lexington-Fayette (1990: 19) describes the destruction of drystone walls in Kentucky as land became more expensive. The Kentucky study quotes a 1967 study which estimated '…*for every 1000 feet of stone fence removed… approximately one acre of ground is gained.*'

Sometimes stone fences are an inconvenience to modern farming practices. Smith (1837: 163) recognised some land had been subdivided '…in so irregular and unsuitable a manner…' and recommended replacing the old walls to create more regular fields – 'Where stone fences

Figure 107: This nineteenth-century dyke ended because the supply of stone ran out. A post and wire fence was built to carry on the duty, but has since rusted away. The dyke continues to give shelter.

exist, it will, in many cases, pay to use the stones for draining, replacing the fences with hedges.'

Smith was an advocate of deep ploughing and draining. He farmed near Deanston in Perthshire, Scotland.

Poorly built walls are not a good advertisement for the craft. Drystone walls had competition in the earlier days of enclosure. Hawthorn or gorse hedges were considered as an alternative because drystone walls, at that time, were unreliable. Innovations such as regular throughbands, secure coping and construction by skilled teams increased landowners' confidence. Accounts of early drystone dyking include the necessity to properly supervise the dykers. If the landowner does not know what to look for, the wallers often cut corners and produced an inferior wall, with a limited life.

GVA West Riding of Yorkshire (1799: 75) acknowledged *stone walls were a good investment and, because of the high cost, it was one for the landowner.* Some of the cost was passed on to the tenants:

> As to the manner of inclosing, we know of no fence equal to a good quick-set hedge of white thorn, when it is properly trained up… Perhaps stone walls are more eligible where sheep are kept. These we would recommend to be built, or rather lipped with lime, and to be six quarters in height, with an additional quarter by way of caping. Probably this, at the

long run, is the cheapest fence; but, being very expensive at first, it should in every case be executed by the proprietor, the tenant paying legal interest upon the outlays.

Readers should note that a 'quarter' is nine inches, one quarter of a yard. The total height of the wall described above is thus 63 inches (7 x 9 inches) or 5 feet 3 inches. 'Lipped with lime' means the drystone wall was pointed up with lime mortar. This was thought to extend the life of the wall.

Small damage to a drystone wall can escalate to major damage if not immediately attended to. Trees, children and livestock can cause damage to walls, fortunately most of it is easily repaired. Mackintosh (1729), a keen advocate for hedges, like many others in the early eighteenth century, writes in his postscript — 'For your Wall, from the Minute it is finished, is every Day growing worse; very often the very first Year, after much Rain or Frost, whole Roods come down that cannot be made up till Summer…'

GVA Berwick (1813: 179-180) notes the frailty of walls then acknowledges their necessity:

> *The great evil of a stone wall is its continual deterioration*, but no other fence is at all eligible in such situations.' His opinion had been clouded by the ineffectual attempts at enclosing land some time before when the dykes '…either from… having been imperfectly executed, or from utter subsequent neglect, they have now all become entirely useless.

By the mid 1800s, drystone walls were much more reliable. Stephens (1853: 592) also thought dykes deteriorated as soon as they were completed. He recommended selecting the best materials and the best craftmanship for their construction and an annual inspection before livestock returns to the fields.

Repairs to drystone walls can be expensive and the repairer unreliable. Lexington-Fayette (1990: 19) refers to the frequent vehicular damage to Kentucky walls and the difficulty of recovering the cost of repairs:

> All too frequently, even if funds are recovered from a motorist's insurance company, the fence owner will either apply the money to other projects, or will employ unskilled workmen who will make inharmonious repairs.

Drystone structures, even in ruins, are a refuge for wildlife and plants. This has a downside if mice, rats, snakes or the like take advantage of the shelter. The islanders on remote Tristan da Cunha in the South Atlantic, hold an annual Ratting Day, a rat catching competition, mainly around the drystone walls in the cultivated area known as the 'Potato Patches'.

A drystone wall is not necessarily attractive to everyone. 'Beauty is in the eye of the beholder'. Incidentally this expression comes from the 1878 novel Molly Bawn, written by Irish (County Cork) writer Margaret Wolfe Hungerford. In 1822 Baines (1969: 278) remarked – 'The situation of Settle [north Yorkshire] is picturesque and singular, but its beauty would be much increased if the high and gloomy stone fences round the fields were displaced by hawthorn hedges.'

4.3 The shelter value of a drystone wall

A drystone wall reduces the force of wind, it percolates through the gaps in the stones and breaks up. When wind hits the solid barrier of a mortared wall it creates a vortex on the sheltered side. Livestock finds this uncomfortable and it can be destructive to growing crops.

Shelter is important for raising any livestock. The old Scots saying 'dry bed, half fed' succinctly describes how livestock can still thrive in cold weather if they have access to shelter, preferably on straw, indoors.

MacWeeney and Conniff (1998: 62) describe the shelter value of drystone walls in The Mourne, south of Belfast, where they are known as 'ditches':

> And they [the walls] are built for shelter, to provide what Philim Doran [the wall builder] calls 'easier living' for the animals. The gappy dry stone ditches are better at this, oddly, than a solid wall. The wind 'plops over' the mortared wall, Doran says, but filters through the dry ditch. A visitor who shelters against one of these walls on a windy afternoon may not perceive this; the wind seems to emerge from every gap with its bite intact. But a yard or so out, the needling rivulets of wind subside, and the air pauses instead of rushing breathlessly past. In this still pocket, the grass grows richer from the long sheltering of the animals.

In Borchgrevink (1982: 89) Charlie Jardine agrees the shelter produces a microclimate. Lambs can lie in the comparative warmth of the lea side and, because it is warmer there, the first grass will appear a little earlier than in the rest of the field:

> On a stormy day if a dyke's four foot six high and you come out about two yard out from the dyke, you'll get more shelter… from the dyke as what you'll get close to the back o' the dyke because the wind's always blowing through some gaps in the dyke.

Rainsford-Hannay (1957: 54-55) tells us a 4½ foot high wall '…is impervious to wind at three feet above the ground, and there is always shelter on one of the sides of the field.' His next statement is worthy of scientific testing:

> The influence of shelter may be felt up to twenty-four times the height of the shelter itself. Thus a dyke five feet high would give its last faint shelter at a distance of forty yards.

GVA Stirlingshire (1812: 121) thought walls interrupted the flow of air to the disadvantage of growing crops:

> Stone walls of about four and a half feet high, with a coppice of sods of earth and turf, sometimes laid on in two rows, are also very frequent. They have the *advantage* of forming a fence of great strength and duration all at once. They have the *disadvantage* of interrupting the current of the winds with such sudden violence as to occasion gusts or eddies upon the surface of the ground, which rush on with increasing impetuosity over the fields, and disturb the stratum of air which lies next to the surface of the earth, and is most conducive to vegetation.

Figure 108: A section of 'double' dyke, built within yards of the section shown in Figure 109. The build style altered to suit the available stone. Tayside, Scotland.

The writer goes on to repeat the, then current, belief that still air in fields was '…highly advantageous to the growth of vegetables, that the stratum of air which lies in contact with the earth, should be as much as possible in a state of rest.'

GVA Perth (1794: 117) also speaks against dykes for shelter value, preferring hedges and trees – 'As far as hedges are superior to stone-walls, in respect of warmth…'

Dickson (1805: 110) gives his opinion that a 'dead fence', one made from wood or stone, cannot be built high enough to provide shelter. He advocates hedges.

Extreme winds can blow down a solid wall, or a drystone wall. D'Arcy (2006: 30) explains this point:

> The logic is simple: a perforated wall is less of a barrier to elemental forces than a solid one – therefore, in theory, more stable. This assumes, of course, that the cumulative frictional characteristic of the wall outweighs the strength of the wind. Usually it does.

A wind speed of 177 mph (285 km/h), was recorded in 1962 at RAF Saxa Vord in Shetland, the island group to the north of Scotland. On 1 January 1992 a gust of 197 mph (317 km/h) was recorded, an unofficial record. The island of Foula lies 20 miles west of Shetland, exposed to fierce winds at all seasons of the year. Manson (1958) built walls which resisted the wind:

On Foula the wind can demolish a stone dyke. Peter Manson built one with holes in it ten or twelve years previously and it is still standing. Previous efforts had been laid to the very foundation every year. He also rebuilt a snaa buil [snow shelter] after the first, vertical, construction had been laid by the wind. The second wall was wider at the base and tapered to the top, and thus resisted the wind. The snaa buil was for sheltering sheep during snow; they could be kept together and fed in the enclosure.

4.4 Stone structures incorporating turf and earth

Figure 109: A section of 'single' dyke, close to the one in Figure 108.
This section was built from right to left using irregular boulders.

Turf, or alternate courses of turf and stone, was once a common construction material. It had advantages and disadvantages. The construction of a wall of stone and turf is easier than building a wall of vertical stone and it went part way to the challenge of building a permanent stone wall.

Turf, in this context, is a layer of vegetation and topsoil, cut from the ground. More detailed, historic, definitions for the different types, including 'sod' and 'turf', are given below, under **fail and divot**.

Some stone/earth combinations have been detailed in Section 4.1, including **Auchenhalrig work**, **backit dyke**, **clawdd**, **Cumberland bank**, **ditch dyke**, **ha-ha**, **half-dyke**, **Galloway**

hedge and **sunk dyke**.

Variations of coping systems using turf are described in Section 3.2 (iv).

Prior to the Enclosure Acts in England and Wales, and the reorganisation of British agriculture in the eighteenth century, communities generally held their land in two forms. Open fields were cultivated near the community's houses. A boundary wall, the head dyke, surrounded the arable ground. Beyond the head dyke was the common grazing where animals wandered freely. The old head dykes were roughly built, not expected to last for more than a season. They had to be stockproof in the summer, to keep livestock off the growing crops; some collapsed over the winter. These gaps were repaired before cattle were put out to the common grazings in the spring.

McParlan (1802: 59) described the poor walls in Donegal, Ireland and realised field walls (called 'ditches' in that part of Ireland) were less necessary when the owners keep their cattle indoors:

> In the mountain region the fences are very bad, small turf ditches, and of those very few; as either a person stands out to watch the cattle from damage, or the cattle are tied by ropes to some of the legs, so as to be able to range about a certain periphery, but to prevent their reaching the potatoe [sic] or corn sowings; the care also, that the common people take in keeping their cattle in the house, to prevent the cruppan and collect manure, renders fences less necessary.

Cruppan was an illness suffered by cattle after eating poor quality grass, probably a mineral deficiency.

Loudon (1825: 444) recognised the limitations of turf walls:

> For a fold, this fence answers extremely well; but for enclosing a field, or indeed any other use where durability is required, it should had never be had recourse to, as, from the moment it is finished, its decay commences, and no pain or attention will be able to keep it in repair after it has stood two or three years.

The construction of turf walls can be quite sophisticated. They were often built on a stone foundation to strengthen the lower part of the walls against the effects of water, frost or impact. The Antonine wall, built circa 140AD as part of the frontier between the Roman Empire and Scotland, was constructed of turf with a stone foundation. Turf construction in Iceland and islands across the North Atlantic goes back to Norse times. Houses in Scotland were built with this system well into the eighteenth century. One big problem with turf construction is the damage caused to grazing. If an area is stripped of turf it means less grass for at least a year.

One type of turf boundary wall was built by piling up the turves, digging a ditch in front of them and throwing the soil behind the pile. Posts were driven into the wall to hold a two-rail fence. The total height might be up to six and a half feet, enough to prevent sheep jumping over. This type of structure was more durable than a simple low turf wall.

The logical progression from turf walls was to walls made of stone and turf, sometimes called 'turf and bool' (turf and boulder) dykes. These still had limitations, recognised by

GVA Roxburgh (1794: 22):

> In some places, dykes are built of alternate layers of stone and turf; they are commonly about five feet high, three feet broad at the bottom, and eighteen inches at the top, and generally cost fourteen-pence the six yards; but all the stones are laid down by the employer, and the turf is cut by the undertaker. Dykes of this construction are not very durable; but as they are speedily raised at no great expence, they serve the present exigency; and will probably in time be superseded by thorn hedges, which, for durability and sufficient fences, are generally acknowledged to be the best.'

For more detailed information on turf and earth construction the reader is directed to Walker and McGregor (1996) and Walker (2006), full details in the Bibliography.

The following terms were extracted from accounts of early drystone construction

Berm a raised mound or ridge of earth. In ancient times berms formed part of a defensive structure; nowadays it is landscaping feature. The term is also used for a mound of earth placed against a wall to provide insulation.

Bofi / boffi / buffie (Shetland) a large thick piece of turf used as a seat in a house, or as a bed or couch.

Cassin dyke (Scots) a turf wall, according to Pride (1996: 22). DSL has 'cassin' as a variation of 'cast', an old Scottish verb with several meanings; one is relevant – 'To take (peats or turf) out of or off the ground by digging.'

Clap dyke (Scots) an earthen dyke or fence so called because they were 'clapped' or compacted with a spade, to make them solid. DSL defines these dykes as being 4 foot wide at the base and from 3 to 5 feet high. These structures were formed from the earth thrown up from a ditch and considered an ideal fence for places where there were no stones.

If ploughing turned up many small stones, they were used to further strengthen the clap dyke, as detailed in GVA Dumfries (1794: 44). Both sides of the earth bank were faced up with stones, one stone thick, with a thin layer of turf between each course of stones. A thorn hedge was planted into the stones on one side of the dyke.

Clay and mott a building style recorded in the parishes of Cromarty and Nigg in the district of Ross and Cromarty, north of Inverness, Scotland by DSL — '…there are houses with internal partitions made of what is locally called 'clay and mott', consisting of a mixture of clay, stones, chopped bent grass, hair, etc.…'

Delf (Scots) a sod, removed by digging. Another meaning, 'a quarry', was common in Middle English, the language used between the eleventh and fifteenth centuries. The meaning of 'delve' in Middle English is closer to 'dig' or 'plough.'

Divot dyke (Scots) a dyke made of divots or turf.

Fail dyke / faily dyke / feal dyke / fealie dik / felly dyke a wall built entirely of grass divots or alternate layers of stone and divots (Figure 110). Omand (1974) describes feal walls in Caithness, northern Scotland, and the progression to more permanent field boundaries:

> The typical old wall constructed to keep animals out of the crops was made of feal (turf),

built 4 ft–5 ft. high, 3 ft–4 ft. at the base and narrowing to 2 ft at the top. Such walls were easily broken by animals and each year after the crops were sown dykes had to be repaired. Shortly after the middle of the eighteenth century stone dykes were constructed on Thurso East Farm. Initially the tops of these dykes were covered with sods but later tough stone coping was used. The planting of hedges, many of them hawthorn, towards the end of the eighteenth century compartmented and visually improved the landscape.

Mackintosh (1729: 71) was an enthusiastic advocate of hedges. He had a low opinion of feal dikes:

> …Earthen or Feal Dikes, as they call them, is the Devil: always a waste, a daily Labour, and never can be kept up: yea, can never be made, the very first Day, a Fence against the heaviest or dullest Sort of domestick Fowl or Beast.

Fail and divot / fale and divet (Scots) a wall built of alternate layers of divot and fail. In the twenty-first century we have fewer words for a piece of turf, unlike previous centuries when it was a valuable building material and fuel source. JSD, under 'fail', details the subtle differences between the various forms, and how they were used in the county of Angus, now part of Tayside, Scotland. The alternative spelling, 'fale', is introduced at the end of the extract:

> *Fail* is used for building the walls of an earthen house, and *divot* for covering it. The *fail* is much thicker than the *divot*, and differs in shape. The *divot* differs also from *tour* or turf, as strictly used; the *divot* being of grass and earth, and the *turf* either of a mossy or heathy substance, or partly of both. *Sod* is properly a thick turf, resembling the *fail*, not so directly used for fuel, as for keeping in the fire kindled on a hearth and casting forward the heat… In building a wall or dyke of *fale and divet*, it is often the custom to set the *fale* on edge, and lay the *divet* flat over the *fale*.

Fald-dyke JSD defines this term, quoting a 1437 reference from Aberdeenshire which refers to a '…wall of turf, surrounding the place appropriated for a fold.'

Falla fóidín (Irish Gaelic, from Dinneen (1904)) a sod wall, from the Irish *fóidín*, a small sod, and *falla*, a wall.

Feally-daek (Shetland) a dyke made of turfs.

Galloway hedge / sunk fence see entry for this combination of hedge, stone wall and earth bank in Section 4.1. There were several variations of this type of structure before the development of reliable freestanding drystone walls.

Nebraska Marble a colloquial term for sod, when used for building a **sod house**, detailed in Section 4.4.

Raik (Scots) a turf dyke which enclosed a drove road where it passed through lowland arable areas, it fenced the route of the animals as they journeyed to the marketplace. The term derives from a meaning of 'raik' as a journey, or wandering aimlessly.

Shirrot (Banffshire, northeast Scotland) a turf or divot.

Sod-dyke / sod fence a turf wall. A term recorded in Scotland and elsewhere.

Figure 110: 'Feal dyke', Highland Folk Museum, Newtonmore, Inverness-shire, Scotland. The vertical white lines are flagpoles in the distance.

Corangamite (1999: 46) describes a sod fence in Victoria, Australia:

> Thomas Russell undertook experiments with sod walling at his Barunah Plains station near Rokewood in the 1860s. On 4 May 1864 Russell entered into an agreement with Patrick Ryan and John Wingrove to construct a 136 chain (2.67 kilometres) sod fence at the rate of 12 shillings per chain. Russell specified that – The said fence to measure 5'6' wide in the bottom and 2'4' on the top tapering gradually to a height of 3'6' and then capped with a large sod on each side leaving a space for furze seed. The sods for said fence to be taken from each side of the wall forming a ditch at least one foot deep.

This specification used a lot of turf, valuable grass for livestock, one of the disadvantages of this type of wall. 'Furze' is another name for gorse *(Ulex europaeus),* also called 'whin' in Scotland.

Sod house Muilenburg and Swineford (1975: 25) detail how pioneers in 1870s Kansas built houses of '...prairie sod, cut into blocks of about twelve inches by two inches by twenty or twenty-four inches and laid like brick to form walls twenty to twenty-four inches thick.'

The walls were coated with rough plaster made from local rock.

Sod, as a building material, was called 'Nebraska Marble'.

Sod, or turf, was also used for building Scottish houses; it required frequent maintenance. In the Statistical Account of Banffshire, 1842, even the poorest people were building their houses of stone and mortar to replace the old drystone blackhouses or turf houses – '… there are few or no black or feal houses.'

Swear dyke / swaird-dyke (Moray Banff, northeast Scotland) a wall of 'sward', an alternative term for turf.

Taft / tafft a homestead and the land attached to it. 'Taft' is common in place names in Orkney and Shetland. 'Toft' is the English form of the word. It also applies to a small patch of enclosed ground set aside for the rearing cabbages etc.

Before drystone walls the wall round the taft (the 'taff dyke' or 'taft dyke') was made of turf. This subsequently morphed to mean a turf wall in general.

Prevost (1957: 100) defines 'taff dyke' as a Galloway (southwest Scotland) term for 'A fence made of turf.'

Tinker fence (Scots) a turf wall.

GVA Berwick (1794: 116) details the construction:

> …the feal are cast [cut] thick, all in the form of the diamond of a pack of cards, and as thick as the length of the side; they are every one set upon edge, the long angles running into one another, and are set perfectly perpendicular; four rows complete the height wanted, and every row, as the dyke advances, is placed two inches within the row beneath. This kind of fence, if it chance to face the north or east exposure, will last a long time, seven years and upwards: but the south exposure is unfavourable. These kind of fences are much used. The name… is a tinker-fence, and the operation is called tinkering.

This method of construction was necessary in that part of the county of Berwick because there were few stone quarries. Ditches and hedges were the main method of enclosure.

Toure dyke (Scots) an alternative name for a dyke made of turves.

Treb / treve / treb-dyke an earthen mound or dyke, many of which are found in Orkney. They are thought to date back to Pictish times (approximately 1500 years ago) as some sort of boundary wall or settlement enclosure.

Truff dyke (Scots) a turf wall.

Turf and bool / turf and boulder dykes built of alternating courses of stone and turf.

Turf cope a grassy top course on a wall, best set in two layers. The first layer is laid grass down, the top layer laid grass up. Stones, or a layer of soil, were often laid on top to secure the turves.

DSL includes an entry, under 'clap', recorded in Shetland in 1897 which defines 'clapper' as '…a pounder or mallet made of stone.' The clapper '…is known to have been used in recent times for beating down and forcing into position the turf coping of dry-stone dykes, and that it is believed to have been made for that purpose.'

Turf-dyke / toure dyke (Scots) a wall of sods.

4.5 Lunkies, smoos, cheeks, posts and alcoves

This section includes special features which need skills beyond that required to build a straight, or curved, wall. Most features are straightforward; they use lintols, steps or some variation of a wall end.

Aran gap / bearna (Aran Islands, Galway, Ireland) a short length of rough walling, within properly built wallends. It can be dismantled so the opening acts as a temporary gateway (Figure 111).

Gaps are filled with anything from iron bedheads to oil drums, not just stone. Many openings have been modernised with metal gates set in concrete, a definite loss to the romantic eye.

Dutton (1808: 152) was describing County Clare and the Burren, the mainland landscape, very similar to the Aran Islands, when he wrote:

> In many parts of this country a gate is a rarity; when cattle are to be moved, a man takes down a yard or two of wall, and, when the cattle are in, builds it up again; this, even with men of property, is the general practice twice or oftener every day …

This type of gate can also be found in the Scottish Hebridean islands and throughout Europe, wherever drystone walling is common.

According to Laheen (2010: 115) most Aran gaps bordering a road have a stile (*staighre*) close by for foot traffic. Stiles are formed by projecting throughstones which lead the walker over the wall, and through openings in the topmost stones which allow easy access over the wall.

Arch a structure which spans an opening and is strong enough to support a load. It is a development from corbelling and the use of lintols. In their most refined form arches use especially cut stones (voussoirs) and a keystone to produce a symmetrical structure. Arches are common in bridges and culverts and within large buildings.

An Arabic proverb tells us 'An arch never sleeps'.

Leonardo da Vinci (1452-1519) defined an arch as two weaknesses leaning against each other, to make a strength.

Barway (New England) a simple gateway with a pair of **cheekends** at either side of the opening. Long poles or rails, slung across the opening, act as a gate.

Battit (Scots) fixed with lead into a hole drilled in rock, for example gate fixings or metal posts. The hole need only be a few inches deep. The post is set in the hole and molten lead is poured in, or strips of lead are punched in with a point, to produce a strong fix. Molten sulphur was sometimes used instead of lead.

Bee alcove / bee bole / bee box / bee garth / bee hole / bee house / bee kep / bee niche / bee shell / bee wall a niche, or groups of niches in a stone wall for straw bee skeps (simple bell-shaped beehives made of straw) or a more conventional beehive. The skep was set on a flat round 'beestane'. Throughout Europe, structures to protect individual hives and groups of hives were an important feature of many religious institutions, providing honey and wax for candles.

Bissing / bissy / bizzie (Shetland, Orkney and Caithness) an individual cattle stall in a byre, often formed from large thin flagstones, set on edge. These flagstones were also called 'buisen stanes', 'buissen stones', 'bowsin-stones' or 'busing stones'. Further south in the west of Scotland these stones were described as 'stallin stanes', named for the job they did – forming the stalls in the byre.

A 'shed' is another name for a byre partition, from the use of the word as a division or dividing act – to shed, to separate, to divide, as recorded in DSL. Grose (1787) records the word 'shed' as meaning – 'Difference. No shed, no difference between things.'

Figure 113 illustrates how flagstones can be used for a partition.

Bole / boal (Scots) DSL includes these definitions:

'A small recess or cupboard in the wall…' — a term used generally in Scotland.

Another definition, for 'bole-hole', comes from southwest Scotland – an 'air hole in wall

Figure 111: An Aran Gap, a stone-filled gateway. Inis Oírr, Aran Islands, Ireland.

of barn or stable; a square hole in the wa' near the fire, where the salt box is kept.'

Boot / bowt / bout (Scots) a hole in the stone wall of a mill lade (the channel guiding water towards the mill) which allows water through to the millwheel and drives the millstones.

Bridge as a verb in drystone work, this means to lay a flat stone spanning an obstacle such as a root, pipe or watercourse. In effect a miniature version of the normal interpretation, an arched or lintoled thoroughfare across a gap or water.

Figure 112: A simple arrangement of the local flagstone creates a niche in a stone shed. The principles are easily adapted for other stone types and many other purposes, including bee boles. Spittal, Caithness, Scotland.

Butt ends (New Hampshire, USA) a standard term for squared-off ends to a freestanding wall. Known in Scotland as a **cheekend** (Figure 114) or in England as a 'wall end'.

Buttress a right angles extension from a wall, propping the wall against a slope or pressure.

Charter bole DSL has this definition, under 'bole':

> …a bole provided for in a feu-charter, which normally prescribes the middle line of a wall as a boundary but may expressly permit a recess on one side of the other to go deeper. Such boles are often founded on to prove that a wall is mutual.

Cheek Rainsford-Hannay (1957: 76) refers to Raistrick (1946) and notes Pennine wallers in Yorkshire build **heads**, conventional wallends, when building up steep slopes and adds – 'In Scotland they would be called 'cheeks' or solid pieces of single walling against which the rest of the work can lean.'

Cheek o' da grind (Shetland) a **cheekend**, the edges of a gateway. From the Scots term 'cheek', the 'edge' or 'face' of something.

Cheekend / cheek end (Scots) the vertical end of a wall, for example at a gate opening, with large stones (headers and stretchers) tied in (Figure 114). The cheek end is normally vertical; it can be battered back, especially if no gate or gateposts are required. Also called **staunchion, sconcheon, wall head, gate head** or **quoin end**. Figure 82 shows a good example from Kentucky.

Cheekstane / cheekstone (Scots) one of the upright stones which supports the grate in a fireplace.

Clachan cinn (SG) end stones.

Cow creeps (Derbyshire, England) as described in Lund (1992: 15):

> ...the lintel can be removed to let a cow pass from one pasture to the next and the wall stones above the lintel are crudely piled together and must be removed and replaced each time that the beasts are moved.

The lintel supports an opening big enough for sheep. Cows can use the passage if the lintol, and the stones piled on it, are removed.

Cow slip / cow stile / open stile / grike (New England, USA) a narrow slit in a wall which allow humans through and exclude cows.

Cradle / craiddle / cradle / credill (Scots) the stonework in a well or shaft. A well shaft is lined by excavating down to the source of water then building up with stone and backfilling as the work proceeds, very like constructing a tight round retaining wall.

Crannie (Scots) a recess in the wall. Old stone houses often had a recess near the fire to keep particular items, such as salt and tea, dry. The Irish phrase *poill agus prochóga* (nooks and crannies) and the Scots word 'boal' for a recess in the wall, may have a common origin.

Evans (1979: 65) notes a 'keeping hole' is found on each side of the fireplace in old Irish cottages — 'Known as boles in Antrim and coves in Cavan... There is an old belief that the left-hand keeping hole belongs to the woman and the right-hand one to the man of the house.'

Creep / creep hole (England) a bridged opening in a wall to allow small animals through, a way of controlling the movement of livestock from one field to another.

Criggie / krigi / kriggi (Shetland) a bend or angle in a dyke or fence.

Cripple hole / cripple gate / cripper gap / cripple gap (Scots) a lintoled opening in a wall which allows sheep through and stops larger livestock. These openings were built at frequent intervals along a head dyke. The shepherds could move sheep without travelling long distances to a gate. The word 'cripple', in one of its older meanings, is apt because the openings are low in the wall and someone must bend right down to get through.

Cundy / cundie / conduit / condy / kundie / cundie-hole a small passage or drain under a dyke, Scots forms of the English word 'conduit'.

Mactaggart (1824) defines 'cundies' as 'hare holes through dykes.'

Cairns (1975: 12) describes the construction of a conduit, apparently in the period before World War I. One shilling (1/-) is five pence in modern British decimal currency:

> ...conduits are made here and there where the water crosses a dyke, this being allowed for, to the tune of 1/- a time in my day, a large sum for a couple of hours' work. Large flat stones have to be found and sometimes carried a long way if the conduit is new.

Culvert a passage for water or animals.

Cwter (W) translated as 'a gully or gutter', a hole through a wall.

Daek-end (Shetland) the end of a drystone wall where it meets the edge of the sea. This stops animals from walking down the beach on a low tide, beyond the wall end and into other fields. Outside Shetland the term appears as 'dyke end' or 'dike-end'.

Dook (Scots) a wooden wedge forced into a hole or crack in stonework, as a fixing point for a nail or hinge.

Dyke-slouch / dyke-sheuch / dykesheugh (Scots) a ditch or open drain at the bottom of a dyke. A 'sheugh' is a drainage ditch in Northern Ireland; they often mark a boundary line.

Fence posts these have been manufactured from rock, a good example of adopting local materials. 'Post-stone', a term from Fife, Scotland, is a hard fine-grained sandstone. Arkell and Tomkeieff (1953: 88) detail how the word 'post' is synonymous with 'bed', the layers of stone in the quarry.

'Post rock', as described in Muilenburg and Swineford (1975), is now a regional trademark

Figure 113: Caithness flagstone partition, showing brackets to fabricate a corner. The principles are easily adapted for other stone types and many other purposes, including bee boles. Spittal, Caithness, Scotland.

for north central Kansas where 'fencepost limestone' has been cut into posts since around 1880 (Figure 115). The posts are typically five or six feet long and 8 or 10 inches in cross section, weighing 350 to 400 pounds. Newly cut posts were seasoned in the sun to dry out and harden the stone. In the field posts were set 30 feet apart with three strands of barbed wire ('the devil's rope'), to make a stock-proof fence. A nail or wire wrap attached the barbed wire to the post.

Conry (2000: 73-74) describes granite fence posts in County Carlow and County Wicklow, Ireland. A typical example measured 1 metre to 1.34 metre tall, 12 to 16 cm wide on one face and 16 to 23 cm on the other. Five holes in the post hold the wire. The posts are set 2.5 to 4.5 metres apart with their longest edge across the line of the fence.

Hubbell (2006: 54) has an image of granite posts 'resembling Easter Island statues' from Higganum, Connecticut. The posts supported a wooden fence. From detail on the image, it seems two sets of wooden rails, one near the bottom and one near the top, were attached to the stone posts with bolts.

In southern India stone fence posts are made from local granite.

In Wales, and the English Lake district, posts are made from slate and schist slabs, with holes punched through for the wire. See also Figures 54 and 120 for Scottish and Irish examples of stone posts supporting wire.

Munday (2012: 49) includes an illustration, from South Australia, of a 'slate slab fence post [at] Willunga' with four holes drilled for fencing wire. The slab's width (estimated 8–10

inches) lies across the line of the fence.

Fence shadow subtracting stonework from the landscape can leave an impression. Murray-Wooley and Raitz (1992: 146) include an image of a double line of old trees growing beside a road. Trees had grown up on either side of an old stone fence. When the fence was removed it left two rows of trees, a yard or so apart. The authors call this a 'fence shadow'.

Figure 114: Cheekends at a gateway in a hill dyke.

Figure 115: Limestone fenceposts from Kansas, recycled for use in a vineyard. Walla Walla, Washington, USA.

Gap a hole in a wall caused by defect or damage or, in the case of **Aran gap**, a simple form of gate.

Gap stead / gap way an opening in a wall. Probably derived from an older English meaning of the word 'stead' meaning 'a place'. The term nicely translates to 'a place with a gap'.

Gate head / gate end where a wall ends at a gate. A gatepost should be installed before the wall is built up to it. This makes sure the stonework is not disturbed. Another term for the Scots **cheekend**.

Gate-slap (Scots) an opening in a dyke for a gateway.

Ginnel (Lancashire, North England) another term for a lintolled opening, known

Figure 116: Pawl stones. Prague, Czech Republic.

elsewhere as a 'lunkie' or 'hogg hole'.

Glenter stone (Durham and Northumberland, northern England) stones or stone posts set against the base of stone walls to keep cart wheels from hitting the base of the wall, particularly on corners or gateways. DSL has 'glent' as a 'glancing blow, a slap'. Also known as **pall** or 'pawl stanes' (Figure 116).

Gloystane (Orkney and Shetland) a projecting stone shelf on a barn wall. Oats were threshed against it. 'Gloy' is oat straw.

Glip / glippik / glipek / glipik / glippik (Scots) a small opening or broken part in the top of a dyke. DSL notes the term derives from Norwegian dialect *glip*: meaning '… a small channel-shaped opening, a crack.'

Greep / groop (Scots) a small drain, often running alongside a dyke.

Greesh (Scots) a stone abutment built against the gable wall of a cottage forming the back of the fireplace. DSL explains the need for this refinement '…so that the fire might be thrown forward a foot or so into the room…'

Grind (Shetland) a gateway through a wall.

Groop / greep a small drain running alongside a dyke.

Hallan / halland (Lancashire) a partition wall in a cottage, between the door and the fire, to keep out draughts. In Caithness, northern Scotland, the terms 'hallan stey'n', 'hallan stane' and 'halland stane' are used for a flagstone division in a byre, a division between stalls. Typically, it was one flake of flagstone set on edge.

Hawg hole (North Carolina, USA) a hole in a wall to allow pigs through.

Head (Yorkshire and Cumbria, England) the vertical end of a wall, at a gateway or formal opening, equivalent to the Scots **cheek** or **cheekend** built with headers and stretchers.

Raistrick (1973: 11), with reference to Yorkshire walls, describes abutting wall heads built at intervals along a wall to indicate divisions of the ownership of the walls. This was part of the obligations under the terms of enclosure in 1788.

Raistrick (1973: 2) includes a drawing of abutting wall 'heads' (conventionally built with headers and stretchers) with symbols carved into the stones at ground level to mark the change of ownership.

Rollinson (1998: 29) includes an image of a wall head, looking like a running joint, beneath a coped section of wall, with the caption — 'The visible line demarcates ownership when a common wall divides land owned by two parties. If the wall *rushes* (collapses), responsibility for repair can be determined.'

Hogg gap / hogg hole (English Lake District) an opening in a wall, maybe two feet high, to allow 'hoggs' (yearling sheep) through and hold back larger animals. The gap can be closed off with a boulder or flagstone.

Jostle stones (County Wicklow, Ireland) a local name for **pawl stones**. Described in Ó Maitiú and O'Reilly (1997: 76) as:

> … cut stones positioned to avoid damage by the hubs of cart wheels to gate piers or the corners of buildings. They are often conical or tapered in shape, although others are low and dome-shaped.

Keeping holes (County Wicklow, Ireland) nicely described in Ó Maitiú and O'Reilly

(1997: 78):

> Lintelled recesses built into boundary walls provided a convenient place for keeping tools or other items and occasionally, housed hens or ducks or perhaps a beeskep. These recesses are sometimes called 'keeping holes' or 'boles'.

Lonker / lonket (southwest Scotland) JSD has this as 'a hole built through dykes, to allow sheep to pass' and suggests it may originate from the Welsh words *llwnic* or *llwng* – 'the gullet'.

Mactaggart (1824) includes 'lonkor' with the same definition but admits 'I have no idea where this word is derived from.'

Loop holes (Lancashire, England) ventilation holes in a barn wall.

Lunky / Lunkie / lonky / lounkey / lunky hole / lunkehole / lunker / lonker (Scots) a lintoled opening in a wall to allow small animals or a watercourse through. Typically, it measured two feet wide and two feet six inches high, built slightly narrower at the bottom than at the top. The opening can be blocked with stones or thorns to restrict the movement of livestock (Figure 119).

Local shepherds know where these openings are, particularly in boundary dykes between estates. If a sheep strays into another estate, it is easily returned to its home grazing without having to search out a gate.

Constant traffic of sheep through a lunkie wears out a shallow depression. Derbyshire farmers, according to local wallers, put lime in this natural basin to keep sheep's feet healthy.

GVA Inverness (1813: 114) emphasises the need for walls to separate the winter grazing from the summer pasture, and provide openings to let sheep though when necessary:

> This fence… ought to be made of stone, and of sufficient height to prevent the leap of sheep, with openings of three feet high and two feet wide, at the bottom of the wall, in every 200 yards, and in a dry place, to let the sheep out and in, when it is needful. When once the sheep are accustomed to these avenues, they will not make such violent efforts to overleap and destroy the walls. These openings may be shut at the season in which the sheep ought to be excluded.

Borchgrevink (1982: 87) quotes Charles Scott Jardine, a Dumfriesshire drystone dyker, who mentions small lunky holes for hares, rabbits and drains. He says these openings are always called a 'lunky hole' in Dumfriesshire though others call them a **run hole** or **running holes.**

Lunt stanes Beaton (1999: 28) describes these as a flat projecting slab close to the door of a mill, high enough to assist when lifting sacks on or off someone's back. Beaton sources the word 'lunt' as a derivation of the Old Norse words *lint* and *lin*, meaning to sit down, rest upon or lean against. The feature is quoted as specific to two horizontal mills in Troswick, Shetland.

Beaton (2000: 70) writes of similar projections called 'porterages' brought to her attention on a property in Whitchurch, Shropshire. They are described by Madge Moran of Shropshire as 'two porterages: external shelves at waist height upon which porters could rest sacks of

Figure 117: Lintoled opening, temporarily blocked. Inis Oírr, Aran Islands, Ireland.

flour, corn and similar goods while loading or unloading.'

Ledges for resting loads exist worldwide, including on Himalayan trekking trails in Nepal.

Oddle-hole (Caithness, Scotland) a hole in the wall of a byre to allow drainage from the gutter. Despite its lowly purpose these holes and associated channelling can be a masterful combination of lintols and paving. Another Scots term for this is 'scuttle hole'.

Ogghole (Yorkshire, England) a lintoled opening in a wall for the passage of hoggs (year-old sheep) or other small animals, while keeping back larger or older animals. The word is clearly derived from 'hogg hole' by dropping the 'h'. This is a good example of local pronunciation being written down.

Pall / pawl stane a post or stone at the foot of a wall, at corners, to protect the wall by deflecting vehicle wheels. These can be quite ornate in formal streetscapes (Figure 116).

Also known as **glenter stones**.

Passthrough (New England, USA) a passage through a wall, with **cheekends** and maybe steps. A type of **stile**.

Pen / pend (Scots) the arch of a bridge. 'Pendstones', 'pend-stanes' and 'ring stones' are the stones of an arch. This clearly defines their function as stones lying along the curve of the arch.

'Pend' is also a Scots name for a small one arch stone bridge. Another common Scots usage for 'pend' is a covered alleyway or tunnel (usually arched), passing through or underneath a building to connect the street frontage to an enclosed courtyard behind.

Pen hole (Scots) a small opening at ground level in a dyke, for rabbits and other small creatures.

Phantom gates these are the 'ghosts' of gates. They exist in any landscape where the land use has changed over the years and the need for a gateway has passed. The wall ends are still there, the gap between is filled in with stone. (Figure 118). Unlike the **Aran gap**, detailed above, the infill is meant to be permanent.

Pier an Irish gate pillar. It can be round or square, made of concrete or cut stone. The author has seen gate pillars formed out of 45-gallon oil drums filled with concrete. The drums rust away, leaving the concrete core.

Pig holes openings found in the drystone walls in the Kona district of the Big Island in Hawaii. The openings allow the pigs a throughway; otherwise, they will tear into the wall.

Pinhead (New England) Thorson (2005: 80) refers to a pinhead as the widening of a wall

Figure 118: Permanently closed opening. St Kilda, Scottish Hebrides.

at an end – 'If well built, the pinhead suggests the intentional strengthening of the end for a barway or gate.' Sometimes this widening is a way of using surplus stone.

Pop holes (Durham Dales, north England) Adair (1998: 37) defines these as small rectangular holes at the foot of a wall, to let rabbits, hares or surface water through. He gives this as an alternative name for **rabbit smoots** and goes on to describe a 20 yard (18.28m) long stone tunnel, 8 inches (0.20m) wide, running alongside a Durham Dales wall, built as a shelter for rabbits in the winter months. The term 'pop holes' is also used in the Mendips, Southern England, for small openings for rabbits. These holes also provide ideal spots for snaring.

Poreen / puicket (Burren, County Clare, Ireland) a sheep pass, like a lunky. Most were undoubtedly built, notes D'Arcy (2006: 28-30), in the eighteenth and nineteenth centuries when large flocks of sheep were reared on the Burren.

Quoin end / quoinend a **cheekend**, where squared stones form up the vertical end. This is a term more likely associated with masonry when cut stone is laid on a mortar bed.

Rabbit smoot (England) a small hole to allow passage of rabbits, known as a **pen hole** in Scotland, a **pop hole** in England and **smoot** throughout Scotland and Northern England.

Run hole / running holes a small lintolled opening in a stone wall.

Sconcheon / scuncheon (Scottish Borders) another term for a cheekend.

Cairns (1975: 9) describes construction – 'A sconcheon has to be very carefully built and tied with the longest stones procurable and no 'wedging' in any direction.'

Scuttle hole (Scots) a hole in the byre wall through which dung is passed.

In Orkney and Shetland, the 'oddle-hole' is the hole where the drainage channel in a byre passes through a wall. 'Sharn hole' is another Scots term for this opening.

Shard / shard way 'shard' is an English dialect term for a gap or broken place in a hedge or wall, hence the Derbyshire term 'shard way' for an opening in a wall. Grose (1787) records the word as meaning 'a gap or notch'.

Sheep creep / sheep gap / sheep run / sheep hole / sheep smoose various names for openings in walls to allow sheep through (Figure 119). They can be closed at any time with a flagstone, a boulder or a pile of small stones. In modern times a wooden pallet, an old bed end or a sheet of corrugated iron serves the purpose.

The term 'sheep's nose' was collected from Vermont, USA. After discussion we realised this was a corruption of 'sheep smoose' or 'sheep smoot'. The term crossed the Atlantic, went from mouth to mouth, was probably never written down, and eventually emerged as something entirely different.

Slap / slape / slappie / slop / slaip (Scots) a hole or breach in a dyke or, as a verb, to make a hole in a dyke. A 'slap' can also be a piece of loose walling which is taken down for entry or exit of livestock, and immediately rebuilt. **Winter-slap**, a Shetland term for the seasonal opening of a dyke, is defined below.

> 'Saturday's slap' was once a general legal requirement in Scotland; salmon traps were opened from a set time on Saturday to a set time on the Sunday or Monday. This allowed free passage of fish and ensured a significant number of fish reproduced.

Slop stone (English Lake District) derived from a variation of 'slap' – a gap or hole.

These are vertically set slabs standing at either side of an opening in a wall. They have holes ready to receive wooden rails which form a gate. They are also known as **stoop** or **stoup** stones.

Smoo (Orkney, Shetland) a gap in a wall to allow the passage of sheep or a narrow passage between two fences. It can also mean a small sheepfold. The word can also mean to 'slink, sneak or prowl'.

Smoot / smoose / smout / smout hole (Scotland and north England) a small rectangular opening at the base of a wall. 'Rabbit smoots' are designed to permit the passage of rabbits and hares and water smoots allow water through. Smoots are smaller than **hogg holes** and ideal for trapping rabbits as they pass through the opening.

A 'sheep smout' or 'sheep smoot' is, of course, built bigger than a **rabbit smoot** – approximately two feet wide by two feet high.

Winchester (2016: 71) tells how these openings were used to control the seasonal grazing.

A record from the manor court at Ousby on the edge of the north Pennines in 1683 '… required the 'smoughts' to be opened yearly on 20 November and to be closed again on 1 March.'

Smugga / smoga (Shetland) an opening or passageway, a path between two fences.

Smyoo-hole (Orkney) a covered drain.

Snoot (Yorkshire, England) a passage in a wall for rabbits, used for trapping.

Squinch Adcock (2011: 30) includes a definition from the *Shorter Oxford English Dictionary* – '[a] straight or arched support constructed across the interior angle of a square tower to carry a superstructure' and notes the superstructure is usually a dome. Adcock also describes a squinch as a diagonal lintol or long stone, built across the corner in a field wall. It supports an overhanging section of cope and prevents sheep from escaping up and over the corner.

Staunchion (Scots) a **cheekend**, the vertical end of a wall for example at a gate opening.

Stiggy (Shetland) a stone stile. Graham (1979: 87) has this definition – 'a stile; a set of steps in a dry-stone dyke for use in climbing over.' The term has alternative spellings – 'stiggie', 'stiggi', 'stigga', 'steggie', 'stuggie' throughout Scotland for an arrangement of steps to allow passage over a wall.

Stile an easy way for pedestrians to get over or through a wall, using steps or a gap, while restricting the movement of livestock. A 'step stile' allows passage over a series of steps. 'Squeeze stiles' allow passage through a narrow gap in wall. A 'step through stile' allows passage over a section of wall built lower than the rest. In Kentucky, USA, the term 'climbing stones' is used for a series of projecting stones which act as steps over a wall.

Stillion / stillín / stellan (Ireland) a slate shelf in a house or dairy, used to keep milk cool or as a safe place to keep a bucket of water.

Stone crib / rock crib (Oregon, USA) in areas where good fenceposts are scarce this construction of stone, within a wire or timber frame (basically a gabion basket), is a secure, solid, replacement for the end or corner of a wire fence.

Stoop / stoup (English Lake District) an upright slab of stone set into the ground against the **wall heads** of a gate or stile (Figure 122). These are an ideal substitute when there are insufficient large stones to build a normal **cheekend**. The wall terminates up against the vertical slabs at each side of the gate opening. The stoop often has holes cut in it to accommodate wooden rails, which act as the gate. Metal hinges, bolted through the slab, are another way of supporting the gate.

In the Lake District, according to LDSPB (1977: 6) – 'Gateways are occasionally bounded by **slop stones** consisting of stoops having five or six holes spaced evenly down their

Figure 119: A lunky, a lintolled passageway for sheep. Fersit, Lochaber, Scotland.

length. Quite often the holes are round in one slop stone and square in the other. This allows the posts, which slot in to make a 'post gate', to be wedged firm and not revolve.' The North American equivalent is a **barway**.

Stop end a **cheekend**. A strongly built vertical end to a freestanding wall.

Thawl / thirl hole (Yorkshire Dales and Caithness, Scotland) a small hole to allow passage for water or small animals.

Thrip (Yorkshire Dales) a passage hole for sheep.

Trow (Shetland) a hole or passage through a wall or fence, a local dialect form of the word 'through'.

Twll defaid (W) literally translated as 'hole sheep' — a lintoled opening for sheep.

Wall head / wallhead an English term for a vertical wall end, known in Scotland as a **cheekend**. Figure 82 shows a good example from Kentucky. A pair of cheekends, butted together, and incorporated into a long stretch of boundary wall, indicate a change of ownership or responsibility for repairs. Wall heads, built occasionally on a slope, will strengthen the wall, and, like a throughband, limit the extent of any collapse.

According to Williamson (2002: 108), railways facilitated the import of regular shaped stones for wallheads. In his area of the Pennines the limestone tends to be small and irregular.

Wall-knocking (Ireland) the deliberate creation of a hole in a wall to allow animals into someone else's field, perhaps out of badness or to poach grass.

Water gap a gap in a dyke at a stream or river. Rails or gates can be slung from pillars or wall ends at either bank.

Water lane a wide space between two parallel walls, leading cattle down to water.

Water pen / waterpend an opening in a wall to allow for the passage of water, very like a passage for sheep or smaller animals. In Scots a 'pend' or 'pen' is an arch or vault, hence an opening with corbelling or a lintol.

Water troughs an important feature of the Aran Islands and the limestone country to the west of Ireland. Large slabs of rock (or several pieces mortared together) catch the rain in a trough and make it available for livestock before it disappeared into the porous limestone. This type of water collection system is also found in the Spanish Balearic Islands, another limestone region.

Weep hole / weeper an opening in a wall to allow water through. A small hole to allow drainage from, for example, behind a retaining wall. Drystone walls are porous, the natural gaps between stones should deal with all but the heaviest water flow. A drain is easy to incorporate, if required.

Figure 120: A wide wall with stone posts on one side, to support fence wires, and lintoled alcoves on the other side, for sheep shelters. Ballyknockan, County Wicklow, Ireland.

Figure 121 (below): A closer view of an alcove seen in the centre of Figure 120.

Figure 122 (opposite): Stoop stone, with signs of previous fittings. Cumbria, England.

Whigmaleerie (Scots) a derogatory term for a piece of ornamentation in stonework.

Winter-slap (Shetland) a gap made in the **head dyke** to allow livestock free access to the arable land after a summer on the common grazings. In the spring crops were sown and livestock returned to the common. Hibbert (1822: 203) gives some detail of the system in Shetland, which had existed to protect the arable ground:

> One of the greatest detriments to the agriculture of Shetland, is the wretched state of the fences. Nevertheless, very ancient laws have existed in the country for the protection on the inclosures destined for cultivated land. It was ordered that all dikes should be in

GRV

sufficient repair by 1 March, … and should any live stock between this day and the 1st of May tread upon the ground of others, the penalty for each swine should be 10s Scots; for each sheep 2s; and for each sheltie 6s besides payment of damages. After the 1st of May, these penalties were doubled: at the same time, every winter-slap left open, every neglect of closing a grind or wilful act of breaking down, or even scaling a dyke, was liable to a fine of 40s Scots. The penalty for swine observed to actually pasture upon the lands or commonty of others, was £10 Scots, beside damages. All these good regulations, which would be highly valued, when agriculture was less oppressed by feudal duties, are now much disregarded, and the imperfect dikes that are constructed of turf or stones, easily yield to the repeated assaults of shelties, sheep or swine.

The reader will note the heavier fines for shelties (Shetland ponies) and pigs; they caused much more damage than sheep. Leaving a gate (a grind) open was punishable by a fine. It was important to keep the head dyke secure over the summer to protect the growing crops.

Cattle are not mentioned; presumably they were herded by small boys over the summer to ensure neither boy, nor cow, got into mischief.

Wirlie (Shetland) a place where a field wall crosses a stream, an opening in a wall to let water through.

4.6 Cairns, boundary stones, heaps and markers

Boundary markers range from piles of stones to finely carved monoliths. The simplest form of stone marker is a heap, as noted in Corangamite (1995: 44): '…boundaries were often indistinctly marked by a plough furrow …or by notches on tree trunks …astute men like Neil Black took to marking their boundaries with 'heaps of stone' along the plough furrows.'

Barp (Scottish Hebridean Islands) immense heaps of stones which were burial cairns, barrows or chambered cairns.

Borestone / boarstone / bore–stane / bore-stone (Scots) a stone drilled to take a flagstaff, one of a series of boundary stones marking the edges of a burgh or town.

Bunker (northeast Scotland) a large heap of material, for example stones or clay.

Bourach / bowrock / boorick (Scots) a small heap of stones, a confused heap of anything. Imagine a 'bourach of hens' at feeding time.

Cairn / carn / carne / cyarn / kairn a pile of stones, shaped like a cone, pyramid or with vertical sides, built as a memorial or to mark a summit or path.

On the Kennebeck River in Maine, USA, cairns were built midstream to support chains, which held back logs until they were gathered and rafted downstream.

In Newfoundland fires were lit inside cairns as a guide to ships.

Cairns on the Caithness coast, in northern Scotland, acted as navigation aids. If a pair were aligned the seamen knew the safe angle to take when approaching a harbour entrance.

Gaelic legends include references to malefactors and suicides buried under cairns.

Mactaggart (1824) defines 'cairns' thus:

Hillocks of stones; whiles they are built with care, but oftener not; they are common on

Figure 123: The urge to create small cairns is irresistible.
These overlook Loch Loyne, northwest Scotland.

the tops of hills; a number of them were built prior to dykes, as marks, to shew *lairds* the *marches* or boundaries of their possessions. In ancient burial grounds they are met with too; our ancestors, like the natives yet of every uncivilised country, huddled a number of stones on the graves of their dead.

'I'll throw a stane on your Cairn,' was a phrase used in the days of yore, and expressive of a kindness that would be shewn persons even when dead – by throwing a stone now and then on to the *cairns* on their graves. Chieftains of course then, and people of note, would be honoured with the largest *cairns*.

The idea of the cairn as a pile of stones representing the fatigue of travellers is explained in Williams (2012: 11, 82) – 'In many Native cultures, tired travelers spit on a small stone and added it to a trailside cairn as a way to transfer and eliminate fatigue.'

Scottish and Irish cairns are sometimes associated with the supernatural and stories of fairy folk living underground. It is bad luck to upset the 'wee folk'. With proper precautions a human can go underground to enjoy their hospitality, and escape.

A 'funeral cairn' was one erected in memory of the deceased by the mourners at a funeral. It might be at the burial site or at a favourite viewpoint, like one erected by Queen Victoria in memory of her husband, Prince Albert, near their Scottish summer home at Balmoral, Aberdeenshire.

After a wall is finished, the left-over stone was often carted to the next section or was built into low cairns for use later.

Cairn circle an archaeological term for the ring of stones surrounding a cairn.

Cairn-master in the early-twentieth century climbing clubs had a 'cairn master' to mark the route up a hill with heaps of stones.

Cairney (Scots) to pile up stone, perhaps at the side of a dyke or a stream.

The term also applies to stony ground, as detailed in an 1804 reference in DSL — '…it was very difficult to labour; it was wild, balky, stony, cairney, and the furrow ill to clear.'

JSD has 'kairney' as a small heap of stones, a diminutive of 'cairn'.

Cairnfields groups of small cairns, sometimes dozens in a group. Most likely the result of agricultural improvements. Some sites have been attributed to ancient religious practices.

Carnell (Scots) a heap, a diminutive of 'cairn'.

Claddach / cladach (Scots) a small heap of stones. To cover with gravel or stones, as a river does during a spate.

Clawin post / clawin steen a rubbing post for cattle. Farmers often instal one, or allow cattle access to an ancient monolith conveniently standing in the field. From a Scots meaning of 'claw', to scratch or rub (Figure 125).

Clearance cairns / consumption piles a collection of stones, often of considerable size, with nicely built sides, to dispose of stone cleared from the land, like a **consumption dyke**, described in Section 4.1.

Dutton (1808: 152) writes of these in County Clare, Ireland:

> Great exertions are made by some gentlemen, but more by cottiers, in clearing their land from stones; for this purpose walls ten feet thick are not uncommon, and pyramids of stones of a large size remain as monuments of Irish industry, and not, like the pyramids of Egypt, everlasting proofs of human folly.

Clearing stones to build cairns can have adverse effects on flora and fauna because this removes shelter and habitat from the landscape, although, according to Williams (2012: 45-46), the cairns themselves create a microclimate which is to the advantage of some plant and animal species.

Man's enthusiasm for cairn building cannot be curbed; Williams (2012: 16) quotes a ranger in the Acadia National Park in Maine – 'Most hikers think they are seeing a natural landscape, but everything is now in the cairns.'

Corangamite (1995: 35, 68, 115) describes how Australian paddocks were cleared:

Originally the stones were gathered into heaps for the purpose of cleaning up the paddocks and acting as fences but the coming of the rabbit meant that great care had to be exercised in the construction of well-built walls.

Löbbecke (2012: 42) details similar stone clearance in Europe: 'The overabundance of stones often necessitated not only the building of useful structures, but also the creation of supplementary stone depositories. They were also surrounded with boundary walls or piled on large rock slabs that could not be removed.'

Löbbecke (2012: 17) also details how derelict corbelled structures found a new use — 'The now superfluous shelters have often been covered up with additional stones in order to clear rocks from the fertile soil.'

Löbbecke (2012: 282) further describes a more sophisticated way of stacking 'superfluous stone', in two *muragghi* near Trebastoni in Sicily. These are tall, well built, solid towers, with external steps, allowing them to be used as viewpoints – 'The carefully assembled hewn stones of the high outer walls surround large amounts of loose stone, thus forming a space-saving depository.'

Alomar et al (2002: 303) details two attractive features in Mallorca, Spanish Balearic Islands, labelled 'stone storage structures'. The outer faces of both structures are well built of dressed rough stone. The inside is filled with stone. The *galera* has typical dimensions of 47.50 metres long by 3.05 metres wide by 1.15 metres high. The smaller *claper* is a round structure with typical measurements of 4.90 metres diameter, total height of 2.15 metres. A rabbit trap was sometimes built into the structure.

Clien (Galloway, southwest Scotland) a small heap of stones.

Coffin road a right of way along which a coffin was carried to the burial ground. The 'resting stane' was a conspicuous stone or cairn on the route where the coffin was laid while the bearers rested.

In northeast Scotland this was called a 'leckerstane' or 'lickerstane'.

Cooling stone (Angus, Tayside, Scotland) a stone in or near a school; a boy who had been whipped was sent there to cool himself. This is mentioned in J M Barrie's 1891 book *The Little Minister.*

Cross (Tweeddale, Scottish Borders) a pile of stones on a hilltop. DSL tells us 'These piles of stones are also termed cairn, pike, currough, cross etc'. Shetland has a varietal spelling of 'kross'. 'Currough' may be a different spelling for **currick**. DSL tells us a 'currough' is an old Scots word for 'a coracle, a small wickerwork boat covered with hides.'

Curricks tall cairns found in the North Pennines, northern England, reputedly used as guidance markers by shepherds. 'Carrick' is the anglicised version of 'creag' or 'carraig', a Scots Gaelic words for 'rock'. A currick is also a term for a small drystone enclosure used by shepherds as lookout points, presumably when guarding their sheep.

Dool / doel (Suffolk and Scotland) a boundary mark in an open field. Conspicuous rocks, ditches or trees in the landscape were important boundary markers before the widespread building of walls.

Garsty / gersty / gorsta / gorsty (Orkney) a ridge of stone and earth, perhaps the remains of an old dyke, left unploughed, to confirm a boundary.

Gauk stane / gowk stane (northern Scotland) a solitary stone. 'Gowk' is Scots for a foolish person, an idiot. The 'gowk stane' is therefore a stone which stands alone, looking awkward in the landscape. It may be a glacial erratic or, perhaps, an old Pictish standing stone (Figure 125).

Gravestone an erect inscribed stone which stands at the head of a grave.

A small stone is left on a gravestone is a message for the next visitor, to let them know someone has been there. Gravestones can be ornate with family crests etc; simple stones have been used in place of the ornate.

Haggle (Shetland) a stone set up as a boundary marker, perhaps of quartz, something easily seen.

Hagmark / hagmet (Shetland) a mark or boundary stone marking between two hill pastures.

Hairstane (Aberdeenshire) a large grey moss-covered stone, a conspicuous stone or stones used as a boundary mark. Also recorded as 'hoar-stone' in England.

Hatter (Scots) as in a 'hatter of stones', a heap of stones, a confused heap of things.

Hot / hott (Scots) a small heap of stones, gathered carelessly into small heaps, for carting away. JSD records **haut** from Ettrick Forest in the Scottish Borders, as – 'properly, to gather with the fingers, as one collects stones with a garden rake.'

Humplock (Renfrewshire, Scotland) a small heap of earth or stones.

Hurrock a north English term for a heap of loose stones or rubbish.

Inuksuk (plural *inuksuit*) originally an Inuit name for stone structures found in northern Canada, used as markers for sites of spiritual interest, as indicators or guides. In the simplest form they are a tall pillar of stone, a pile of several flat pieces or a flat slab set up on smaller boulders.

Figure 124: Memorial cairns. Cille Choirill, Roy Bridge, Scotland.

Readers will be familiar with 'Michelin Man' figures made of flat and rounded stones which appear on many modern Canadian logos. These are cartoons of Inuksuk which are, in reality, more complex, ancient structures. The subject is well covered by Norman Hallendy, see bibliography for full details.

Keuss (Orkney and Shetland) a pile, a heap, a mass. JSD has this as a derivative from an Icelandic word — '…*kös* – a heap, a pile, as of stones, blubber &c, from *kasa*, to heap earth or stones upon, to earth, as was done to witches, miscreants and the bodies of outlaws.'

Kellach (Scots) anything built high and narrow, from the name of a large conical wicker basket.

Kelp grids (Northern Ireland) boulders set in rows at the sea edge as a growing platform for seaweed which was harvested on a three year cycle. After the seaweed was cut the rocks were turned, ready for new growth. Seaweed was used as a manure or burnt to produce ash used in several industrial processes. Also known as a **plan** in nineteenth-century County Down, Ireland.

Kyerro (Orkney) a heap of stones, for example a collapsed ruin.

Lowping-on stanes defined by Mactaggart (1824) as 'Stones common about country inns for getting into carts, or mounting horses from.'

JSD has 'Uplie-stane' – 'the uplifting or leaping on stone. A stone in the form of a step to assist getting on horseback.'

March stane / mairch stane / marchstane (Scots) a boundary stone.

'To fitch a marchstane' meant 'to make a slight change in the situation of a landmark'.

'Mairch-stane' is a Shetland variation. *Clach-chriche* is Scots Gaelic for a marchstone or landmark. See also **wrang dyking**.

Markstane (Galloway, southwest Scotland) stones set up on end to mark the marches, the boundary line, of farms and estates. The stones often have marks or initials cut into them to indicate ownership.

Menhir a Breton or Cornish word for a solitary standing stone, an upended boulder, a monolith. They can appear in groups or singly, mostly in western Europe. One of the most outstanding alignments of menhirs is at Carnac, Brittany, France, where more than 3,000 stand over a 4 kilometres (2½ miles) distance. The various alignments are possibly over 6,000 years old.

Mett / meith (Scots) a boundary stone or, as a verb, to measure or mark a boundary.

'Land meithing' was an old Scots term for the fixing and marking of boundaries.

Milestone a stone marker, used to indicate miles to or

Figure 125: St Demhan's Cross, a Pictish stone. Creich, Sutherland, Scotland.

from another place.

Monolith a slab of rock set vertically into the ground, or a large stone standing alone.

Obelisk a stone pillar with a square or rectangular cross-section, a pointed pillar. The Washington Memorial in Washington DC, USA, is a modern example.

Orthostat an archaeological term for a large stone or slab set vertically.

Wildgoose (1991: 215) identifies 'double orthostat walls', a double row of vertical slabs, two metres apart with a filling of small boulders and rubble, at Roystone Grange, Derbyshire, England.

Parch an old Scots term for a pile of rocks at the side of a river or harbour entrance, used as a navigation guide.

Pike (southern Scotland and Northern England) DSL provides a definition – 'A pointed pile of stones, a cairn.'

Pillar stone (Ireland) a standing stone.

Pirrack (Perthshire, Scotland) a heap or pile of stones, a cairn built on top of a hill.

Pit-stane / puttstane (Scots) a boundary stone, a march stone.

Plan (nineteenth century County Down, Ireland) a seashore site where stones were laid out as a growing surface for seaweed. Also called **kelp grids**.

Prap / prop (Scots) a heap of stones placed as a boundary marker, memorial or grave marker. The stones are 'propped' up against each other.

Rangel a nineteenth-century Scots term for a heap of stones.

Raughle (Scots) a rough heap of stones, a wall loosely built without mortar and appearing ready to fall, a **rickle dyke**, as described in Section 4.1.

Reassurance markers a term used by The US Department of Agriculture Forest Service in their publication 'Trail Construction and Maintenance Notebook' for confirmations of a trail route. Various methods are recommended, including 900 mm (35.5 inches) high stone cairns. They go on to say that rock ducks (a small pile of two or three stones with one pointing the way) '…are no substitute for cairns and should not be built.'

Ronnie / runi (Shetland) a heap of stones or prominent rock on a hillside.

Roo (Shetland) to pile up into a heap, a heap of any kind.

Ruckle (Yorkshire) a loose heap of stones.

Rubbing stones large earth-fast stones, naturally occurring or installed by farmers, for horses and cattle to rub on. Called a **clawin post** in Scotland.

Sodger (Fife, Scotland) a soldier, a standing stone or monolith.

Stake and stones / post and stones corner markers (Vermont, USA) a post embedded in a cairn of stones to mark the corner of a property line.

Standin'-stane a Scots form of the English term 'standing stone'. A **monolith** or **menhir**. The term also appears as 'standand-stane' in southwest Scotland.

Steeve / stieve (Scots) immobile, as a rock in the ground. From first sight the word suggests a connection with the christian name 'Steven'. DSL makes no such link and suggests a connection with the word 'stiff'. Saint Stephen, apparently the first Christian martyr, was stoned to death.

Sterlich DSL has this as an Aberdeenshire term from 1932 meaning 'a small amount, heap, etc of anything.'

Stone Johnnies tall narrow cairns built by shepherds, many of them Basques, in sheep

Figure 126: An old style of walling. Orthostats with smaller stone infill. St Kilda, Scottish Hebrides.

herding parts of the USA. Williams (2012: 116-117) quotes a sign which advises – 'These stone columns were probably built to indicate distance and direction to waterholes and provide the sheepherder with a pastime while tending his flock.' The name, according to Williams, may originate from the Basque *harri mutilak* meaning 'stone boys'.

Tappietourie / tappin / toppin / toppeen (Roxburgh, Dumfries, Galloway, Scottish Borders) a cairn on a hilltop. Scots also has 'tappiloorie' for anything raised high on a slight or tottering foundation.

Tor a rock formation, notable in Dartmoor, southwest England, looking like a stack of weathered slabs or a pile of biscuits.

Tourock / toorackie (Scots) anything rising high or to a point, for example a pile of sand or dust.

Trip (Orkney and Caithness) a pile of stones set up as a division on a beach in the days when access to seaweed, fish and shellfish was important. This follows the sense of 'trip' as a break or obstruction along a route.

Tryst-stanes (Scottish Borders) a stone or stones used as a mark for a rendezvous point.

Tsunami warnings along the Japanese coast there are hundreds of stone markers, some 600 years old, warning to build above the flood line. One stone marker in the coastal city of Kesennuma, northeast Japan, warns 'Always be prepared for unexpected tsunamis. Choose life over your possessions and valuables.' Another typical warning asserts 'Do not build your homes below this point.'

These ancient warnings were given much publicity in the media after the tragic earthquake and tsunami off the east coast of Japan on 11 March 2011.

Vauch / wauch (Scots) a bank or mound of stones, set up in a river as a kind of salmon trap.

Winter-steen (Shetland) DSL defines this as:

> …a standing stone or large boulder visited on the first winter moon by an unmarried person wishing to divine their future spouse by walking round it several times clockwise and anticlockwise while reciting the names of desirable candidates.

Witter hole / witterstane a mark or depression made in a witter-stone, the hole distinguishes the stone from ordinary big boulders or rocks. A 'witterstane' is a boundary stone. This follows a meaning of 'witter', according to DSL, as something serving as identification, an identifying mark or token.

Wrang dyking (Selkirk, Scottish Borders) literally 'wrong dyking' — to falsify boundary markings or put them out of alignment. DSL quotes an old reference in The Border Magazine, (1939: 46-47) referring to the 'riding of the marches' (an annual circuit of the boundary) when any encroachment or illegal building on common ground was dealt with, immediately and thoroughly. The piece mentions a man having had the temerity to build a house on the town's common and '...the whole community pulled it down.'

Youffing-stone (Scottish Borders) a boundary stone, a stone on which participants in a Riding of the Marches ceremony were 'bumped' (a kind of ritualised assault) as an emphatic reminder of where the boundary stones were. This ensured the whereabouts of the stones were known to succeeding generations.

The term 'perambulation' was used in Britain and New England to describe travelling round the boundary of an area, along existing walls or paths, to confirm ownership rights and responsibilities.

4.7 Paving, steps and cobbles

There is a rich heritage of stonework ensuring paths and roads are weatherproof and permanent. Most of this stonework dates from before the introduction of concrete and asphalt paving.

Bishop (Scots) a heavy piece of wood used by pavers to level the slabs, an instrument for ramming down stone or earth. A 'priest', a lower rank of clergy, is a smaller club-like instrument used by anglers to kill fish with a sharp blow on the head.

Blin (Scots) to pack smaller material between larger stones forming the bed of a road or path. The equivalent English term is 'to blind'. The term comes from an old meaning – 'to close up'. 'Blinding' is the act of applying smaller material, for example, dry sand brushed into cobblestones.

'Filled to refusal' describes when no more material, perhaps sand, can be added to larger material. The spaces between the larger stone is filled.

Bluestone (New York and Pennsylvania, USA) a type of hard sandstone used for paving. It comes in shades of blue, grey and buff. Cows and rock are sometimes called 'blue'. This means a shade of green/grey rather than the colour of the summer sky.

Border stone a flat stone, used on edge, to establish the side of a path and hold cobblestones or paving in place.

Boulder stane (Scots) a paving stone. The use of rounded stones seems counterintuitive. They are laid tightly together, length into the ground, to form a solid, though uneven, surface. The production of regularly cut stones, either as flagstones or granite setts, made it easier to lay a good flat surface.

Bouldering (USA) paving with cobblestones, naturally occurring rounded stones.

Brig stane / briggsten / brig-steen / brigstane / briggiestane(s) / briggiesteen (Scots) a pavement of flagstones; a platt of flagstones laid in front of the door of a house; flagstones laid over a gutter; a stepping stone.

Broad-stone(s) Arkell and Tomkeieff (1953: 13) define these as paving slabs, in an extract from a 1703 Builders' Dictionary, 'so called because they are raised broad and thin out of the Quarries… not above 2 or 3 inches in thickness'.

Bull nose the rounded front edge of, for example, step treads.

Casey-bool (Scots) a rounded boulder used in paving.

Cassie / cassy / cassay / cassey / casey / casy / cassa / causey / calsy / cawsey a cobbled street or pavement, the cobbled part of a cow byre, the paved portion around a house door. Cobbled areas have a rougher surface than paved areas made of flagstones. We can see how the word 'causeway' is connected. The term can also apply to individual cobblestones. The variations are a good example of local spelling, in the days before standardised English.

'Cobblestones' outside the Dundee College of Art and Technology in Reform Street Dundee were made of wood, so that Victorian-era students were not disturbed by the noise of horseshoes on stone.

Cacicr is an old Scots term for a paver, one who set cassies, and a more modern name for a maker of 'cassies' or cobblestones. DSL defines 'cassie man' as a stone mason who specializes in shaping cobblestone for roadmaking and includes an extract from *The Scotsman* newspaper of 19 January 1991, page 13 – 'His father, James Bissett, master cassie man at Fyfe's quarry, had gone to Odessa to teach the Russian quarrymen the refinements of modern granite dressing.'

Cassie dunter (Scots) a heavy piece of wood used to 'dunt' or level paving stones by

Figure 127: Stone pitched, long edge into the ground, to protect shoreline. Sarclet, Caithness.

knocking them down. The wood will cause far less damage than impact with a hammer or another stone. A rubber mallet is a modern equivalent.

Causey-stane / causey-steen (Scots) cobble stone, a granite sett. As a verb – to pave with small stones, driven with their long edge into the ground. Strictly speaking a 'cobblestone' is rounded and a 'sett' is cut to a shape.

Channel stane (Scots) a paving stone.

Clach-urlair (SG) paving stone.

Clachans / clauchans (Argyll and Bute, Scotland) stepping stones. The Irish word *clochan* can also mean a stepping stone, as well as a corbelled **beehive hut** structure, detailed in Section 4.1.

Clacharan / clacharain (SG) a pavement, stepping stones across a river or boggy ground.

Cobble a rounded stone used for paving, set with its length into the ground.

Cobble (New England, USA) Thorson (2002: 117) identifies this as a piece of ground, often at the corner of a field, where horses gathered; a piece of compacted ground where the topsoil has been washed away, leaving the stones exposed.

Coggle / cogglestone a rounded waterworn stone, especially of the size suitable for paving, a cobble, also known as a 'copple-stone'. In Scots something which is 'coggly' is unstable and easily overturned.

Cornizer (Scots) described in Pride (1996: 26) as a 'thin offcut from whin setts inserted vertically in the ground to form a non-slip pathway'. 'Setts' are hard basalt-type rock cut into rectangular paving blocks. 'Horonising', from McMillan (1997: 60), is a similar process. The image on the page shows a mason preparing squared granite setts with the narrative that the offcuts could be used for pinning walls or '…in the practice known as horonising, where fragments are set tightly on edge…'

Corrying the cobbling or paving on the actual river bed, under the arch of a bridge. This reduces damage from fast flowing water. This term was used by Norman Haddow, a dyker from Crieff, Perthshire, Scotland, describing a bridge he built.

Munday (2012: 46) details 150 metres of stream bed in South Australia which had been stabilised in the early 1900s, with offcuts laid with their edges into the stream bed and their long edge along the direction of flow.

Culver (Somerset, southwest England) Arkell and Tomkeieff (1953) note this as 'A blue stone used for steps…' They do not identify the rock type; it is probably a hard limestone. A 'culver' is a woodpigeon, a grey bird. The term 'blue' is often used for shades of grey.

Curbing stone used for edging pavements, plant beds or a kerb. From 'curb' in the sense of enclosing or holding back.

Door stane / dorestane (Scots) the stone or flagstone in front of the threshold.

Dumplings / Norfolk dumplings big flint cobblestones used for paving streets. Norfolk is an eastern county of England.

Flag flair (Scots) a flagstone floor.

Flagging a collective term for flagstones, a surface paved with flagstones or the process of setting flagstones.

Flagstone a flat stone, a general name for any flat stone used for paving, typically from limestone or sandstone areas where sedimentary rocks break into sheets and require the minimum of finishing to get a flat surface.

Flatwork paving and such work done on the flat, with no vertical plane to it. The skills of breaking the joints and proper placement of stones still apply, perhaps even more so, bearing in mind the pressures the paving will be subject to.

Floorstane (Scots) literally a 'floor stone', a hearth, a stone on the floor to support or contain a fire.

Gutter stane (Scots) a kerbstone.

Holloway (Northern England) a walled track, an ancient trade route.

Horonising this is detailed under **cornizer**.

Kaffer stein (Switzerland) upright stones set every two or three metres to mark the edge of a path when covered with snow. According to Collins's *German Dictionary,* second edition, 1991, *'kaffer'* is a thickhead or a duffer. These stones might be so named because they warn the simpleton of the edge, or because only an idiot would stand there.

Knapper / stane-knapper one whose job was to break down large stones to a size suitable for roads. To some this was a form of punishment. To others it was a recognised way of making an income. Stone was dumped by farmers in designated places by the side of the road and the knapper sat there, at work.

See **knapper** in Section 5.1 (stoneworkers) for a full description.

Knap of the causey (Scots) the middle stones of a street, the crown of the road.

Lippin' stane (Scots) a stepping stone.

Mitchels / Purbeck Michels (Isle of Purbeck, Dorset, England) noted by Arkell and Tomkeieff (1953: 76) as limestone paving stones about 18 inches square, probably named after Humphrey Mitchells, Clerk of Works at Windsor Castle in the sixteenth century under Elizabeth I.

Nidger Ryan (1992: 18) differentiates between 'nidgers' which were produced in Welsh quarries and 'nudgers' which were produced in Scottish quarries. Both were large sized granite blocks with smooth and regular faces. They were treated carefully because 'They were used to pave the more important streets in the cities.' Setts (smaller squared off granite blocks) on the other hand, were roughly handled and dumped in the ships' holds.

Padstone the top layer of stones on a house wall, the ones upon which the roof timbers lie. The term is also applied to stones at ground level beside the first step of a stile, a flagstone set into a lawn or the first step inside a gate. These places are subject to wear from foot traffic; the padstone helps keep the surface dry and clean.

Palladiana paving flat paving made from rough slabs of stone, pieced together, usually with very tight joints. Commonly known as 'crazy paving'.

Pament / payment (old Scots) pavement or paving.

Panged (Scots) describing something tightly side by side, for example cobblestones.

Parpoint (Yorkshire and Lancashire, England) a thin stone, used for interior walls and paving.

Paver (Scots) a paving stone.

Paver (Colonial USA) a person who laid paving stones or one who maintained pavements.

Paving a generally accepted term for laying a smooth flat surface such as flagstones, cut stones or brick. **Pitching** is different: it produces an uneven surface because it uses rougher material.

Pentil steen / pentilsteen / pendle stone (northeast Scotland) a flagstone, a Scots version of the English 'pendle stone', a name for shaley limestone.

Peters (West Yorkshire) Arkell and Tomkeieff (1953: 84) define these as 'Flagstones of unusual width, either wider or narrower.'

Pitching the process of taking natural stones and packing them tightly together in a trench or channel to form up a path or roadway (Figure 128). The stones are laid, length into the ground, and tightened up with shims or rough sand. Pitching takes a path up a slope by raising the level of each succeeding line of stones a fraction. A formal step is sometimes necessary every few yards, to achieve the gradient. A pitched surface is never completely flat, unlike a paved surface.

Pitcher defined in Arkell and Tomkeieff (1953: 86) as '…a stone used for paving; a sett. To pitch is to pave, especially with small stones set up edgeways. From the Middle English – *piche*, to fix, erect, place, &c.'

Plainstane / plain-stane / plennies / plettie stanes (Scots) a paving stone, a flat flagstone or the pavement itself.

Plank / planking Arkell and Tomkeieff (1953: 86) define this as any '…coarse flaggy stone that can be used for paving…' with the note that 'planking' is '…a verbal substantive meaning 'flooring' known since 1495, from [Middle English] *planke*, a slab.'

Quarrel (Scots) to pave with square shaped slabs. The word has other meanings associated with quarrying and roughly shaped stone.

Figure 128: Pitched stone ramp. Deia, Mallorca, Spain.

Road-stone a common name for inferior stone used for repairing roads, in the days when roads were made of compacted stone. Many of these old roads are still viable although they suffer under the weight of modern vehicles.

Ruckly a Shetland term describing an uneven surface, as in paving.

Saddle a strip of stone at the door, raised slightly above the floor level to improve weatherproofing and discourage water ingress.

Setts hand-cut rectangular stones used for the top surface of roads. A typical sized sett is 8x4x5 inches (20x10x13 cm). Each was cut from granite, basalt or similar hard rock, using hammer and chisel. 'Setts' and 'cobblestones' are regarded as synonymous. Strictly speaking setts are cut and cobbles are uncut, or naturally rounded.

In a letter to *The Scotsman,* on Friday, 6 April 1906, a cabman wished the Society for the Protection of Cruelty to Animals had spread sand on the streets to stop horses slipping on the cobblestones during the long dry season.

Stanner-steps stepping stones in a river.

Stap stane / step (Scots) a stepping stone. 'Step' is another Scots term applied to a stepping stone in a river.

Steps an 'average' formula for steps emerges from various sources. The minimum width (from front to back) of a step should be nine inches (23 cm). Twice the height of the risers (the vertical distance between each tread), plus the width of the step should be around 26

Figure 129: Whaligoe Steps, Caithness, Scotland.
Approximately 330 stone steps wind down to the old fishing station.

inches (66 cm). Those who build with natural stone will realise such a formula is only a guide. In the drystone world the mason will have to work with whatever slabs are available (Figure 129). Superstition says a flight should have no more than 12 steps.

Terrazzo a type of ornate stone mosaic flooring where small pieces of tile or stone are laid into mortar and polished.

Thrashel-stane (Scots) a stone threshold.

Tread a flat stone, the horizontal walking surface of a step.

Trink (Caithness, northern Scotland) a long narrow flagstone.

Trod (England) an ancient footpath with stone flags or occasional paving to ease travel across boggy ground. Trods existed in medieval times along trading and pilgrimage routes at a time when most goods were carried on peoples' backs or on pack horses.

Urrs (Leicestershire, England) particular beds of limestone used for paving floors.

Wading stone / riding stone DSL provides us with a definition from Perthshire, Scotland, for 'riding stone' – 'a stone in a river at a fordable place serving as an indicator of the depth of the water for those intending to cross on horseback.' DSL goes on to mention 'two natural stones in the middle of the river [Isla], called the riding stone and the wading stone.'

Warped a description of a rough flagstone which requires tool work to level it off.

Yorkshire stane / York stone a fine-grained type of sandstone especially suitable for flat paving stone. It is a similar material to Caithness flagstone.

5. WALLERS AND DYKERS, OUTPUT AND INCOME

5.1 Stoneworkers

Dykers were an important part of the agricultural community, and remained so into the twenty-first century when the bulk of their work was maintenance of field boundaries built a hundred and fifty years before. Dykers were granted an extra cheese ration in 1942 because they worked away from canteen facilities. In the 1950s, they were allowed deferment from National Service, along with other farmworkers. Their skills were needed to increase food production.

Cairns (1975: 21) describes the hard life of dykers before World War I:

> They got to work about 11 or 12 years of age and really couldn't lift a heavy stone, so by the time they got to 60 or 70 you will find they had a pretty rough passage, and no wonder they got a bit bent with the wettings they got in the back. Then a lot of them like my father lost their eyesight through stooping and heavy lifts. Then again they all drank as much as they could afford and had great fun at the local inns as long as the money lasted. There is a dry dyke opposite the Crook Inn, Tweedsmuir, opposite side of the road, that was built by drunk dykers, each doing a bit when on the 'skite' ...and it cost the landlord nothing, and it's a lovely piece of work... On the whole they could have made their lot a bit happier if they had looked better after their money and their health, but that wasn't the way in those days.

Apprenticeship and training for drystone dyking was not as formal as in other trades. Traditionally most tradesmen had to 'serve their time', beginning as a general labourer. Over time apprentices gained experience and were given more responsibility, until they were judged to be 'time served' and qualified in all aspects of their chosen trade.

Cairns (1975: 22) thought there was no definite 'time' a lad had to serve as an apprentice dyker in the Scottish Borders; it was up to him to learn. When he was good enough, he worked in a pair for a full share of the pay, otherwise he worked by himself and was paid by the yard.

A dyker's apprenticeship was short by comparison to other stone trades. In the Ballachulish slate quarries, according to Fairweather (1988: 8), – 'It was considered necessary to serve a seven year apprenticeship before qualifying as a slate worker, dresser and splitter.'

In the nineteenth century, a stonemasonry apprenticeship might take fifteen years; it took a long time to learn all the intricate elements of the trade. After serving a full apprenticeship the mason would be classified as an 'improver' for three years. Nowadays a stonemasonry apprenticeship is three years for standard, four for an advanced craft qualification.

Borchgrevink (1982: 80-82) quotes Charlie Jardine, a dyker with 50 years' experience:

> If you're with a good dyker it takes three years for tae learn the job and be able for to do as much as what'll make a day's work pay — so much a yard.

Charlie goes on to detail how the young lads started with stripping the dyke down, prior to repair, graduated to placing the hearting, then worked up to building – always ahead of the main men who kept up the pace, kept the youngsters interested and corrected any error.

Corangamite (1995: 91) includes a dissertation by Rod McLellan on the nineteenth-century stone walls in the volcanic area west of Melbourne, Australia. Trainees started as labourers:

> Wallers worked variously in teams, pairs or alone, with apprentices serving at least two years collecting stones before they would be allowed to work by themselves.

Margaret Brooks (1977: 97) gives more insight into young dykers:

> People who have grown up amongst wallers where wall building and rebuilding is merely part of the normal cycle of activity, naturally find it difficult to remember and describe how they first learned the craft. Certainly they were not actually taught, in any formal sense of the word, but they did watch fathers, brothers and hired hands… Farm children might actually help with a few stones, but they probably would not be called upon to do the work themselves until they were perhaps fifteen years of age. Myers Johnson [a waller from the West Riding of Yorkshire] says that before that they would lack strength, patience and judgement. Sometimes he has been tempted to speed his walling task by using a boy as a helper but has found he actually makes more work for himself correcting the boy's mistakes.

Margaret Brooks quotes a general observation from wallers — 'Best education to this buildin' a wall is to take three yards down and put it up again. Anybody, they'll learn.'

Rainsford-Hannay (1957: 46) gets to the very essence of a good drystone dyker by referring to the work of John MacAdam of Craigengullich (inventor of the 'sneck' and the 'locked top'), describing him as '[an] observant fellow, indeed, who really studied how to use the varying shapes of stone to the best advantage for every separate dyke which he built.'

The skills of a drystone waller are commonplace to some, amazing to others. D'Arcy (2006: 33) quotes from Lord Chief Barron Edward Willes' letters to the Earl of Warwick wherein Willes marvels at the walls in County Clare, Ireland during his visit in 1761. He thinks the walling '…very nearly resembles a net set to catch fish… I wou'd defy a London mason to build one of them without serving an apprenticeship to the trade, and yet every cottager makes them in this manner.'

5.1 (i) A brief note on the international nature of drystone walling

When Europeans emigrated, they took their agricultural practices, including fencing, with them. For many it was a hard lesson adapting to foreign climates. They frequently met local cultures with their own stone-based architecture, but usually gave those peoples scant credit for their achievements. These include the drystone walling systems of the Koni people in southeast Africa (described in Delius et al (2014)), the ruins of Great Zimbabwe, the stone constructions of the indigenous peoples of Australia (described by Memmott (2022)) and, of course, the stonework in South America and southeast Asia.

Indigenous peoples' ability to adapt and learn was often exploited, or traded, to suit the needs of the colonists.

Daniel Gookin (1612-1687) wrote two books on his experiences with Native Americans in New England. They were not published until much later. Gookin (1792: 22) tells how, after periods of unrest between themselves, the different Native American groups learned to cooperate with the English settlers and worked to earn supplies:

> But this good effect the war had upon some of them, namely, to turn them from idleness; for now necessity forced them to labour with the English in hoeing, reaping, picking hops, cutting wood, making hay, and making stone fences, and like necessary employments, whereby they got vićtuals and clothes.

It is not clear whether these Native Americans already knew how to build walls or learned from the settlers.

Russell (1976: 188) tells us, from the early 1700s:

> In Rhode Island and southeastern Massachusetts walls were sometimes laid up by Indians, not always willingly. If the Indians did not work according to their contract they were 'summarily flagellated'.

Australia also provides examples where the indigenous population assisted the European settlers. Munday (2012: 67), after giving details of a wall, 65 kilometres long, built through the Camel's Hump and Browne Hill Ranges in South Australia in the 1860's, quotes historian Nancy Robinson who:

> …garnered verbal reports indicating that Aboriginal people also played a role: 'As they were employed in almost every other station task, this is quite possibly true, though the construction of anything as permanent as the walls seems a little contrary to their nomadic culture.'

Immigrants also helped build walls. Munday (2012: 107), cites a case from the mid 1800s:

> The old-timers suggest that some of the walls in this region [of South Australia] were built by Chinese immigrants… who proved very good workers... But many shepherds also came out from Scotland and they might well have brought walling skills with them.

Dykers working overseas on contract is not a new thing. Corangamite (1995: 59, 61, 78, 119, 138) includes references to professional wallers, mainly Scottish and Irish, who were contracted by Australians to build walls in the late 1800s.

Ballard (2021:23) describes how economic refugees, from modern-day Croatia, often called 'Dalmatians', built drystone walls in North Island New Zealand prior to 1914. Many were conscripted to do labouring work during the Great War – 'Stonewall building was an ideal occupation for them since many had stoneworking skills brought from their homeland.'

Their effect was considerable. Ballard (2021: 24) records the effort of walling gangs made up of these Dalmatians/Croatians, Māori and the unemployed, on government relief schemes from the 1920s to the 1940s:

> Whether they were experienced in stonewalling or not, they built a lasting memorial to those times of severe financial hardship. Their labours are still contributing to the economic efficiency of the farms they helped develop.

Prisoners of war are often a source of cheap skilled labour. Throughout Britain many miles of wall were reputedly built by Napoleonic prisoners of war in the early 19th century. This is difficult to prove, although there were probably individual cases when skilled craftsmen were glad to resume their trade, even as prisoners in a foreign land.

Munday (2012: 59, 64, 73) refers to Yugoslav and Italian prisoners of war working on South Australian walls in the 1940s. They were well respected; many were welcomed as immigrants after the war.

The following terms relate to drystone wallers, dykers, associated stone trades, and others

Banker mason one who spends most of their working day at a banker, a raised bench where a stone is held safe and secure while it is shaped.

Barraman / barrowman (Scots) a mason's labourer.

Bigger (Scots) a builder, from the verb 'big' or 'bigg', to build.

Something which is 'biggit' is built and to 'bigg up' means to rebuild or repair.

Boarders / splitters (Caithness, Scotland) men who hand-split the flagstones. Each splitter works with a hammerman who hammers the steel wedge selected by the splitter. The wedges are made from old truck springs, or similar very hard steel.

Bristol pinners a derisory term used by Cotswold wallers for those who use pinnings in their walling, as noted by Rainsford-Hannay (1957: 80). Cotswold wallers prided themselves in tight work; there was less need for pinning up gaps with small stones.

Brock (Scots) a description of an unskilled workman or work badly done. The term is also used for small pieces of spoiled rock, or fragments of wool when sheep are clipped.

Builder-mason (Scots) a mason who builds with prepared stones, also known as a 'fixer mason' in North America. Traditionally Scots stonemasons were trained to cut and set stone, in England these were two different jobs.

Clay davie (Aberdeenshire) a farm worker, an agricultural labourer.

Connach / conach (northeast Scotland) an unskilled worker, a bungler.

Cowan / cowen / cowane / cowaner Scots terms for a dyker or a semi-skilled mason, one who has not served a formal apprenticeship or been initiated into the Freemasons.

According to Stevenson (1990: 42-43) the Freemasons' statutes of the seventeenth century did not allow cowans to work with masons, or restricted cowans to stonework which did not involve the use of lime mortar, cutting or carving. These regulations were aimed at preserving the exclusivity of the masons and separate their training, knowledge and practices from ' ...the semi-skilled and uninitiated.'

'Cowan' has been used as a derogatory term for a mason with no apprenticeship and 'cowan-hearted' may have arisen from the presumption that a cowan feared initiation into freemasonry.

Mactaggart (1824) includes a definition:

> Cowans — those who would wish to know the mysteries of free masonry, without being regularly initiated, a thing both mean and foolish to attempt; mean, because they are cowards, and dare not to go boldly forward like men; foolish, for unless they do so, masonry they never, never can know, no not the slightest thing about it; they may rail against the ancient and glorious institution, because they are held in ignorance, a thing which harms it nothing, for masonry, while time remains, will always find patrons in warm-hearted, social, and independent men. All other arts but it, of man's formation, sometime or other decays, and are reeled rotten into the bastile of oblivion, it alone remains the oracle of ages, and stands on a foundation so deeply grounded in nature, that no storm, though ever so rude, can make it in the smallest totter.

Figure 130: Drystone walling beside the river at Langdale using stone from the river bed. Cumbria, England, 1953. © Illustrated London News Ltd/Mary Evans.

'Cowan' is a common Scottish surname and was used in Australia for a drystone waller, as quoted several times in Corangamite (1995).

Diggers a puritan group who destroyed enclosing fences in the early 1600s in midland England. They believed the land belonged to the people and they were being robbed by the squires. Also recorded in southwest Scotland in the early eighteenth century as **levellers** or **dyke-breakers**.

Dorbie / darbie (Aberdeenshire) a stonemason.

Dressers (Caithness, Scotland) those who finish off the edging of the flagstones.

Drywalleros (Mexican Spanish) a term recorded in California for drywall fixers. They are not drystone wallers, but skilled erectors of 'drywall', known elsewhere as 'plasterboard'.

Dust (Ireland) 'The Dust' was a slang term used by Irish stonecutters to describe themselves. 'Stour' or 'stoor' is a Scots version of the word 'dust'.

Dyke loper / dyke louper (Scots) literally 'a wall jumper', an animal which jumps dykes, an immoral person. The natural urges of male mammals can cause havoc with breeding plans.

In old Cumberland dialect, from northwest England, a 'lowpy-dyke' was a wayward beast or an unfaithful wife or husband. DSL includes this entry, from Roxburgh, Scottish Borders, in 1825, for 'Dyke-louper':

> I am informed, that the old Session records of the parish of Hobkirk take notice of a female who was commonly known by the *soubriquet* of Bessy *Loup-the-Dykes*; and who is said to have been brought before the Session for having been guilty of *dyke-loupin*.

Dyke sider one who skulks around, a secretive person.

Dyke-priser / dykepriser (Scots) an official at a march-riding ceremony who inspects the state of the walls and fences of the common land. The boundary records of a town or county are often recorded in detail, listed as a series of landmarks and prominent stones. Travelling round the boundary on horseback is still an annual ceremonial event in the Scottish Borders. These Common Ridings helps preserve tradition and reaffirm ownership.

Dyker / dry-dyker / dry-diker / drystane dyker / drystone dyker (Scots and northern England) a builder of drystone walls.

Dyker one of a gang digging the Fenland drainage dykes in south-eastern England.

'Diking' in Lancashire, northern England, means cleaning out drainage courses.

Dyker a nickname for an inhabitant of Cellardyke, a village in Fife, Scotland.

Edge mason (Spanish Balearic Islands) from Trias and Arrom (2002: XLII). One who edges the terraces, a builder of retaining walls.

Fence mason (Kentucky, USA) a stonemason who specialises in rock fence construction. Equivalent to a Scots dyker or an English waller. From Murray-Wooley & Raitz (1992: 192).

Fence viewer (New England, USA) an official, from mid-seventeenth century onwards, who mediates in disputes and ensured local fences, including stone walls, were kept in good order. The office still exists in some states.

Foggy-Toddlers (Forfarshire, Tayside, Scotland) mentioned in an interesting extract from DSL, from 'Up Glenesk' by J. Paul, published in 1894:

> The men from Forfar, who were free-stone masons, appear to have been looked on as in the nature of intruders in the Glen. They were locally known as 'Foggy-Toddlers', the soubriquet arising from the circumstance that their trousers bore patches of the hue contracted by working among free-stone.

The masons had been working in Glenesk, presumably in sandstone or similar sedimentary rock because it is referred to as 'free-stone'. Their trousers were stained from the rock and the mosses and lichens thereon. The nickname likely stems from the stained trousers and the sight of the masons scrambling across the rocks, like bees.

'Foggy-Toddlers' is also a Scottish term for a bumblebee (*Bombus terrestris).*

Freemason originally the term applied to a mason who worked in 'freestone', stone which can be cut in any direction without meeting any grain or bedding plane.

Gapper one who repairs gaps in stone walls. In drystone walling competitions younger competitors may be given a gap to repair. This involves less work than a full stint, which is dismantled down to the foundations. Gapping takes skill, it is important to replicate the style of the walling to each side of the repair. *Cau bwlch* (literally 'to close the gap') is the Welsh for 'gapping'.

Goniel a Scots term for an apprentice mason. DSL has this as a 1923 term from Roxburgh in the Scottish Borders to describing both 'a loud-spoken, empty headed person' and an 'apprentice (mason, or gardener).'

Hayward an officer in charge of fences and enclosures with the added responsibility for impounding stray animals. An obsolete term from colonial USA and England.

Homer (Scots) a job done 'on the side', **nixer** is the Irish equivalent.

Journeyman originally a term applicable to a craftsman paid by the day (*jour*), subsequently applied to a fully qualified craftsman who travelled in search of work. They were not apprentices or assistants. Irish terms for these itinerant stonemasons include 'stamp stonies' or 'journey-men'.

Karoo (South Africa) while in conversation with a South African visitor it was interesting to learn drystone walls were introduced to southern Africa by Scottish Presbyterians who ministered to the Boers because the Dutch clergy would not go to South Africa.

Low drystone walls were built as overnight defensive shelters by the Welsh regiments serving in southern Africa during the Anglo-Zulu War (1879) and Boer Wars (1880-81 and 1899-1902)

Knapper / stane-knapper a person who breaks up rocks to create material for roads, from the Scots verb 'knap' meaning to 'strike sharply' or to 'break down to smaller pieces'. Stone was dumped in designated places, by local farmers. The 'knapper' or 'stane-knapper' sat there, breaking the pile, using a 'knap hammer' or 'knapping hammer', a light hammer with a long shaft, for maximum impact. The task was sometimes allocated as a form of punishment for minor offences but, for some, it was a source of income. Machinery eventually replaced the knapper.

Many knapping sites became an overgrown jumble of different rock types.

This backbreaking work has been depicted by several artists including Gustave Courbet (*The Stone Breakers* (1849)) and Edwin Landseer (*The Stone Breaker and his Daughter* (1830)).

DSL, under 'knapper', quotes from Donald MacLeod's *Past Worthies of Lennox* (1894), a piece called '*A Hebridean Stane-knapper and his Cuddy*'. The first line captures the imagination — 'Once upon a time there dwelt within the confines of the parish of Cardross an Irish stone-breaker, who kept a donkey.'

'Flint knapping' is another familiar term; lumps of flint are struck together to produce sharp flakes.

The term 'knapper' also refers to a Yorkshire walling hammer.

Layar an obsolete Scots term for a person who lays stones, from the old verb 'lay', to

Figure 131: This elevated wall end prevents sheep from walking round. Modern wire, and pallet, add to its efficiency. Galtee Mountains, Ireland.

build.

Levellers / dyke breakers in 1724 gypsy travellers and smugglers in Galloway, southwest Scotland, found their tracks and pack-trails blocked by new dykes and demolished them. The demolition of the walls was also a protest against the eviction of tenants and the erection of dykes as part of the landlords' wish to make more profits from raising cattle. The military were called in and the demolition stopped. The term was also applied to English protesters who opposed the hedging or walling of common land.

A similar group called 'Whiteboys' (named for their white smocks) carried out disruptive action in eighteenth-century Munster, Ireland, to protect the rights of tenant farmers.

Lowker (Yorkshire, England) a term for a gatherer of stones.

Masher (Angus, Tayside, Scotland) a quarryman who breaks large blocks of stone into smaller fragments with a mash hammer.

Mason in North America the term is in general use for anyone who works with brick, block or stone. The purist prefers the term is restricted to those who work with stone.

The term 'stonemason' applies to a mason who works cutting, carving or laying stone.

Originally, going back to the eleventh and twelfth-century cathedral building heyday, the term 'master mason' meant a man who combined the functions of designer and builder and could perform any task on site.

Terms such as 'block mason', 'brick mason' and 'stone wall mason' have emerged to more fully describe the craftsman and the material he/she works with. The term 'drystone mason' has emerged in modern times, for drystone wallers who produce very accurate work with tight-fitting joints.

Mechanic an old term from North America, applied to someone skilled in installation of dimensional stone – stone cut to specific shapes or sizes. 'Mechanic' was used as a generic term for a journeyman of any trade in the eighteenth and nineteenth century Statistical Accounts of Scotland.

Nixer an Irish equivalent to the Scots term **homer** – a piece of work done outside, or inside, normal working hours. It is done for personal monetary benefit rather than for the employer, often using the employer's tools and facilities.

Pickieman / pickman / pikeman (Scots) a flour miller's assistant who dressed millstones with a mill pick to keep the grinding surfaces rough. This job had its hazards. It was alleged that a pickieman with two good eyes was not doing the job right.

Pluggers Corangamite (1995: 91) includes this term for those who followed the wallers, filling small holes by tapping in chips of stone. In nineteenth-century Australia it was important to make the walls rabbit proof.

Quareor / Quereour an old term for a mason, from the Latin word *quadrare,* to square up. The word morphs into 'quarrier / quarryman' — a quarry worker. 'Quarry' is also a term for a square or diamond shaped pane of glass, from the French word *carre* meaning 'square'.

Rickler (old Scots) an unskilled stoneworker. He produces 'ricklie' work.

River (Cumbria, northwest England) pronounced 'r-*ee*-ver', one who splits slate in a quarry.

Roch dyker (Scots) a drystone waller, a dyker. 'Roch' is pronounced with the same ending as the Scots word 'loch'. 'Roch' in this case derives from the use of 'rough' undressed rock, rather than any suggestion the work is rough, bad or substandard.

Rock musicians late in the eighteenth century 'rock music' was developed in Keswick, in the English Lake District, when Peter Crosthwaite discovered striking the local stones (a unique mix of hornblende, slate and gneiss) produced a musical note. The Richardson, Abraham and Till families perfected the rock xylophone and were much admired by John Ruskin. They toured Britain and the USA until the late 1890s.

In more recent times John Lennon formed a skiffle group called 'The Quarrymen', an early formation of the Beatles.

Ruch layer / ruch mason / ruch dyker (Scots) masons and dykers who work with rough undressed stone.

Saint Stephen the patron saint of stonemasons, bricklayers, deacons, headaches, coffin makers and horses, with a feast day of 26 December. Stephen was the first martyr, stoned to death in Jerusalem for blasphemy around 35 CE. Images show him carrying a pile of stones.

Scuffer (Caithness, Scotland) men who level off the surface of a slab by chipping off high spots.

Seasonal work Stephens (1891: 231), confirms the seasonal nature of drystone work:

> A dyker cannot work in wet or very cold weather, as handling rough wet stones is hurtful to bare hands. On this account, dry-stone dykes are usually built in summer.

The seasonal nature of stone working is recorded in the Gaelic proverb — *clachair Samhraidh, diol deire Geamhraidh* — 'a mason in summer, a beggar in winter'.

DSL, under 'dorbie', quotes old Banffshire weather lore, saying of the first of March, the start of spring – 'Corbie an' dorbie baith a-biggin wid be.' With the return of warm weather, the crow and the mason will start building. 'Mason's ghost' is a nickname for the European robin, a small red breasted bird (*Erithacus rubecula*). They should not be confused with the North American robin (*Turdus migratorius*), a larger bird, more like the European thrush. All summer robins followed the mason around, hoping for crumbs, and remained on site after masons left for the winter. Lime mortars were commonly used in the days before Portland cements. The slow-maturing lime mortars are prone to frost damage. Building work usually stopped until spring.

Drystone wallers were not only men. In the early days of enclosure whole families worked together to improve their tenancy by building drystone walls. The manual nature of the work meant all the family participated, especially during the summer months. Munday (2012: 57-59) confirms this for South Australia:

> Whereas the pastoralists employed tradesmen to build the long epic dry-stone walls… many of the shorter stone fences were built by farming families, the women played a role we too often think was the domain of men.

Prevost (1957: 92) gives some insight into how large-scale enclosure was achieved in southwest Scotland:

> The hill dykes '…were built during the summer months, whole families migrated to the moors with tents, food and the minimum of equipment to do the work. Stones were gathered and rock was quarried and hauled on slypes to the site, or carried on hand barrows across ground impassable to horses. Preparations for the erection of park dykes were carried on during the winter months when stones were hauled on the farmer's stone-carts and dumped conveniently, to lie till spring or the coming of warmer weather. Both male and female, old and young, worked at the dyking. The young boys were given the job of packing and the women laboured for the men.

Rainsford-Hannay (1944: 24) supports this point, he writes:

> Dry stone dyking in a comprehensive way, began in about 1712, when the brothers M'Kie of Palgown, in the extreme west of the Stewartry of Kirkcudbright, set the example by

Figure 132: A mix of 'single' and Galloway dyking.
Such dykes can only be built by teams. Southwest Scotland.

starting to enclose many square miles of otherwise useless moors. They allowed free-rented pieces of land to people who would help them in the summer. Early in the summer these people with their women and children took to the hills with poles, tents, and blankets. In two or three days they had built themselves turf huts, and heather formed their beds. The brothers themselves superintended, and in a very few years the work, which had first looked like the labours of Hercules, was done.

There is anecdotal evidence for seasonal workers from Ireland who came over to Scotland for harvest work, there is little doubt many will have worked on the new sheep dykes in the nineteenth century and taken the skill back to Ireland.

Sinclair (1793: 229) refers to the parish of Tarland in Aberdeenshire and the difficulty of finding farm labourers in the summer because the men went south to build dykes:

...for some years back... a great number of able-bodied winter servants engage to go out of the country in summer, for building dykes and enclosures to those gentlemen who are carrying on improvements in the south country; and by these means they make double, sometimes triple, the wages in the summer half year that they could get or expect by engaging with a farmer at home. The Dyker, as he is called, gets from L.2 to L.3 sterling, and sometimes more, for 3 months in summer…

These seasonal migrations to the south of Scotland were a forerunner of longer trips to the British Empire, particularly Australia, where Scots and Irish were recruited to build drystone walls in and around the sheep stations as told in Munday (2012)

Sisyphus (Greek mythology) King Sisyphus tricked the gods. His punishment was to push a boulder to the top of a hill. Every time he reached the summit the massive stone rolled back down, forcing him to begin again. His punishment has no time limit.

'Sisyphus Syndrome' is defined in the McGraw-Hill *Concise Dictionary of Modern Medicine* as 'a mindset typical of a stress-driven type 'A' person, who obtains no gratification from accomplishing the difficult goals he or she places upon himself or herself.'

This stubborn acceptance will be familiar to many dykers who have worked through stormy days in a muddy field with rough stone, no shelter, and a broken thermos flask. Negative feelings are normally short-lived; the dyker's efforts are not futile. There are bad days but, unlike Sisyphus's disheartening, pointless, slog, the dyker will complete the task and can return, hopefully on a sunny day, to appreciate the result.

The author is grateful to Sebastian Rico for bringing an update to our attention. In the 2020 video game 'Hades', developed by Supergiant Games of San Francisco, the Sisyphus character accepts his never-ending punishment with good humour and has established a friendly relationship with the boulder, which he affectionately calls 'Bouldy'.

Snod dykit feedle a letter in the *Aberdeen Weekly Journal* newspaper for 24 December 1942, by A R Torphins, explains this term, and describes the dyker:

> 'Feedle' is a field, and snod dykit means a tidy built stone wall round said field or 'feedle.' I can well remember the 'snod dyker' in the Howe of Alford at that time. His name was William Pirie and he was a bit of a character. He always wore a brown velvet jacket going to work, and right on the back was a large patch of darker brown.

Stone cleaver (County Carlow, Ireland) a local term for those who cleaved the stone in 'the rocks' – places where granite boulders had been deposited by glaciation or erosion. These rocks were easier to work and more convenient than quarrying.

Stone cutter (Colonial USA and Ireland) one who cuts and dresses stone, and passes it to the masons who install it.

Stonepicker (Colonial USA) a person hired to remove stones from fields before sowing a crop.

Stonemason a craftsman skilled in cutting, carving or setting stone. Ryan (2005: 31), quoting medieval sources, tells us 'Stonemasons were required to be of exemplary character and of legitimate birth. They were not allowed to have concubines.'

Stonie (Ireland) a stonemason, a stoneworker.

Tradesman (Scots) one who practises a skilled handicraft, an artisan, a craftsman.

DSL explains the different between a tradesman in Scotland and one in England:

'In Scotland a craftsman was a 'tradesman' and the crafts were the 'trades', whereas in England a tradesman was and is a retail dealer – baker, butcher, etc – who in Scotland was and is a merchant.'

Craftsmen have always been respected. According to Fitchen (1999: 50) the Mogol Prince Tamburlaine (1336-1405), ruthlessly massacred populations after capturing cities during his campaigns across western Asia, but gave strict orders to spare all craftsmen. They were employed in reconstruction. By way of contrast, De Camp (1963: 293) tells of demanding times for masons working on the Great Wall of China – '[Emperor] Tsin ruled that a workman who left a crack between stones wide enough to stick a nail into should be instantly beheaded.'

Transferable skills stonemasons and drystone wallers share some terminology; their skills are not always directly transferable. Cairns (1975: 23) apprenticed with stonemasons before working on drystone dykes:

'I served my time in Edinburgh and thought I could build anything, but I got a shock when I went to work on dry dykes. Nothing would lie at all for the first few days. I laid them the wrong way always and built too plumb.'

Cairns (1975: 22) relates how his father, on mason jobs in Edinburgh '…could build as much hammer rubble (rough stone dressed with catchie hammer only, such as for inside walls) as any two men…' These dykers were proud of their ability to keep up with masons in the city.

Trias and Arrom (2002: XLII) gives some insight into the decline and merging of drystone with masonry work in the Spanish Balearic Islands after 1945:

'In his book *Pedres, Parets i Paredadors*, Jaume Portell discusses the decline of the trade in the post-war period, parallel to the decline of the profitability of agriculture and livestock breeding. He also mentions how the demand for walls in new housing complexes and along roads (the latter sponsored by government agencies) spurred the recovery of the trade. By the end of the twentieth century, the master dry stone masons' tasks had diversified and as in other places, masons lost their links to the rural world to become increasingly integrated in the construction industry.'

Morton (1855: 386) compares the skills required for stonework and brickwork:

> Building a wall with stones differs from brickwork in this, that, while from the regular form of the bricks, the workman can, from mere habit or exertion of memory, form a perfect piece of work; in stone work, on the contrary, from the varying sizes and forms of the materials generally used, a constant exertion of skill is necessary to secure the best disposition in the wall.

A man skilled in using randomly shaped stone may not find it so easy to work with regular building brick. A Working Man (1908: 101) tells of an old stonemason employed to lay bricks. The boss noticed he was not breaking the joint and told him so. The old mason replied – 'I know, and how the **** can I, when the bricks are all one length?'

Stonework is not easy but masons are not always appreciated or valued. Muilenburg and Swineford (1975: 59) detail how, in the limestone districts of Kansas, with so much stone available, the main cost of constructing a house was labour. Settlers learned the skills because '…few could afford to hire a trained stonemason, a craftsman entitled to top wages — $1.50 to $3.00 a day in 1880…'

Norma Campbell (1990: 67) quotes a dyker from the Edinburgh area in her unpublished

Figure 133: This advertisement, from *The Inverness Courier* of 10 November 1853, led to the construction of the dyke in Figure 134.

CONTRACTORS WANTED
FOR BUILDING a MARCH DYKE between the Lands of STRONE and BALLACHROAN, in the parish of Kingussie, according to Specifications to be seen on applying to Duncan Macpherson, banker, Kingussie; or to Thomas Falconer, solicitor, Inverness.
9th November 1853.

Figure 134: This dyke was built, probably in 1854, following the advertisement shown in Figure 133.

thesis:

> Why is a dyker's skill so undervalued? I charge £5 an hour in the village, £7 out and about. I'm cheaper than a plumber. But the rates are even cheaper in the south.

Waar / waaer (Scots) a waller, one who builds walls, a mason, a dry-stone dyker.

Wailer (Argyll, Scotland) DSL defines this as 'a surface-worker at a coalmine who picks out stones and other rubbish from the coal, a picker.'

Waller (England) a drystone wall builder.

In sixteenth-century England a 'salt waller' was a person who boiled down brine to produce salt.

The author walked the length of the dyke referred to in Figures 132 and 133 on 9 November 2016, the 163rd anniversary of the advertisement. It is reasonable to assume it was built in 1854. There are collapsed sections, mainly due to the boggy ground and tree felling work. The tight build and properly placed throughbands have added to the strength and longevity of the parts that survive. This dyke could be restored to a stockproof boundary, but that function is now performed by post and wire fencing. The dyke is approximately 1¼ miles long, and is consistent in its width, height and placing of throughbands, covers and copes. The contractors clearly worked to a tight specification, and to a high standard.

5.2 Output and income

Manual work in the 1700s and 1800s was brutal, with small reward for massive effort. According to Cameron (1978: 229), Peter Still the Aberdeenshire poet (1814-1848) '…went out stone-breaking for a pay of 4d. (2p) a ton. His working day stretched from five in the morning… until seven at night. He died at 34, broken by work.' Note that, prrior to decimalisation of the British currency on 15 February 1971, the pound (£) was made up of twenty shillings (abbreviated to 's'), each of twelve pennies (abbreviated to 'd'). After decimalisation the pound was made up of one hundred pence (abbreviated to 'p').

Historic comparisons are difficult, we shall leave it to the reader to follow their own lines of enquiry. Suffice to say Peter Still's wage was low. Using the currency convertor available online at https://www.nationalarchives.gov.uk/currency-converter a wage of four pennies a ton, in 1820, converts to approximately £1 in 2017.

Employee, and employer, expectations have changed dramatically over the years. Coldstream (2004: 20) gives a impressive illustration of how regulation, administration and the building processes have changed over the centuries. Only four clerks administered the 1,000 workers sent to build Harlech Castle in 1286.

The following is presented with minimum commentary; most speak for themselves. Rods, roods etc, are fully described in Section 6.2, Linear measurement.

Working for 'dog's wages' in Scotland meant food was the only payment. This might be offered to labourers, not to skilled men.

Bowie (1855: 421) includes an agreement with a dyker for maintenance of extensive feal (turf) dykes signed on 27 January 1630. A merk was two thirds of a Scots pound. This lifetime contract suggests stability with no expectation of the great changes ahead:

> Adame Hadden dyker, agrees to mend and uphold yearly, during his lifetime, the park dykes of Sir Duncane Campbell of Glenurquhay… for which the said Sir Duncane binds himself to pay the said Adame Hadden yearly the sum of twenty merks Scots, and twenty shillings for his expenses in travelling, a stone of cheese and a pair of shoes yearly, and to give him meat and drink for his said labour…

Winchester (2016: 34) details wages for walling work on an estate in Rydal, Cumbria, England, in the 1690's – 'Typically, a fortnight's work each year at 4d per day (plus tobacco) for each workman was enough to keep the walls in good repair.'

Prevost (1957: 86-91) quotes costs and wages in the Scottish Borders with an extract from the Drumlanrig Estate book between 1740 and 1745. 'John Tait, a dyker, was paid £4 8s 6d for raising and leading stones and building 59 roods of snap dyke at 1s 6d a rood.'

By the end of the eighteenth century dyking had, to quote Prevost, '...developed into a very considerable business.' In the 1800s prices increased:

> The cost of building a stone dyke in 1812 varied from 7s 6d or 8s per rood of 18 feet to 12s or more if the materials were very remote, and the price included the whole cost of quarrying and erection. The cost of repairing a march dyke between Craigbeck and Dumcrieff in 1818 amounted to 2s 6d a rood...

In the early 1900s, Prevost, referring to the rebuilding of 15 miles of stone dyke on the farm of Kinnelhead, tells us the price was '...was 4s 6d a rood and wages at that time varied from 4s 3d to 5s a day.' Eighteen dykers were employed on the project. In these extracts from Prevost a rood is equivalent to six yards. The rood had different values in other areas.

The cost of rebuilding and repairing was cheaper than building a new wall because there was no cost for winning or transporting stone.

Rollinson (1998: 6) gives rates for wallers in the English Lake District:

> Wages were low; in 1774 a rood, or seven yards (6.5m), of wall five and a half feet (1.5m) high could be built for between 1s 6d and 1s 8d per rood, but by 1877 this had increased to 6s 6d per rood.

GVA Argyll (1798: 63-64) quotes a very low price for dyking, using abundant stone and cheap labour, 'In most places, they may be built from 4 to 5 feet high, at 1s. per yard.'

Rainsford-Hannay (1944: 34) advises us – 'An old estimate gives in 1814: 2s. a rood quarrying, 2s.6d. a rood carting, 2s. a rood building.'

The following is an extract from an anonymous 'Pitlochry Area' estate account book found at the Highland Folk Park in Newtonmore, Inverness-shire, Scotland. On the same page the price of a bottle of whisky is given as two shillings:

> April 27, 1840 – Paid Alexander Stewart, Strathtummel, to building 56 Roods of stone Dyke on the Farm of Chapletown Borenick 1/8 per Rood = £4–13–4.

Munday (2012: 35) gives rates for walling in South Australia:

> The going rate for stone fencing in 1870 was four shillings and sixpence per rod, providing the stone was close at hand... by 1875 Keynes [the landowner] was paying five shillings per rod, which became the customary 'pound a chain.' By 1878 the payment was five shillings and six pence: 'If stone was close to hand or carted to the waller his team would complete five chains per week (a mile in sixteen weeks). A typical team consisted of two men and two boys with basic rations provided and pasture for three horses.' In a footnote Munday adds: 'There are 4 rods to a chain and 80 chains to a mile, making this rate equivalent to 18 shillings (18/-) per chain or £72 per mile.

Corangamite (1995: 91) gives prices for the Western District of Victoria, Australia:

> Wallers' charging methods and rates varied upwards from working for 'baccy and tucker'. The general rate was 120 pounds per mile in 1880, 3s 6d per day at the turn of the century, and is about $25 per metre today, collection of stone not included.

The Australian currency was decimalised in 1966.

Readers will note the general price of £120 per mile in Victoria, quoted by Corangamite and the price of £72 per mile in South Australia, quoted by Munday for around the same period. This might be because of a difference in the wall's specification, or due to the reliability of the reporting.

Some New England drystone walls were reputedly built for a ridiculously low rate. Henry Matthews, part Mohican Indian, built fine walls around 1860-70 at Raymond Hill, Connecticut. One wall, possibly a couple of years' work, was reputedly built for a barrel of rum. The build quality is so good it is more likely he received a fair wage for his effort.

Hart (1980: 23) provides comparative costings from Yorkshire:

> Before World War I, a waller made about 1s. 3d. (6p) per yard, and expected to build a rood a day (the rood being six or seven yards according to the district and type of stone). By World War II, labour costs had approximately tripled.
>
> To build a yard of 4½ feet (1.37m) wall in 1976 costs about £6 for labour, including digging foundations, setting out and clearing up. Stone and haulage cost a further £13, making a total of £19 per yard.

Rainsford-Hannay (1944: 34-35) describes hard work:

> Wages at this time, March 1944, are difficult to assess, but dykers should get more per hour than the agricultural labourer. They can do the latter's work and have their own special skill. Two skilled dykers, with stones put to them, can build two roods (18 ft. to a rood) of 4 ft. 9 in. dyke in a nine hour day. So the two men would build sixty-six yards in a fifty hour week. That is fast work by men who 'never luik at a stane twice,' and, in practice, less will be done if the men have to divert water, dig out roots, or have much dressing of stone to do with the hammer.

Mundell (1981: 119-121) records the work of Charlie Jardine, a fourth generation dyker who started dyking in 1924 and spent over 50 years in the profession:

> Family records show prices in the old days working out at from 6d to 9d a yard – and a man's average length a day is still around six yards. In these old days dykers spun out their working year by rabbit-catching and draining.

Charlie Jardine reckoned a man could average about six yards a day for a 4ft 6in high dyke:

> If he builds this he is turning out six tons of stone a day. On some dykes it may be more and could be less. But a good dyker doing six yards is not wasting any time.' The cost of a dyke had increased with the years – 'What had been 1s. 8d a yard for rebuilding dykes

Figure 135: Charlie Jardine after a dyking competition at Mid-Kelton Farm, Castle Douglas, southwest Scotland, on 6th October 1945. The original caption tells us — 'Mr C S Jardine standing alongside the excellent dyke he built in four hours to win both a special award for speed and the Challenge Cup for workmanship.' © Illustrated London News Ltd/Mary Evans.

when Charlie started to follow the trade and build the first of his seventy-two miles of dyking was around £2 a yard in 1975.

Charlie Jardine's life and work are also recorded by Borchgrevink (1982). See Figure 135 for more on Mr Jardine.

Prevost (1957: 93) writes:

> Two dykers with stones ready to hand can build 12 yards of 4 feet 9 inch dyke in a nine-hour day, but the speed is considerably reduced if a man is working single-handed. In fact if there are any heavy stones to lift into a dyke it is neither a one-man job nor an economical one. For building a new dyke two men are almost essential.

This point is confirmed by other writers who note that two men working, one each side, eliminates the need for jumping from side to side; they can take on a longer section and finish it in the same day.

Nowadays dykers often work alone. 'Working overhand' or 'working overarm' means one person builds both sides of the wall, from one side; the side furthest away is the 'overhand' side. This can result in a lower quality finish on the far side because the builder cannot constantly review the work as it progresses.

In a cutting from the *Colac Herald* (Colac is a town near Melbourne) dated 'approximately 1990', Corangamite (1995: 121) quotes 'Striker' Roy Gardner:

> He said the going rate for pay for stone walling building in the 1950's was 2 pound per foot and although he could build only twelve to twenty feet of wall a day it was still very lucrative work while it lasted… The rate is $30 a foot now and too expensive for most people to afford…

Some landowners commissioned construction of walls during times of economic depression, for example the Irish Potato Famine (1845-49). These were called 'penny walls' or 'famine walls'. The daily wage was low; literally a penny or two. In Golspie, Sutherland, northern Scotland, anecdotal evidence says the landowner paid fisherman a very basic rate to build walls when it was too stormy to go to sea.

Some idea the phenomenal manual effort which went into building drystone walls is given by Stilgoe (1982: 190), describing field boundaries in New England in the first half of the nineteenth century:

> Most walls stood four or four-and-one-half feet tall, and many were topped with one or two split rails. Wall building consumed vast quantities of time and human energy. An experienced man set between twenty-four and sixty-four feet of wall a day but only if the fieldstones were already heaped along the line the wall was to follow.

Stilgoe also includes an extract from an 1858 *Southern Planter* article which was trying to convince farmers in the southern United States to build stone walls instead of fences from wooden rails:

> About one hundred and eighty common-sized stones will build a rod of fence… work nine hours in the day, and handle one stone the minute, and you will have made three rods of fence; and surely a man can do this.

The calculations following do not allow for the time taken to gather the stone and transport it to site.

The 'rod' was a measure of length, equivalent to 5½ yards or 16½ feet. It should not be confused with the 'rood' which was an unstandardised measurement of output. (Weights and measures are discussed in Section 6.)

Callander (1986: 80-81) estimates there are over 3,000 miles of drystane dyke in Deeside, Aberdeenshire. He further calculates these dykes contain more than 5 million tons of stone and, if stone was on site '…on the basis of a dyker building a rood a day (i.e. 6yds.), and working an average of 200 days a year, then the work would have taken fifty dykers about 100 years or 100 dykers 50 years.' Although he recognises these figures are rather arbitrary but they give some idea of the scale of activity when '…nearly all the dykes we see in Deeside were built within 100 years and the majority of them in fifty years, 1820-1870.' Müller (2013, V1: 80-81), with explanatory notes, estimates the effort required to build drystone walls and how many have been lost (Figure 136). Comparative figures for most European countries are included in Müller's magnificent publication.

	Length of field walls in 1880 (km)	Weight in tonnes (see note A)	Man hours required to build (see note B)	Workers required to build the walls in one year (see note C)	Length of field walls in 2012 (km)	Percentage of walls lost between 1880 and 2012
Great Britain and Northern Ireland	269,750	286,395,598	572,791,196	238,663	115,500	57%
Ireland (not including the six counties of Northern Ireland)	568,000	669,309,900	1,338,619,800	557,758	400,000	30%
France	490,000	697,760,000	1,395,520,000	581,467	170,000	65%
Spain	400,000	352,800,000	705,600,000	294,000	200,000	50%
Italy	357,000	646,174,686	1,292,349,371	538,479	150,000	58%
Greece	333,300	497,323,596	994,647,192	414,436	200,000	57%
Croatia	375,000	663,796,875	1,327,593,750	553,164	300,000	20%
Sweden	228,572	784,403,390	1,568,806,779	653,669	160,000	30%

Figure 136: The effort required to build drystone walls. Extracted from Müller (2013: Vol 1: 80-81). Note A: On page 80, Müller gives average dimensions for each nation's walls to produce an estimated cubic capacity, which, at 2.5 tonnes per cubic metre, calculates the weight. Some walls are taller or wider than others. This produces apparent inconsistencies as exampled by the United Kingdom and Swedish figures. Note B: Müller estimates: 'On average a drystone waller manages to erect 0.5 tonnes = 0.2 m^3 = 1 m of field wall per hour or 1.6 m^3 per day of 8 hours. Field stones weigh 2.5 tonnes per m^3.' Note C: Müller calculates this by taking the number of hours in column D and assuming an 8 hour working day and a 300 day working year. A kilometre of 1000 metres is equal to 3,280 feet, 1,093 yards or 0.62 miles. A metric tonne of 1000kg is equal to 2,204 pounds or 0.98 Imperial tons (as used in the British Isles), or 1.1 'short tons', each of 2,000 pounds (as used in the USA).

5.2 (i) Checking and measurement of completed work

Once the work was done it was measured.

'Mettage', according to DSL is an old Scots term, defined as the '…measurement of the extent of (a pieceworker's work) in order to meet the due payment.' It stems from an old Scots word 'met' meaning 'to measure or verify the accuracy of'.

On big contracts the dykers were guaranteed a payment per unit of work satisfactorily completed. A good team often influenced the measurement, as described in Cairns (1975: 8-10) during the period before and after World War I:

'The method of measuring a dyke was usually by chain and this pleased the dyker best, as it sank into all the hollows of the ground and gave better measurement than the tape, which if drawn tight missed the hollows.'

Cairns also reports 'I have seen some of the tough old 'uns with the tape rolled round their fist a few times…' when they held the end of the tape for the person checking the yardage.

This could add half a yard to every 20 yard length of the measuring tape, thus increase the dykers' income.

The estate factor, the laird's manager, often reduced the agreed payment if he decided the dyke had not been built to the agreed standard, or the dykers were suspected of poaching.

Dykers charged extra for special features such as gate openings, stiles or arches. Each of these takes more time to build and are further confirmation of the craftsman's skill. Cairns (1975: 9) wrote:

> The gates were measured right through as this allowed for the extra work in building the ends, 'sconcheons', and the measurement went right round a single sconcheon, at the end of a dyke, to allow for the extra there.

Cairns measured the distance between the wall ends in a gateway as though they were built dyke.

At the end of a dyke the yardage included the width of the vertical face. The price for these 'extras' was agreed as part of the dyking contract. Another accepted way of charging for a sconcheon (also known as a cheekend) is, and was, to consider the last yard, before the cheekend, as two yards, to allow for the extra work.

Curves took more time and were charged for, as per Stephens (1891: 234) — 'In building curves in dykes, builders charge a half more per rood than for plain work.'

Daily targets were not always measured with a tape. Adair (1998: 40) describes how the old-timers picked the heaviest hammer and threw it up the hill as far as possible to set the intended output for the day.

In Corangamite (1995: 118) describes one way of checking work:

> Jack Boyd tells the story of one works boss who would inspect a day's labor with a crowbar, which he would put into a crevice and bounce up and down on it. If anything moved the section might have to come down.

Another oft-quoted way of checking a dyke was to give it a solid kick. If the centre was hollow the face stones gave way, or the clink of falling hearting was heard.

According to GVA Galloway (1813: 86) the cope was tested by pushing a wheelbarrow along the top.

Prevost (1957: 101) describes a measuring device, called 'Tumbling Tom':

> The name given to an instrument used for measuring stone-dykes. It consisted of a kind of an adjustable frame, with two arms which dropped down on both sides of the dyke which it was required to measure. A similar kind of instrument is described by Dr Singer, but a plummet and line takes the place of one of the arms.

GVA Dumfries (1812: 152) includes Dr Singer's more detailed description of the instrument, used for measuring a Galloway dyke:

> A rectangular instrument, with a common three-feet rule may answer this purpose. The instrument has two legs, firmly put together in an angle of ninety degrees; one leg six feet long, the other thirty-four inches long, with a plummet and line, moveable in the short leg. The long leg is placed upright, the foot close to the base of the dyke on one side; the short leg projects over the dyke, and the plumb-line descends at the other side, having been set corresponding to the thickness at the grass. A three-foot rule applied at the first lift, and at the single work, horizontally, ascertains the thickness; and placed vertically at the top it shows the height.

In simple terms, Singer's description is of a plumb line hung over the dyke to reach down to the foundation or the top of the 'double'. Measurements taken from the vertical plumb line to the face of the dyke confirmed the dyke's width. Dr Singer goes on to describe how to check the strength – 'As to the solidity, it is common for the inspector to ascertain it by pressure, firmly striking against the dyke with the sole of his foot.'

'Catch a' (literally 'catch all') was a system used by Scottish Border dykers, if they quoted for the repair of a long dyke. They quoted a price per yard and hoped there were not too many sections where a big repair was necessary.

'Slump' was a Scots term for a quick, rough estimate.

Modern wallers and dykers rely less on field walls as a source of income and are more likely to do garden or ornamental work, measuring the daily output in terms of square feet rather than linear yards. Payment is often for a complete project rather than for yardage done. Repairs are usually agreed on an hourly or daily rate.

5.3 Some statistics

This section gives information about the lengths of walls in various countries. For an initial, overall impression, the reader is referred to Figure 136 and 137.

Estimates for the lengths of drystone wall in a particular area vary significantly. These estimates are affected by what is defined as a wall, the survey method and when the estimate was made. Whatever the length, we cannot fail to recognise the vast tonnages of stone used, and that every stone was placed by human hand.

The existence of a wall and its inclusion in an overall mileage figure does not mean it is stockproof.

Brooks and Adcock (1999: 13) include a graphic from CCP482, the 1996 Countryside Commission Report. Only 4% (2,795 miles or 4,500 km) of English dry stone walls are 'stockproof and in excellent condition'. Another 9% (6,086 miles or 9,800 km) have minor defects. A further 38% (26,517 miles or 42,700 km) show 'major signs of advancing deterioration'. A total of 49% (34,483 miles or 55,600 km) are not stockproof and becoming, or are, derelict, or exist as remnants.

The total for walls, in all conditions, in England, comes to 69,926 miles or 112,600 km. This is about 2.5 percent more than the Report's estimate of 68,204 miles for English drystone walls (see under the heading **England** below). We presume that this is because of mathematical quirk in the calculation.

The high percentage of ruined walls is disappointing. It very much reflects the change in agricultural practices. These piles of stones may have no value as a stockproof fence but retain value as wildlife habitat, and a record of human endeavour.

Ireland

McAfee (2011: 100) writes — 'There are reckoned to be about 400,000 kilometres (250,000 miles) of stone walling, both dry and mortared, in Ireland. Much of this is only about 150 years old…'

This figure is for the whole of the island of Ireland, including the six counties of Northern Ireland.

The Burren area of western Ireland and the offshore Aran Islands have remarkable arrays of drystone walls. McAfee (2011: 99) estimates 'The total amount of dry stone walls in the three islands is roughly 1,500 kilometres.'

Cook (2011: 167) includes an essay by Brendan Dunford which refers to the '…quite extraordinary heritage of dry limestone walls — over 2,400 kilometres in an area of under 44 square kilometres – which dominate the Aran Islands.' These figures, translated out of metric, mean 1,491 miles of wall on 10,872 acres.

The Céide Fields near Ballycastle, County Mayo, Ireland, hold the remains of a significant Neolithic farming site, with buried sandstone walls which may be over 5,000 years old. They were discovered by probing into the bog. According the McAfee (2011: 17,18) 'The overall area of the site is about 10km^2 – the largest such site in Europe. Some of the walls are 2 km long and the total amount of stone is estimated at 250,000 tonnes, although the individual stones are not very large.'

Indications from excavated walls suggests the walls were not built to any pattern. This is certainly evidence of drystone walls although not to the sophisticated specification of those built from the early eighteenth century. The name of the site comes from the Irish *Achaidh Chéide*, which means 'flat topped hill fields'.

The United Kingdom

The UK is made up of Scotland, England, Wales and the six counties of Northern Ireland and has '…about 200,000 kms (125,000 miles)…' of dry stone walls, according to the Dry

Figure 137: The density of walls in Britain. Image courtesy of the Centre for Ecology and Hydrology, based on data from the Countryside Information System (CIS).

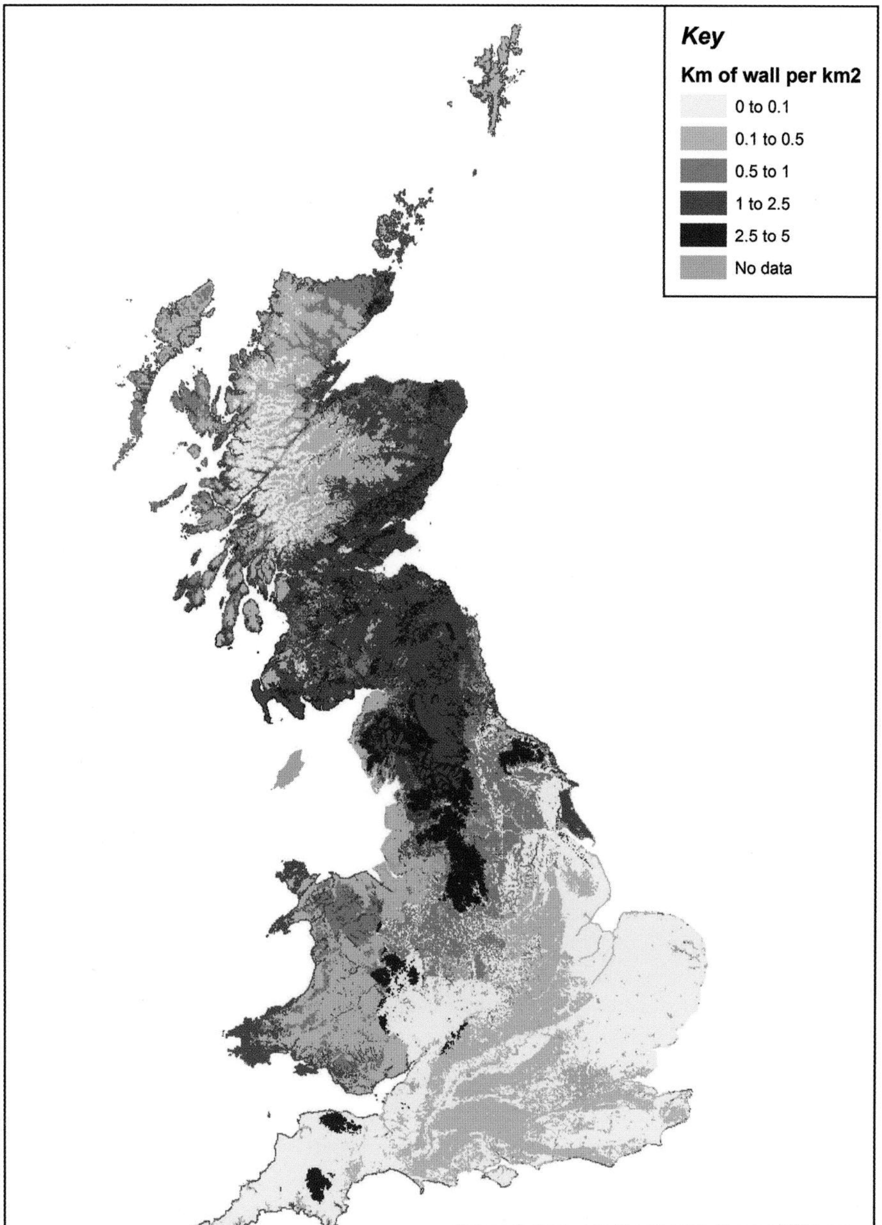

Stone Walling Association of Great Britain in their 2006 booklet *Walls and the Landscape, a guide to British walls.*

The density of walls in Britain is shown in Figure 137. The walls are clearly associated with areas of sheep farming, in the hill ground, and arable farming, in the more fertile valleys.

Note also how the density of walls is less in the south of England, where hedges are the most common field boundary.

England

England has an estimated '…68,204 miles (109,830 km)…' of walls according to Brooks and Adcock (1999: 13). Half of the walls are reckoned to be in the four counties of North Yorkshire, Cumbria, Cornwall and Derbyshire. These figures are extracted from CCP482, a Countryside Commission report from 1996.

DEFRA (2007: 8) includes a table breaking down estimated lengths of wall and density in English counties. North Yorkshire is reckoned to have 20,900 km (12,986 miles) of walls, Cumbria 15,050 km (9,351 miles), Cornwall 14,460 km (8,985 miles), and Derbyshire 9,420 km (5,853 miles). The figure for Cornwall is declared to include 'Cornish hedge banks', a stone-faced earth bank.

DEFRA (2007: iii) reviews various studies and survey methods. In the report's conclusion it states:

> A number of studies have provided estimates of the total length of stone wall in England. However, there is significant variation in these estimates probably due to different sampling techniques used, and the way in which walls were defined. These estimates vary between 80,000 km and 112,600 km. Very roughly this is equivalent to 10% of the total length of field boundaries in England.

Different sources are consistent in quoting a length of about 4,000 miles (6,437 km) of drystone walls in the Cotswolds, a famous limestone area in south central England. This was compared to the Great Wall of China before modern reserach increased its total, combined, length to over 13,000 miles.

Scotland

DEFRA (2007: 5,6), gives a figure of 87,100 km (54,121 miles) for stone walls in Scotland, from '…the Countryside Survey 2000 (CS 2000) which was based on sample field surveys that took place in 1998.'

This is said to be equivalent to 22.8% of field boundary length for Scotland. The figure may be an overstatement; the definition of 'stone wall' for the survey was 'a built structure of natural stone or manufactured blocks, mostly of traditional dry stone wall construction but including mortared walls. Includes walls with fences and lines of trees or shrubs'.

Callander (1986: 80-81) estimates approximately 3,000 miles of dyke in Deeside, an area in Aberdeenshire of about 800 square miles or half a million acres. Callander makes comparisons with the Rainsford-Hannay (1948: 80) estimate of 7,000 miles of dyke in Kirkcudbrightshire, southwest Scotland, an area of 900 square miles.

Deeside and Kirkcudbrightshire are approximately the same area but there are more walls in Kirkcudbrightshire because, thinks Callander, there is '...a much greater extent of dykes running back and forth over its green hills than are to be found amongst the heather of Deeside.'

The Dumfries and Galloway Standard and Advertiser for 24 June 1939, in a piece promoting an upcoming dyking competition, under the title 'An Old Industry', states: 'It is computed that in the Stewartry there are 25,000 miles of such dykes.'

The 'Stewartry' refers to the county of Kirkcudbrightshire, which is now, in 2018, part of the local government area known as Dumfries and Galloway. There is a considerable difference between the newspaper's 1939 figure of 25,000 and Rainsford-Hannay's 1957 estimate of 7,000 miles. They cannot be referring to the same ground; perhaps the newspaper article refers to all of Dumfries and Galloway.

North Ronaldsay, in the Orkney Islands group, has a dyke about 13 miles (20 km) long and 6 feet (1.8 metres) high, built around the coast to keep the sheep on the beach. They have adapted to graze on the seaweed and only occasionally get access to the grassland inside the dyke.

Northern Ireland

According to McCann et al (2012: 2) there were 5,973 km of dry stone wall in Northern Ireland in 2007 with a further 3,084 km of 'ruined dry stone walls and earth banks'. These total 9,057 km or 5,627 miles.

The longest drystone wall in Northern Ireland is the Mourne Wall which surrounds the watershed of Belfast's water supply. It is 22 miles (35 km) long and was built between 1904 and 1922.

Dry stone walls are known as 'walls', 'ditches' or 'Danish fences' in different locations throughout Northern Ireland.

Wales

According to Smart, et al, (2009):

'In 2007, the estimated total length of walls in Wales was 13,000 km [approximately 8,100 miles], evenly split between the upland and lowland zones.'

The definition of 'wall' for that survey was '...[a] built structure of natural stone or manufactured blocks, mostly of traditional dry stone wall construction but including mortared walls. Includes walls with fences or banks/grass strips and/or lines of trees or shrubs.'

A full version of the Countryside Survey, UK results for 2007 is available from Carey, et al (2009).

New England, USA

New England consists of six states – Maine, New Hampshire, Vermont, Massachusetts, Rhode Island and Connecticut. New England sources are consistent, each quoting about 250,000 miles (402,336 km) of drystone wall. It is difficult to break this figure down between viable walls, and those in various stages of disrepair. If we make a reasonable guess, based on the figures for English walls, from Brooks and Adcock (1999: 13), quoted at the top of this section, and bear in mind the changes in farming systems, and the abandonment of New

England farming ground, it is likely that only a fraction of the 250,000 miles are still viable walls. Much of the rest is lost in the woods.

Allport (1990: 17) quotes detail from a Department of Agriculture report for 1871, *Statistics of Fences in the Unites States*, published in 1872:

> In 1871 approximately one-third of Connecticut's fences were made of stone, amounting to 20,505 miles of stone wall – enough to extend almost one time around the equator. Most of Rhode Island's 14,030 miles of fencing were stone, as were nearly half of Massachusetts: 32,960 miles. In New York, eighteen percent of the fences were made of stone, staggering 95,364 miles, more miles than there are in the coastline of the entire United States. … The work that went into them, according to one estimate, would have built the pyramids of Egypt one hundred times over.

Allport (1990: 18) goes on to illustrate the farmers' achievement — 'It would have taken …15,000 men 243 years to build the 252,539 miles of walls in New England and New York.' Readers should note this figure includes New York state, in addition to the six states which make up New England.

Thorson (2002: 1) refers to a 1939 calculation by Oliver Bowles, which drew on the same 1872 Department of Agriculture report as Allport, above, to estimate approximately 240,000 miles of stone walls in New England and adds 'That's longer than the US coastline, or even the distance to the moon at perigee.'

'Perigee' is when the moon's orbit is closest to Earth.

Hubbell (2006: 66) tell us 'A scientist once theorised that the stone walls built in central New England between 1810 and 1840 held a mass 'greater than that of the great pyramids of Egypt'.'

The extent of walling in New England deserves comparison with other regions. The land area of the six New England states is approximately 72,000 square miles. The land area of Ireland (excluding the six counties of Northern Ireland), is approximately 27,000 square miles. Figure 136 tells us Ireland had 568,000 kilometres (353,000 miles) of stone walls in 1880. The accepted equivalent for New England is 250,000 miles. Let us accept this figure also applies to 1880 – a reasonable assumption. Based on these figures, Ireland has 40% more walls than New England but on only 38% of the land mass. The reasons for that come down to the nature of the land and how it was used.

The timescale for the build is also an interesting comparison. It is probable that most of the Irish walls were built in the 1800s, a time of widespread agricultural improvement. The New England walls were more likely built over a longer time period.

Kentucky, USA

Kentucky is famous for its horses and its stone walls. Limestone makes good bones, and good walls. Minerals in the soil make good blood.

Unfortunately, there has been no audit of the length of walls in the state, and much of the nineteenth century drystone heritage has been destroyed. Lexington-Fayette (1990: 18-20) quantifies this destruction – 'Professor Karl Raitz estimates that over 90% of the stone fences which once were a part of the Bluegrass region are gone, and that 'nearly all were victims of 'progress." The report goes on to say '...there are only 38.7 miles of stone fence left in the Fayette County as compared to 49.6 miles found in 1967, a reduction of 11 miles, or a loss of 25% in 23 years.' This report helped stimulate interest in the state's stonework and the restoration of many historically important walls.

Australia

Corangamite (1995: 124) includes a 1991 newspaper cutting. 'There are about 3000 kilometres of drystone walls in Victoria's Western district and, according to Nathan Perkins, only three people working full-time at repairing them.'

Nathan was the first Australian to qualify as a Master Craftsman through the Dry Stone Walling Association of Great Britain.

As at the time of writing (August 2021) drystone walling is thriving in Australia. Many old walls have been restored.

Munday (2012: 65) refers to one wall, over 40 miles long, the '...remarkable wall that runs some sixty-five kilometres along the Camel's Hump and then Brown Hill Ranges, making it 'the longest stone wall in South Australia, and probably the longest in Australia'.'

Europe

Müller (2013) is a good source for the estimated length of drystone walls in European countries. Some comparative figures are given in Figure 136. It is remarkable to see how many drystone walls were built throughout Europe, and how much has survived.

6. WEIGHTS AND MEASURES

Weights and measures and conversion factors are easily accessible from many other sources. This chapter will concentrate mainly on those found in drystone literature.

Local weights and measures have become aligned and standardised over centuries This was essential as trade increased between parishes, counties and countries. Standardisation inevitably meant the loss of antiquated units, for example the North Welsh 'hobbit' of barley (180 pounds) and the British 'pottle', a measure of liquid equal to four pints.

The British system of weights and measures, termed 'Imperial' or 'British Imperial', is based on the Weights and Measures Act of 1824 and subsequent legislation. The Metric system was developed by the French during their revolution (1789-99) and was widely adopted by other nations during the nineteenth century. The system of measurement used in the United States of America was based on the British system; after 1824 several differences emerged.

The older written sources usually define their terms of measurement, thus reducing confusion for the modern reader. For more detail on the development of measuring systems in Scotland, readers are recommended to Simpson and Connor (1996: 104), details in the bibliography, or their book – *Weights and Measures in Scotland: a European Perspective*, edited by A Morrison-Low and published by National Museums of Scotland Enterprises Ltd, Edinburgh in 2004.

6.1 Weight

Imperial

16 ounces = 1 pound (lb).
14 pounds = 1 stone (st).
2 stones or 28 pounds = 1 quarter (qr).
4 qrs (112 lb) = 1 hundredweight (cwt).
20 hundredweight (2,240 lb) = 1 ton (t).

The link between stone walls and sheep has been well documented. Another link between wool and stone is confirmed by Hill and Birch (1994: 14) when referring to sheep farming in the Cotswolds, southern England. The 'stone' was '…the old English weight equivalent of 14 lb = 1 stone was based on a standardised block of stone hoisted inside a barn to weigh fleeces.'

14 lb is equivalent to 6.4 kg.

Metric

1,000 milligrams (mg) = 1 gram (g).
1,000 g = 1 kilogram (kg).
1,000 kg = 1 tonne (t) = 2,205 imperial pounds.

United States customary units

100 pounds = 1 short hundredweight (cwt) (The British equivalent is 112 pounds).
2,000 pounds = 1 short ton (The British 'long ton' is 2,240 pounds).

These differences mean 8 British tons are almost exactly 9 USA tons.

6.2 Linear measurement

Imperial

12 inches = 1 foot.
3 feet = 1 yard.
6 feet = 1 fathom (for depths of water or mines).
5 ½ yards = 1 rod, pole or perch (16½ feet, ¼ chain or 5.029 metres).
4 poles, rods or perches = 1 chain (66 feet or 22 yards). Measured using Gunter's Chain.
10 chains = 1 furlong (660 feet, 1/8th of an Imperial mile).
8 furlongs = 1 Imperial mile (5,280 feet or 1,760 yards).

Miles have had various definitions over the years. Older definitions may have been used in early descriptions of drystone walling.

The English statute mile was introduced by Act of Parliament in 1592. There was further standardisation in 1824 with the introduction of the Imperial mile. It was defined as having 5,280 feet, equal to 1,760 yards or 1,609.344 metres.

The Old Scots country mile consisted of 1500 paces of 5 feet each = 7500 feet = 2500 yards = 1.42 statute miles = 2286 metres, as described by Simpson and Connor (1996: 105):

1 Scots mile = 5928 feet = 1976 yards = 1.12 statute miles = 1807 metres.
This was rendered obsolete by the Weights and Measures Act of 1824.
1 Irish mile = 6720 feet = 2240 yards = 1.27 statute miles = 2048 metres.

The Weights and Measures Act of 1824 abolished the Irish mile. It was still in occasional use in some county road signs until standardisation in 1926. The Irish mile was longer because the Irish 'rod' or 'perch' was 21 feet, compared to 16½ feet in England.

Gunter's Chain is described in detail at the end of this section.

Metric

10 millimetres (mm) = 1 centimetre (cm).
100 centimetres = 1 metre (m).
1,000 metres = 1 kilometre (km) (approx. 0.62 miles).

The spellings 'metre' and 'kilometre' are found in Europe, Australia and Canada and spelt as 'meter' and 'kilometer' in the USA.

The following terms have been associated with measurement of drystone walling

Ell (Scots) an old measurement of 37 inches. It originated from a measurement used for yarn and cloth. Many towns had their particular 'ell' marked out near the site of the local marketplace. The 'Plaiden Ell' measurement in Dornoch churchyard (Sutherland, Scotland) measures 39 inches between two metal studs set into a flat stone (Figure 138).

Rainsford-Hannay (1957: 69) quotes a 'ell' length of 3 feet 1⅕ inches, (37.2 inches) when describing dyking in Aberdeenshire in 1741. This, slightly longer, 'ell' was used in the eighteenth century. An ell measurement of 37.0598 inches is quoted in the early nineteenth century. Simpson and Connor (1996: 104) say this was based on 'a faulty interpretation of the Edinburgh standard ell bed in 1811.'

These intricacies, and their repercussions, are beyond the scope of the normal dyker.

An 'ell-wan', 'ell-wand', 'elvan' or 'ellvand' was a Scots measuring rod, one ell long which, in time, became the 'yardstick'. The ell is now only of historic interest and was replaced by the 'yard'.

Fa / fall an old Scots measurement used for measuring land area and distance. One square 'fall' equalled 36 square 'ells', an area six 'ells' by six 'ells'. 160 square 'falls' equalled one Scots acre. In linear terms a 'fall' was a distance of 6 ells, each of 37 inches, the length of the surveyor's standard rod or rope. The total distance is 222 inches, eighteen and a half feet, or 5.64m.

GVA Argyll (1798: 63-64) defines the 'fall' as 6 ells: 'Nearly 6¼ yards English.'

DSL defines 'Fall' as, 'a lineal measurement of land equal to six ells…' with the quote, 'Ane rod, ane raip, ane lineal fall of measure, are all ane…' which translates as 'a rod, a rope, a linear fall of measure, are all the same.'

Rods / poles / perches this measurement had individual values in different places; in modern measurement systems they are considered the same.

Morton (1855: 1125) gives a straightforward definition: 'Perch, Pole, or Rod, a measure of length equal to 5½ yards = 16½ feet.'

He goes on to provide regional British examples for 'perch' as a measurement of output: '…sometimes 18 feet for rough work' in Berkshire; 'Stone work 16 ½ feet in length, 1 in height and 22 inches in thickness' in Devon; [In Lancashire the perch could be] '…5½, 6, 6½, 7, 7½ or 8 yards in different parts of the county.'

Morton further defines a perch as 6, 7 or 8 yards of labourers' work in parts of Wales and 18¾ feet (6¼ yards) in Scotland.

He also defines a perch as 'a rod of 3 ells, or 9 feet 3 inches' in Dumfriesshire, southwest Scotland, which seems out of step with the other definitions.

It is not clear from some of these definitions whether the measurements of output are square yards or linear yards.

Thorson (2002: 250), with reference to New England walls, equates a 'rod of stone wall' with a length of sixteen feet.

Todd (1859: 462), writing for the New York State Agricultural Society, gives definitions of 'rod' and 'perch':

> Stone walls are, usually, built by the perch or by the rod, running measure, with the foundation and top of a given width and the wall of a given height. A perch of wall is sixteen and a half feet long, and a foot square, or sixteen and a half cubic feet. A rod of

> wall may be two feet wide or ten feet wide, and almost any given height, but always sixteen and a half pickets.

In this case a perch is a measurement of volume (16½ cubic feet) and a rod is a measurement of length — sixteen and a half 'pickets', presumably 16½ feet.

Murray-Wooley and Raitz (1992: 195) writing from Kentucky, USA, confirm 'rod' as a 'Unit of measure, 5.5 yards, 16.5 feet, or 5.029 meters.' 'Pole' is defined as a length of '… sixteen and one-half feet; similar to a rod.'

Murray-Wooley & Raitz (1992: 195), include a definition for 'perch':

> a unit of measurement of twenty-four and three-fourths (or twenty-five) cubic feet; a volume one foot tall and one foot deep by twenty-five feet long; sometimes one foot high by eighteen inches deep by sixteen feet long.

Vivian (1976: 66), from New England, writes:

> If you'd like to sound old-timey, you can order rock by the *perch*, based on an old surveyor's measure. A perch is a stone wall 16½ feet long, 1 foot wide and 1½ feet high or 24¾ cubic feet, something less than a cubic yard.

Conry (2000: 78) confirms one Irish perch is a measure of length, 7 yards or 6.5 metres.

The Weights and Measures Act of 1878 set the pole as 5½ yards or 16½ feet.

Rood modestly defined in modern walling literature as '…6 or 7 yards depending on the type of stone…' This measurement has had an interesting history, including as the term for a Scots quarter acre.

DSL, quoting a 1687 source, gives a definition which ties the rood to surface area of work done, rather than just a straightforward length — 'A rood of work… contains but thirty six ells, that is, if any piece of work be found to be eighteen ells in length and two ells in breadth, it makes a rood. Twelve ells in length and three ells in breadth is a rood…'

DSL goes on to define a rood as 'The quantity of work performed by a mason, slater or other tradesman or craftsman…'

GVA West Lothian (1794: 14) defines rood as 'This kind [of stone wall] is generally built about 5 feet high: the rood is 36 yards in length, and 1 yard in height.'

GVA 1794 Dumfries (1794: 47) defines the rood as a length, the other dimensions to be agreed when the price of the dyke is fixed:

> The rood for all fences in this county is six Scottish ells, or nineteen feet of length, running measure along the dyke or ditch, without measuring height, depth, or breadth. These last dimensions are all fixed at making of the bargain, and particular respect is had to them in fixing the price of the rood of fence.

GVA North Wales (1810: 125) refers to roods of eight yards each, when detailing expenditure of 'about £4,000' on the Plas Gwynn estate walls:

They measured about 4500 roods of eight yards each, at 17s. per rood. The walls are five feet high, all entirely pointed: and the coping, of one foot depth, worked with mortar.

Morton (1855: 1126) gives two local definitions for the rood in Britain: 'Durham, of wall building 7 yards… Berwickshire, of labourers' work, 6 or 7 yards; of masonry — 6 yards square, 2 feet thick.' No height or other dimension is quoted.

Rainsford-Hannay (1944: 34) tells us:

A rood of length is a Scottish measure, and in most counties is 18 ft. In a few it is 19ft., and in one or two 20ft. In one of the old surveys dyking was done by the ell of 37⅕ in. and a man did 'a rood of 6 ells a day of 5 foot dyke.' In the Hebrides and Argyllshire the measure [for dyking] was the 'fall' of 18ft. A rood, therefore, is not a rod, nor is it a pole nor a perch.

Figure 138: The Dornoch Ell. A local standard measurement, set in the grounds of the cathedral. Cloth was measured between two brass studs.

In his later publication Rainsford-Hannay (1957: 34) defines the rood as '…a measure of length: 6 yards in granite districts, 7 yards in limestone districts. It is generally a day's work for one man.'

Murray-Wooley and Raitz (1992: 196) give the linear definition of a rood as '…a British unit equal to seven or eight yards, or sometimes the same as rod.'

Alan Brooks (1989: 5) probably gives the best modern interpretation for the rood:

The rood is unstandardized… whatever the exact distance, it was considered to be a reasonable length for a craftsman dry waller to build in a day, assuming the foundation trench was cut, the stones ready to hand and the wall was not unusually high. Now that the work day has shortened, few wallers would care to attempt this rate except in repair work where the foundation and lower courses are still secure.

Quarter Murray-Wooley and Raitz (1992: 195) define 'the quarter' as an 'English unit of height equalling nine inches (one quarter of a yard).'

This is confirmed by Stephens (1853: 588) regarding Scottish dykes. He defines the term 'five quarter wall' in detail and uses the 'rood' as a measurement of area:

> Dry-stone dykes are measured by quarters – that is, quarters of a yard of 9 inches each. A 5-quarter dyke is the usual measure of a field-fence, that is 45 inches, or three feet 9 inches to the under side of the cover upon which the cope stones stand – the cover and cope-stones usually measuring 12 inches, so that the dyke stands altogether 4 feet 9 inches in height. The dyke, when finished, is measured by the rood of 36 square yards upon its face under the cover, so that every 30 yards of a 5-quarter dyke will be 1 rood in length. The usual thickness of such a dyke is 2 feet at the base, and 15 inches under the cover.

These figures can be used to check the value of a rood:

A quarter is a quarter yard, or 9 inches. A five-quarter dyke is 5 x 9 inches high = 45 inches, to the underside of the cover and cope. 45 inches is 1.25 yards. A length of 30 yards x 1.25 yards high = 37.5 square yards. This differs slightly from the 36 square yards calculated by Stephens.

Local terms of length

Remote areas can retain a standard measurement long after it disappears elsewhere. There is no need to alter what works if everyone in the district understands and accepts what it represents. Currency and spoken language are other good examples.

In personal correspondence (2013) with Patrick McAfee, Dublin stonemason and writer he said 'An Aran Island 'tape' has been described to me as 66 feet long and £2 was paid to build a wall this length up to the 1950's.'

This measurement originated from the days when the Department of Agriculture recorded walls for grant purposes by the number of 'tape' built. The length corresponds to Gunter's chain — 66 feet or 22 yards.

Gunter's surveying chain was an early instrument for land surveying, developed in 1620 by English clergyman and mathematician Edmund Gunter (1581-1626).

The chain is 66 feet long (22 yards or 20.12 metres) made up of 100 links joined by rings. Each set of link and ring measured 7.92 inches, therefore —

7.92 inches = 1 link.
25 links = 1 pole.
4 poles or 100 links = 1 chain of land (66 feet).
80 chains = 1 mile.
100,000 square links or 10 square chains = 1 acre.

Trautwine (1937: 282) advises that Gunter's chain had been 'abandoned' by engineers in favour of a chain of 100 feet with each link measuring one foot or 12 inches. This allowed measurements in feet and square feet — more appropriate for engineering purposes.

Corangamite (1995: 118) quotes an article from *The Age,* a Melbourne newspaper, dated 21 January 1989, entitled 'A country with an art of stone' by Jenny Brown which includes the sentence: 'He and his brother, Keith, working on either side, could pick up £1 a foot in the 1950's and on a good day they might get a chain (seven metres) completed.'

Presumably this was a standard length for the area, perhaps a hangover from older times. Seven metres are just short of 23 feet, approximately one third of a Gunter's chain.

Marshall (1805: 398) in *Provincialisms of West Devonshire* includes the local measurement known as a 'staff' defined as 'a measure of nine feet; half a customary rod, perch, or pole.'

Morton (1855: 1125) lists 'lug' as a measurement with regional definitions:

> '(Dorsets.), of land 15 feet 1 inch; called also Goad, used instead of a pole of 16½. (Herefords.), 20 feet. (Wilts.), a pole or rod of 15, 16½, or 18 feet.'

JSD defines 'a raip' as a sixteenth-century Scots measurement of six ells. Raip is a Scots variation of the word 'rope', the nature of the measuring device is used to describe the measurement itself: 'A measure of six ells in length, a rood; so called, as being measured by a rope, as rood is from the use of a rod, and line…'

Dickson (1805: 112) mentions 'a rope of 20 feet' in Somerset.

Morton (1855: 1126) includes – 'Somerset — a rope of wall building — 20 feet long.'

6.3 The relationship between boundary length and enclosed area

The length of wall required to enclose a field is not directly proportional to the area enclosed.

Martin (1887: 47) includes a table to record the weight of barbed wire required for various lengths of boundaries to prove '…the larger the area enclosed the smaller is the amount of fence required per acre.'

The table of figures is paraphrased thus:

— an area of one acre has a boundary length of 990 feet, (0.1875 of a mile).
— an area of 5 acres has a boundary length of 3/8 mile.
— an area of 10 acres has a boundary length of ½ mile.
— an area of 20 acres has a boundary length of ¾ mile.
— an area of 40 acres has a boundary length of one mile.
— an area of 80 acres has a boundary length of 1½ miles.
— an area of 160 acres has a boundary length of 2 miles.

An acre is 4,840 square yards, 43,560 square feet, around 4,047 square metres, or approximately 0.002 square miles.

These figures confirm bigger fields mean fewer walls, but the stone is carried further. Farmers practicing subsistence agriculture on stony ground exert the least effort for the most benefit. They tend to move stone short distances to create a landscape with many small fields.

BIBLIOGRAPHY

Three resources proved invaluable for preparation of this volume, though not credited for every individual term they assisted with. Full details for them are given in the bibliography:

Dictionaries of the Scots Language (DSL) in its multi-volume format and online presence.
English Rock Terms by Arkell and Tomkeieff (1953).
Jamieson's Scottish Dictionary (JSD), also in its multi-volume format and online presence.

A lot of practical drystone knowledge lies just beyond the horizon of human recollection and will never be properly recorded — lost like tears in the rain. Newspapers are a valuable resource for finding everyday events and practices as they pass quietly into history. Two archives were consulted while preparing this volume. The British Newspaper Archive (www.britishnewspaperarchive.co.uk) gives access to digitalised pages from back issues of British and Irish Newspapers. An Australian equivalent, Trove (https://trove.nla.gov.au), created by the National Library of Australia, gives access to newspapers, and a host of other information relating to Australia.

There are several international drystone organisations which provide information on drystone walling and drystone construction. This short list should provide sufficient links to a wide world of information. Drystone walling is not limited to northern Europe and North America. These website addresses are correct at the time of writing (2022):

Dry Stone Walling Association of Great Britain — https://www.dswa.org.uk
Dry Stone Wall Association of Ireland (DSWAI) – https://www.dswai.ie
Dry Stone Conservancy (DSC, Kentucky, USA) – https://www.drystone.org
The Stone Trust (TST, Vermont, USA) – https://thestonetrust.org
Dry Stone Walls Association of Australia Inc. (DSWAA) – https://dswaa.org.au
L'architecture en Pierre Seche (France) – www.pierreseche.com
Artisans Batisseurs en Pierres Seche (France) – www.pierreseche.fr

Throughout the text 'GVA' has been used as an abbreviation for 'A General View of the Agriculture of _____', the eighteenth and nineteenth century audits of agriculture in the counties of the British Isles, including the counties of Ireland. This abbreviation instantly confirms the regional origin and date of the reference.

The following list creates a link between the abbreviation and the original author, making it easier for the reader to trace the original through the bibliography.

Spellings are as in the original publication.

Aberdeenshire (1811) — George Skene Keith DD, minister of Keith-hall and Kinkell.
Angus or Forfarshire (1813) – Reverend James Headrick, minister of Dunichen.
Argyll (1798) – John Smith DD, one of the ministers of Campelton.
Berwick (1794) – Alexander Lowe.
Berwick (1813) – Robert Kerr, FR and ASSE, farmer at Ayton, in Berwickshire.

Caithness (1815) Sir John Sinclair, Bart.
Derbyshire Vol. I (1811) – John Farey, Senior, mineral surveyor.
Derbyshire Vol. II (1813) – John Farey, Senior, mineral surveyor.
Dorset (1812) – William Stevenson.
Dumbarton (1811) – Reverend Andrew Whyte and Duncan Macfarlan, DD.
Dumfries (1794) – Bryce Johnston, DD, minister at Holywood.
Dumfries (1812) – Dr W Singer.
Fife (1800) – John Thomson, DD, minister of Markinch.
Galloway (1813) – The Reverend Samuel Smith, minister of Borgue.
Hebrides or Western Isles of Scotland (1811) – James Macdonald, AM.
Inverness (1813) – James Robertson, DD, minister at Callander in the county of Perth.
Midlothian (1793) – George Robertson, farmer at Granton near Edinburgh.
North Wales (1810) – Walter Davies, AM, rector of Manafon, in Montgomeryshire.
Northumberland (1813) the third edition – J Bailey and G Culley.
Orkney Islands (1814) – John Sherreff.
Perth (1794) – James Robertson, DD, minister at Callander in the county of Perth.
Ross and Cromarty (1813) – Sir George Steuart Mackenzie, Bart.
Roxburgh (1794) — Reverend David Ure of Glasgow.
Selkirk (1794) – Mr Thomas Johnston.
Stirlingshire (1812) – Patrick Graham, DD, minister of Aberfoyle.
Surrey (1809) – William Stevenson.
Sutherland (1815) – Captain John Henderson.
West Lothian (1794) – James Trotter, farmer at Newton, county of West Lothian.
North Riding of Yorkshire (1800) – John Tuke, land-surveyor.
West Riding of Yorkshire (1799) – Robert Brown, farmer at Markle, near Haddington.

The following publications and websites were consulted (these website addresses were correct at the tim of completion of this text: May, 2024)

Aalen, F H A; Whelan, Kevin; Stout, Matthew. (ed.) *Atlas of the Irish Rural Landscape.* Revised and expanded second edition. Cork University Press 2011.

Adair, Hunter. *A Guide to the Countryside – Stone Walls.* The Abbey Press, Hexham 1998.

Adams, I H *Agrarian Landscape Terms. A glossary for Historical Geography.* Special Publication Number Nine. Institute of British Geographers. London 1976.

Adcock, Sean. *Masterclass: Corbelling III/Roofs.* Stonechat, the branch magazine of the North Wales Branch of the Dry Stone Walling Association, No. 23, Spring 2011, pp 30-31.

Adcock, Sean. *Stone fox traps.* Stonechat, the branch magazine of the North Wales Branch of the Dry Stone Walling Association, No. 25, Autumn 2011, pp 12-17.

Adcock, Sean. *Stonework, a technical guide to standards and identification of common faults in dry stone walling.* North Wales Branch of the Dry Stone Walling Association, 2012.

Adcock, Sean, *Clawdd construction.* North Wales Branch of the Dry Stone Walling Association, 2012.

Aitken N. and Rippingale N. *Shaken, but OK.* The Waller and Dyker, the official Magazine of the Dry Stone Walling Association of Great Britain, Spring 2001, p. 3.

Alexander, William. *Notes and Sketches illustrative of Northern Rural Life in the Eighteenth Century.* David Douglas, Edinburgh 1877.

Alexander, William. *Rural Life in Victorian Aberdeenshire.* The Mercat Press. Edinburgh 1992.

Allen, Edward. *Stone Shelters.* The Massachusetts Institute of Technology, Fifth Printing 1981.

Allport, Susan. *Sermons in Stone: the stone walls of New England and New York.* W.W. Norton & Company, New York 1994.

Alomar, G; Ferrer, I. Grimalt, A; Reynés, A; Rodríguez R. *La pedra en sec. Materials, eines i tècniques tradicionals a les illes mediterrànies. (Dry-stone work. Materials, tools and traditional techniques in the Mediterranean Islands.)* Officine Grafiche Riunite, Palermo 2002.

Ancona, George. *Cutters, Carvers & the Cathedral.* Lee & Shepard Books, New York 1995.

Arbennigol, Craig. *Professional Column, Line and Batter: part 2.* The Waller and Dyker, the official magazine of the Dry Stone Walling Association of Great Britain, Summer 2011, pp 10-11.

Arkell, W J and Tomkeieff, S.I. *English Rock Terms, chiefly as used by miners and quarrymen.* Oxford University Press 1953.

Arnold, Bob. *On Stone, a builder's notebook.* Origin Press, Boston 1988.

Arnold, James. *The Shell Book of Country Crafts.* John Baker, London, 1977 reprint.

Audel, Theo. *Audel's Masons and Builders Guide Vol. 1.* Theo Audel & Co. New York, 1950 edition.

Bailey, J and Culley, G. *General View of the Agriculture of the County of Northumberland.* London 1813.

Bailey, N. *An Universal Etymological English Dictionary etc.,* The Twenty First Edition, London 1675.

Baines, Edward. *Baines's Yorkshire, Volume 1.* Reprint of original 1822 work. David and Charles (Publishers) Limited, Newton Abbot 1969.

Baker, Ira Osborn. *A Treatise on Masonry Construction.* John Wiley & Sons, Inc., New York, tenth edition 1913.

Ball, Philip. *Universe of Stone, a biography of Chartres Cathedral.* Harper Collins, New York 2008.

Beall, Christine. *Masonry Design and Detailing for Architects, Engineers and Builders.* Prentice-Hall, New Jersey 1984.

Beaton, Elizabeth. *Lunt stanes. Resting or lifting stanes at the Clumie Burn horizontal mills, Troswick, South Mainland, Shetland.* Vernacular Building 23, 1999, p. 28. Shetland Special Issue. Scottish Vernacular Buildings Working Group. Edinburgh.

Beaton, Elizabeth. *Porterage: an English equivalent to the Shetland Lunt Stane.* Vernacular Building 24, 2000, p. 70. Scottish Vernacular Buildings Working Group. Edinburgh.

Bianchina, Paul. *Illustrated Dictionary of Building Materials and Techniques.* John Wiley & Sons, Inc., New York 1993.

Blackwood's Edinburgh Magazine. *The Shepherd's Calendar*– storms, p. 75, Vol. V, *April-September,*

1819. Edinburgh 1819.

Blaser, Werner. *The Rock is My Home*. Van Nostrand Reinhold Company, New York 1977.

Borchgrevink, Anne-Berit Ø. *A Dumfriesshire Drystone Dyker*. Extracted from, *Gold Under the Furze: studies in folk tradition*. Edited by Alan Gailey and Daithi O Hogain, pp 76-92. The Glendale Press, Dublin 1982.

The Border Magazine. *Gleanings from the past of the Royal Burgh of Selkirk*. (pp 46-47), A. Walker & Son Ltd, Galashiels 1939.

Bowie, William. *The Black Book of Taymouth with other papers from the Breadalbane charter room*. (edited by Cosmo Innes). T Constable. Edinburgh 1855.

Brewster, David. *The Edinburgh Encyclopaedia*, Volume 1. William Blackwood, Edinburgh 1830.

Brooks, Alan. *Dry Stone Walling, a practical conservation handbook*. BTCV, Wallingford, first published 1977, revised edition 1989.

Brooks, Alan and Adcock, Sean, edited by Elizabeth Agate. *Dry Stone Walling, a practical handbook*. BTCV, Wallingford, second edition 1999.

Brooks, Margaret. *The Craft of a Dry Stone Waller*. Folk Life, a Journal of Ethnological Studies, Volume Fifteen, 1977, pp. 98-105.

Brown, Robert. *General View of the Agriculture of the West Riding of Yorkshire*. Edinburgh 1799.

Brown, Robert E. *The Book of the Landed Estate etc*. William Blackwood and Sons 1869.

Buchan, David. (editor) *Scottish Tradition: a collection of Scottish Folk Literature*. Routledge & Kegan Paul Plc, London 1984.

Burgoyne, General Sir John. *A Treatise on the Blasting and Quarrying of Stone*. Crosby Lockwood and Son, London 1895.

Burn, Michael. *The Age of Slate*. Quarry Tours Ltd, Blaenau Ffestiniog circa1970s.

Bush, J. *Hibernia Curiosa. A Letter from a Gentleman in Dublin to his Friend at Dover in Kent. Giving a general View of the Manners, Customs, dispositions, Etc, of the Inhabitants of Ireland*. London 1769.

CAPEB with ABPS, Muraillers de Provence, CBPS, CMA 84 and ENTPE. *Guide de bonnes pratiques de construction de murs de soutènement en pierre sèche*. Universite de Lyone (ENTPE) 2008.

Cairngorms Partnership. *The Cairngorms Local Biodiversity Action Plan*. Cairngorms Partnership 2002.

Cairns, Robert. *Drystone Dyking*. Biggar Museum Trust, Biggar. First published 1975, reprinted 1986.

Callander, Robin. *Drystane Dyking in Deeside*. Published by author, Haughend, Finzean, Aberdeenshire, first edition 1982, revised edition 1986.

Campbell, Norma. *Rebuilding the Walls, an enquiry into the small-business potential of drystane dyking in Scotland*. Unpublished thesis for BA (Hons) in Business Studies at Napier Polytechnic of Edinburgh, December 1990.

Cameron, David Kerr. *The Ballad & the Plough, a portrait of life in the old Scottish farmtouns*. Victor Gollancz Ltd, London 1978.

Carey, P D , Wallis, S, Chamberlain, P M , Cooper, A , Emmett, B A, Maskell, L C, McCann,

T, Murphy, J, Norton, L R, Reynolds, B, Scott, W A, Simpson, I C, Smart, S M, Ullyett, J M. *Countryside Survey: UK results for 2007.* Centre for Ecology & Hydrology November 2008 (revised June 2009). Accessed online, 24 June 2022, at – file:///C:/Users/stone/Desktop/Documents/Countryside%20Survey%202007,%20British%20walls/N005191CR%20UK%20Results.pdf .

Clifton, Robert T. *Barbs, Prongs, Points, Prickers & Stickers, a complete and illustrated catalogue of antique barbed wire.* University of Oklahoma Press, fifth printing 1984.

Coldstream, Nicola. *Medieval Craftsmen – Masons and Sculptors.* British Museum Press, London, fifth impression 2004.

Conry, Michael J. *The Carlow Fence.* Chapelstown Press Ltd. Carlow 2000.

Cook, Mariana. *Stone Walls: personal boundaries.* Damiani, Bologna 2011.

Corangamite Arts Council. *If these walls could talk, the report of the Corangamite Dry Stone Wall Conservation Project.* Terang, Australia, 1995, reprinted 1999.

Countryside Commission for Scotland. *Drystane Dykes, information sheets 4.10–4.10.15.* Countryside Commission for Scotland 1988.

Cramb, Ian. *The Art of the Stonemason.* Betterway Books, Cincinnati, Ohio 1992.

Cridland, Robert B. *Practical Landscape Gardening.* A T DeLaMare Company, Inc., New York 1922.

Cruickshank, Helen B. *Collected Poems.* Reprographia, Edinburgh 1971.

The Cultivator. *Farming among the Hills*, pp 304–305, The Cultivator, a monthly journal devoted to agriculture, horticulture, floriculture, and to domestic and rural economy. New series – vol III. Albany, New York 1846.

D'Alton, Martina. *The New York Obelisk or how Cleopatra's Needle came to New York and what happened when it got here.* The Metropolitan Museum of Art/Abrams, New York NY 1993.

D'Arcy, Gordon. *The Burren Wall.* Tir Eolas, Co. Galway 2006.

DARDNI (Department of Agriculture and Rural Development, Northern Ireland) *Field Boundaries.* Originally consult online at www.dardni.gov.uk/ruralni/field_boundaries.pdf on 20 January 2013. This link no longer works. In 2016 this government department was renamed the Department of Agriculture, Environment and Rural Affairs (DEARA). The document on field boundaries does not appear to have been carried forward to the new website but has been partly replaced by: https://www.daera-ni.gov.uk/publications/northern-ireland-countryside-survey-2007-field-boundaries-summary. (Accessed 18th August 2021).

DEFRA (Department for Environment, Food and Rural Affairs) *Defining Stone Walls of Historic and Landscape Importance. Final Report produced for Defra and partners by Land Use Consultants with AC Archaeology, April 2007. Incorporating the findings of methodology field testing.* Bristol 2007. The report was consulted online in June 2022: https://assets.publishing.service.gov.uk/government/uploads/system/uploads/attachment_data/file/320858/Defining_stone_walls_of_historic_and_landscape_importance.pdf.

DSL *Dictionaries of the Scots Language,* comprising electronic versions of the two major historical dictionaries of the Scots language – the 12-volume *Dictionary of the Older Scottish Tongue* (DOST) and the 10-volume *Scottish National Dictionary* (SND) with its two supplements (1976 and 2005), frequently consulted online at http://www.dsl.ac.uk/. Original paper copies were consulted at the

Highland Folk Museum, Kingussie.

Dawber, E Guy. *Old cottages, Farm Houses, and other stone buildings in the Cotswold District.* Illustrated by W. Galsworthy Davie. B T Batsford, London 1905.

De Camp, L Sprague. *The ancient engineers. Technology and invention from the earliest times to the renaissance.* Barnes and Noble Books, New York 1993.

DSWA. *Building and Repairing Dry Stone walls.* Dry Stone Walling Association of Great Britain, based on the original 1982 publication of the same name by Lawrence Garner, revised and reprinted 1988.

DSWA. *Dry stone walls, the national collection.* Dry Stone Walling Association of Great Britain, Sutton Coldfield 2002.

DSWA. *Specifications for Simple Retaining Walls.* Dry Stone Walling Association of Great Britain, Milnthorpe, Cumbria 2021. Available online at www.dswa.org.uk/wp-content/uploads/2021/02/Retaining-spec-2020.pdf (consulted June 2022).

DSWA. *Walls and the landscape, a guide to British walls.* Dry Stone Walling Association of Great Britain, Milnthorpe, Cumbria 2006.

Davies, Walter. *General View of the Agriculture and Domestic Economy of North Wales; containing the counties of Anglesey, Flint, Caernarvon, Meirionydd, Denbigh, Montgomery.* London 1810.

Davis, Henry. *The works of Plato. A new and literal version, chiefly from the text of Stallbaum. Vol II containing The Republic, Timaeus and Critias.* Henry G Bohn, London 1849.

Dean, Carolyn. *A Culture of Stone, Inka perspectives on rock.* Duke University Press, North Carolina 2010.

Deane, Samuel. *The New England Farmer or Georgical Dictionary containing a compendious account of the ways and methods in which the important art of husbandry, in all its various branches is, or may be, practised, to the greatest advantage of this country. The second edition, corrected, improved, and enlarged, by the author.* Worcester, Massachusetts 1797.

Delius, Peter; Maggs, Tim; Schoeman, Alex. *Forgotten World. The stone-walled settlements of the Mpumalanga escarpment.* Wits University Press, Johannesburg 2001.

Dickson, R W. *Practical Agriculture; or a complete system of modern husbandry with the methods of planting and the management of livestock.* Vol. I. Richard Philips, London 1805.

Dineen, Reverend Patrick S (editor). *Foclóir Gaedhilge Agus Béarla. An Irish-English Dictionary, being a thesaurus of the words, phrases and idioms of the modern Irish language, with explanations in English.* The Irish Texts Society, Dublin 1904.

Donovan, Molly and Fiske, Tina. *The Andy Goldsworthy Project.* Thames & Hudson, London 2010.

Drake, Robert. *A Solitary Trade, the art and craft of dry stone walling.* Bookcase, Carlisle 2008.

Dry Stone Conservancy. *Building and Repairing Dry Stone Fences and Retaining Walls.* Dry Stone Conservancy, Inc., Lexington, Kentucky. Third printing, revised 2001.

Dutton, Hely. *Statistical Survey of the County of Clare with observations on the means of improvement.* Dublin 1808.

Dutton, Hely. *A Statistical and Agricultural Survey of the County of Galway etc.* The University Press,

Dublin 1824.

Dwelly, Edward. *The Illustrated Gaelic-English Dictionary*. Consulted online at http://www.dwelly.info/.

Edmundson, T. *An Etymological Glossary of the Shetland and Orkney Dialect.* Adam and Charles Black, Edinburgh 1866.

Emerton, Gerald. *The Pattern of Scottish Roofing.* Historic Scotland, Edinburgh 2000.

Erlande-Brandenburg, Alain. *Cathedrals and Castles: Building in the Middle Ages.* Harry N Abrams, Inc., New York 1993.

Eyles, V. Glossary of Colloquial and Dialect Rock Terms. *Nature* 173, 7 (1954). Accessed on 18th August 2012 at https://doi.org/10.1038/173007a0.

Evans, E Estyn. *Irish Folk Ways.* first published by Routledge & Kegan Paul 1957, seventh impression, Routledge & Kegan Paul Ltd, London 1979.

Evans, Thammy. *The Great Wall of China, Beijing and Northern China.* Brandt Travel Guides 2006.

Fairweather, Barbara. *The 300-Year Story of Ballachulish Slate.* Glencoe and North Lorn Folk Museum, Glencoe 1988.

Farey, John. *General View of the Agriculture and Minerals of Derbyshire; with observations on the means of their improvement.* Vol 1. London 1811.

Farey, John. *General View of the Agriculture of Derbyshire with observations on the means of its improvement.* Vol II. London 1813.

Feininger, Andreas. *Stone and Man, a photographic exploration.* Dover Publications, Inc., New York 1979.

Fenton, Alexander with Mulhern, Mark A. *A Swedish Field Trip to the Outer Hebrides, 1934.* NMS Enterprises Limited (National Museums Scotland), Edinburgh 2012.

Ferrucci, Katherine. *Limestone Lives, Voices from the Indiana Stone Belt.* Indiana University Press 2004.

Fields, Curtis P. *The forgotten art of building a stone wall.* Yankee, Inc., New Hampshire 1971.

Fitchen, John. *Building construction before mechanisation.* The MIT Press, Massachusetts, seventh printing 1999.

Fleming, John; Honour, Hugh; Pevsner, Nikolaus. *The Penguin dictionary of Architecture and Landscape Architecture.* Penguin books. London, paperback edition 1999.

Forsyth, Robert. *The Beauties of Scotland: Containing a Clear and Full Account of the Agriculture, Commerce, Mines, and Manufactures; of the Population, cities, Towns, Villages, &c. of each County,* Vol II. Edinburgh 1805.

Freeman, Rich and Freeman, Sue. *Cobblestone Quest, Road Tours of New York's Historic Buildings.* Footprint Press Inc., Florida 2005.

French, Lew. *Stone by Design, the artistry of Lew French.* Gibbs Smith, Salt Lake City 2005.

Gage, Mary and Gage, James. *The Art of Splitting Stone: early rock quarrying methods in pre-industrial New England 1630-1825.* Powwow River Books, Amesbury MA, 2002.

Gallagher, Rob; Malone, Sean; Piazza, Joe. *Building Dry-Stack Stone Walls.* Schiffer Publishing

Ltd, Pennsylvania 2008.

Garner, Lawrence. *Dry Stone Walls, Shire Album 114,* Shire Publications Ltd, Aylesbury 1984.

Garner, Lawrence. *Dry Stone Walls.* Shire Publications Ltd, Princes Risborough 2005.

Gardner, Kevin. *The Granite Kiss: traditions and techniques of building New England stone walls.* The Countryman Press, Vermont 2001.

Gaskell, Janet and Paterson, Nigel. *The Dry Stone Walls of the Chalfont Area.* The Chalfont Village Walls Project 1999.

Glover, Thomas J. *Pocket Ref.* Second edition, 30th printing, September 2001. Sequoia Publishing, Inc Littleton, Colorado 2001.

Goldsworthy, Andy. *Stone.* Viking, London 1994.

Goldsworthy, Andy. *Wall at Storm King.* Thames & Hudson, London 2000.

Goldsworthy, Andy and Craig, David. *Arch.* Thames & Hudson, London 1999.

Gookin, Daniel. *Historical Collections of the Indians in New England.* Originally published by Apollo Press, Boston 1792. Reprinted by Arno Press, New York 1972.

Goulder, Dave. *Notes on Building a Cairn.* Dry Stone Walling Association of Great Britain, revised edition 2013.

Goulder, Dave. *A Short Tale of Floating Stones.* The Waller and Dyker, the official magazine of the Dry Stone Walling Association of Great Britain, Winter 2018, pp 24-25.

Graham, John G. *The Shetland Dictionary.* Shetland Publishing Company, Lerwick 1986.

Graham, Patrick. *General View of the Agriculture of Stirlingshire.* Edinburgh 1812.

Gregor, Reverend Walter MA. *Notes on the Folk-lore of the North-east of Scotland.* The Folk-lore Society, London 1881.

Griffiths, David. *In there somewhere.* Dry Stone Walling Association of Great Britain 1999.

Grose, Francis. *A Provincial Glossary with a collection of local proverbs and popular superstitions.* London 1787.

Hadingham, Evan. *Circles and standing Stones, an illustrated exploration of the megalith mysteries of early Britain.* Anchor Press, New York 1976.

Hallendy, Norman. *Inuksuit, silent messengers of the Arctic.* British Museum Press, London 2000.

Hallendy, Norman. *Tukiliit, the stone people who live in the wind.* University of Alaska Press, Fairbanks 2009.

Halstead, Byron D. (editor) *Barns, Sheds and Outbuildings, placement, design and construction.* Alan C. Hood & company, Inc., Chambersburg, Pennsylvania 1994 reprint of an original 1881 edition.

Hardcastle, Jim and Nisbit, Merryn. *Lifelines – the vital Dry Stone Walls of the Mendip Hills Area of Outstanding Natural Beauty.* Mendip Hills AONB Service, Bristol 2008.

Harris, Cyril M. (editor) *Illustrated Dictionary of Historic Architecture.* Dover Publications, Inc., New York 1983.

Hart, Edward. *The Dry Stone Wall Handbook.* Thorsons Publishers Limited, Wellingborough 1980.

Harvey, John. *The Master Builders. Architecture in the Middle Ages.* Thames and Hudson, London 1971.

Harvey, Nigel. *Fields, hedges and Ditches, Shire Album 21.* Shire Publications Ltd, Aylesbury 1976.

Hazard, Charles. *Finnish Dry Stone Walls, a brief historical background.* The Waller and Dyker, the official magazine of the Dry Stone Walling Association of Great Britain, Spring 2006, pp 4-5.

Hazard, Charles. *Dry stone dykes of the Estonian Islands.* Vernacular Building 31, 2008, pp 67-76. Scottish Vernacular Buildings Working Group. Edinburgh. 2008.

Headrick, Reverend James. *General View of the Agriculture of the County of Angus or Forfarshire.* Edinburgh 1813.

Henderson, Captain John. *General View of the Agriculture of the county of Sutherland.* London 1815.

Henderson, Richard. *The Young Estate Manager's Guide.* William Blackwood and Sons, Edinburgh and London 1898.

Henderson, Richard. *The Estate Manager.* William Green & Sons, Edinburgh and London 1910.

Hibbert, Samuel. *A Description of the Shetland Islands comprising an account of their scenery, antiquities and superstitions.* Published by Archibald Constable and Co., Edinburgh 1822. Reprinted T & J Manson, Lerwick 1891.

Hill, Jack. *The Complete Practical Book of Country Crafts.* David and Charles 1979.

Hill, Michael and Birch, Sally. *Cotswold Stone Homes.* Alan Sutton Publishing Ltd, Stroud 1994.

Hodgson, Fred T. *Hodgson's Estimator and contractor's guide for pricing Builder's work.* Frederick J Drake & Co., Publishers, Chicago 1918 edition.

Hogg, James. (The Ettrick Shepherd). *The Shepherd's Guide: being a practical treatise on the diseases of sheep, their causes and the best means of preventing them; with observations on the most suitable farm-stocking for the various climates of this country.* Archibald Constable and Co., Edinburgh 1807.

Houston, Anne L (editor). *Lest we forget, the parish of Canisbay.* Congregational Board of Canisbay Parish Church, Caithness 1996.

Hubbell, William. *Good Fences. A Pictorial History of New England's Stone Walls.* Down East, Maine 2006.

Hunt, Marjorie. *The Stone Carvers: master craftsman of Washington National Cathedral.* Smithsonian Institute, Washington DC 1999.

Hutton, Arthur Wollaston (editor). *Arthur Young's tour in Ireland (1776-1779),* Vol 1. London 1892.

JSD — Jamieson's Scottish Dictionary, *An Etymological Dictionary of the Scottish language: illustrating the words and their different significances, by examples from ancient and modern writers; …to which is prefixed, A Dissertation on the Origin of the Scottish Language: by John Jamieson DD, fellow of the Royal Society of Edinburgh, and the Society of the Antiquaries of Scotland. A new edition, carefully revised and collated, with the entire supplement incorporated, by John Longmuir, AM, LLD, and David Donaldson, FEIS* Published by Alexander Gardner, Paisley. In four volumes, 1879–1882.

JSD — Jamieson's Scottish Dictionary, *Supplement to Jamieson's Scottish Dictionary with memoir, and introduction* by David Donaldson, FEIS published by Alexander Gardner, Paisley and London 1887.

JSD – *Jamieson's Dictionary of the Scottish Language, abridged by John Johnston. A new revised edition, revised and enlarged by John Longmuir, AM, LLD, 1856* is available, freely online, at https://electricscotland.com/history/jamieson/ (consulted July 2022).

Jae, Yim Seock. *Stone Walls and Paths, a study of Korean architecture.* Ewha Womans University Press. Seoul 2005.

Jekyll, Gertrude. *Wall and water gardens.* Reprint of original 1901 publication, introduced and revised by Graham Stuart Thomas. The Ayer Company, Salem, New Hampshire 1983.

Jenner, Michael. *Ireland through the ages.* Penguin Books, London 1992.

Jerome, John. *Stone Work: reflections on serious play and other aspects of country life.* Viking Penguin, New York 1989.

Johns, Ryan Luke; Wermelinger, Martin; Mascaro, Ruben; Jud, Dominic; Gramazio, Fabio; Kohler, Matthias; Chli, Margarita; Hutter, Marco. *Autonomous dry stone. On-site planning and assembly of stone walls with a robotic excavator.* Construction Robotics (2020) 4:127-140 Volume 4, Issue 3-4, December 2020.

Johnston, Bryce. *General View of the Agriculture of the County of Dumfries.* London 1794.

Johnston, Thomas. *General View of the Agriculture of the County of Selkirk.* London 1794.

Joyce, P. W. *A Social History of Ancient Ireland. Volume I, Second Edition.* Longmans, Green & Co., London 1913.

Juracek, Judy A. (1996) *Surfaces, visual research for artists, architects and designers.* Thames & Hudson, London reprinted 2000.

Juvanec, Borut. *Square Root of Three in Vernacular Use.* Proceedings of the International Dry Stone Walling Congress. Dry Stone Walling Association of Great Britain, Milnthorpe, Cumbria 2010.

Juvanec, Borut. *Corbelled Constructions on Menorca.* Stonechat, the branch magazine of the North Wales Branch of the Dry Stone Walling Association, No. 28, Spring 2013, page 3.

Keith, George Skene. *A General View of the Agriculture of Aberdeenshire.* Aberdeen 1811.

Kenlan, Joseph. *Guidelines for Laying Stone Walls, master the basics and gravity will do the rest.* Article featured in 'Masonry, the best of Fine Homebuilding', pp 8-11, The Taunton Press 1997.

Kerr, Robert. *General View of the Agriculture of the County of Berwick.* London 1813.

King, Ross. *Brunelleschi's Dome.* Penguin Books, London 2000.

Knott, Laura Lynne. *The Historic Rock Fences of Blanco County: Their Past, Their Future.* Thesis presented to the Faculty of the Graduate School of the University of Texas at Austin. December 2004.

Kona Historical Society. *Traditional Cattle Walls of Kona.* Kona Historical Society, Hawaii, USA 2002.

Lambert, Joseph. *Observations on the Rural Affairs of Ireland; or a Practical Treatise on farming, planting and gardening etc.* William Curry, Jun. and Co., Dublin 1829.

Laheen, Mary. *Drystone Walls of the Aran Islands, exploring the cultural landscape.* The Collins Press, Cork 2010.

LDSPB (Lake District National Park Information Service). *Stone Walls and Appendages.* Lake District Special Planning Board, Windermere, Cumbria 1977.

Lassure, Christian. *The new myths of dry stone walling.* http://www.pierreseche.com/new_myths.htm. February 8th, 2004. (Accessed 5th May 2018).

Lassure, Christian. *Building a drystone hut: an instruction manual.* Centre d'études et recherches d'architecture vernaculaire (CERAV), Second edition, Paris 2009.

Lassure, Christian. *Is dry stone work earthquake resistant?* htpp://www.pierreseche.com/dry _stone_and_quakes.htm. March 18th 2006. (Accessed 24th November 2017).

Lathem, Edward Connery. (editor) *The Poetry of Robert Frost.* Henry Holt and Company. New York, NY 1969.

Lavicka, W L. *Masonry, Carpentry, Joinery: The Art of Architecture, Engineering, and Construction in 1899. Selections from the International Library of Technology published by the International Textbook Co., Scranton, Pa., beginning 1899.* Chicago Review Press, Chicago 1980.

Leask, Harold G. *Irish Churches and Monastic Buildings. Volume 1, the first phases and the Romanesque.* Dundalgan Press (W Tempest) Ltd, Dundalk, Ireland 1955.

Levy, Matthys and Salvadori, Mario. *Why buildings fall down: how structures fail.* W W Norton & Company, New York 1992.

Lexington-Fayette Urban County Government. *The Stone Fences of Fayette County.* Division of Planning, Lexington Kentucky 1990.

Linklater, Andro. *Measuring America: how an untamed wilderness shaped the United States and fulfilled the promise of democracy.* Walker Publishing Company, Inc., New York 2002.

Löbbecke, Renate. *Corbelled Domes.* Verlag der Buchhandlung Walter König, Köln. 2012. More detail and information are available on her website – https://www.renateloebbecke.de/corbelled-domes/ (consulted 15th June 2022).

Logan, James. *The Scottish Gael: or Celtic Manners, as preserved among the Highlanders; being an historical and descriptive account of the inhabitants, antiquities, and national peculiarities of Scotland; etc.* (First American Edition) A. Andrus and Son. Hartford 1843.

London, Mark. *Masonry: how to care for old and historic brick and stone.* The Preservation Press, Washington DC 1988.

Long, Charles K. *The Stonebuilder's Primer.* Camden House Publishing Ltd, Ontario 1981.

Lord, Thomas C. *One on Two, and Two on One: Preliminary results from a survey of dry stone walls on the National Trust Estate at Malham.* In White, R & Wilson, P (editors). Archaeology and Historic Landscapes in the Yorkshire Dales. Yorkshire Archaeological Society Occasional Paper No. 2, 2004, pp 173-186.

Lord, Tom. *Guide to the Medieval dry stone walled landscape at Winskill in the Yorkshire Dales National Park.* XII International Dry Stone Walling Conference, University of Cumbria, Ambleside Campus, 4th–6th of September 2010. Consulted at www.lowerwinskill.co.uk/walls.htm on 12th June 2021.

Loudon, J C. *An Encyclopaedia of Agriculture etc.* London 1825.

Low, David FRSE. *Elements of Practical Agriculture; comprehending the cultivation of plants, the husbandry of the domestic animals, and the economy of the farm*. Edinburgh 1834.

Lowe, Alexander. *General View of the Agriculture of the County of Berwick*. London 1794.

Ludlow, Edmund. *Memoirs of Edmund Ludlow Esq*. London 1751.

Lund, Geoff. *Yorkshire Dales Stonewaller.* with text by Richard Muir and Maurice Colbeck. Dalesman Books, Lancaster 1992.

Martin, George A (editor). *Fences, Gates and Bridges, a practical manual.* Republication of 1887 original by Alan C Hood & Co., Inc., Chambersburg, PA 1992.

McAfee, Patrick. *Irish Stone Walls.* The O'Brien Press Ltd, Dublin 1997, first paperback edition 2011.

McAfee, Patrick. *Lime Works, using lime in traditional and new buildings.* The Building Limes Forum of Ireland, Dublin 2009.

McCann, T; Rogers, D; Cooper, A (2012) *Northern Ireland Countryside Survey NICS2007: Field Boundary – Summary Report 1998-2007.* Northern Ireland Environment Agency Research and Development Series No.12/13, Belfast.

McConnell, Jack. *Stone Walls of New England, a postcard book.* The Globe Pequot Press, Guilford, Connecticut 2003.

McCutcheon, Ivan (editor). *Stone Wall Construction in West Cork.* West Cork LEADER Co-operative Society Ltd 2000.

Macdonald, Alexander. *The Improvement of Hill Pasture Without breaking it up.* Transactions of the Highland and Agricultural Society of Scotland. Fourth Series, Vol XIX 1887. Accessed at https://electricscotland.com/agriculture/page72.htm on 23rd March 2021.

Macdonald, James. *General View of the Agriculture of the Hebrides or Western Isles of Scotland.* Edinburgh 1811.

Macfie, Cody. *Getting started with mortared stonework.* Lark Books, New York 2006.

Mackenzie, Sir George Steuart. *General View of the Agriculture of the Counties of Ross and Cromarty.* London 1813.

Mackintosh, William. *An Essay on the Ways and Means for Inclosing, Fallowing, Planting etc* Edinburgh 1729.

McMillan, Andrew A. *Quarries of Scotland.* (Technical Advice Note 12). Historic Scotland, Edinburgh 1977.

McParlan, James, MD. *Statistical Survey of the County of Donegal, with observations on the means of improvement; drawn up in the year 1801, for the consideration, and under the direction of The Dublin Society.* Dublin 1802.

McRaven, Charles. *Building with Stone.* Republished by Storey Books, Vermont 1989, originally published by Lippincott & Crowell, New York, NY. 1980.

McRaven, Charles. *Stonework techniques and projects.* Storey Publishing, Massachusetts 1997.

McRaven, Charles. *Stone Primer.* Storey Publishing. Massachusetts 2007.

MacSween, Ann and Sharp, Mick. *Prehistoric Scotland.* Batsford Ltd, London 1989.

Mactaggart, John. *The Scottish Gallovidian Encyclopedia, or the original, antiquated, and natural curiosities of the South of Scotland etc.* Published by the author, London 1824.

MacWeeney, Alen and Conniff, Richard. *Irish Walls.* Stewart, Tabori & Chang, New York 1986.

MacWeeney, Alen and Conniff, Richard. *Ireland: Stone Walls & Fabled Landscapes.* Frances Lincoln Limited, London 1998.

Mader, Günter and Zimmermann, Elke. English translation, *Walls – elements of garden and landscape architecture.* W.W. Norton & Company Inc., New York, 2011. Original German edition *Mauern, Elemente der Garten und Landschaftsarchitectur* published 2008 by Deutsche Verlags-anstalt, Munich.

Magnusson, Magnus (editor) *Echoes in stone, 100 years of ancient monuments in Scotland.* Scottish Development Department, Edinburgh 1983.

Magnusson, Magnus and Palsson, Hermann. *The Vinland Sagas, the Norse Discovery of America.* Penguin Books Ltd, Middlesex 1965.

Manson, Peter. 1958. *Dyke building to resist the wind: etc.* School of Scottish Studies Archive SA1958.188, Tobar an Dualchais/Kist o Riches, http://www.tobarandualchais.co.uk/en/fullrecord/34931/2 . (Accessed 7th February 2015).

Mavor, James W and Dix, Byron E. *Manitou, the sacred landscape of New England's native civilisation.* Inner Traditions International, Vermont 1989.

Maxwell, Robert. *Select Transactions of the Honourable, The Society of Improvers in the Knowledge of Agriculture in Scotland etc.* Edinburgh 1743.

Marshall, William. *The Rural Economy of the West of England: including minutes of practice, in that department.* Vol. 1, Second edition. London 1805.

Melville, Herman. *Moby-Dick; or, The Whale.* London and New York 1851.

Merrill, George P. *Stones for Building and Decoration.* John Wiley & Sons. New York 1903.

Mitchell, William R. *Drystone Walls of the Yorkshire Dales.* Castleberg, Settle 1992.

Morgan, Morris Hicky (translator). *Vitruvius, the ten books on architecture.* Dover Publications, New York 1960.

Morris, John T and Walsh, Candace. *Stone Designs for the Home.* Gibbs Smith, Utah 2008.

Morton, John C (editor). *A Cyclopedia of Agriculture, practical and scientific, etc. Vol II.* Blackie and Son, Glasgow, Edinburgh and London 1855.

Morton, Tom, with Jenny Anderson, Harriet Lindsay, Rebecca Little, Jane Mackintosh and Elizabeth Parker. *Soft Capping in Scotland. The context and potential of using plants to protect masonry.* A research report in two volumes. Historic Scotland, Edinburgh 2011.

Muilenburg, Grace and Swineford, Ada. *Land of the Post Rock, its Origins, History, and People.* University Press of Kansas 1975.

Müller, Georg. *Europe's Field Boundaries, hedged banks, hedgerows, field walls (stone walls, dry stone walls), dead brushwood hedges, bent hedges, woven hedges, wattle fences and traditional wooden fences.* Neuer Kunstverlag, Stuttgart 2013.

Munday, Bruce. *Those dry-stone walls: stories from South Australia's stone age.* With photographs by Kristin Munday. Wakefield Press, Kent Town, South Australia 2012.

Mundell, Chris. *Large Scale Testing of Drystone Retaining Structures.* Thesis submitted for the degree of Doctor of Philosophy, University of Bath, August 2009. https://researchportal.bath.ac.uk/en/studentTheses/large-scale-testing-of-drystone-retaining-structures (Accessed 14th November 2021).

Mundell, Matt. *Country Diary.* Gordon Wright Publishing, Edinburgh 1981.

Murphy, Glenn. *Properties of Engineering Materials.* International Textbook Company, Scranton Pennsylvania, third printing 1943.

Murphy, Seamus. *Stone Mad.* First published 1966 by Routledge and Kegan Paul Ltd. Republished by Collins Press, Cork 2005.

Murray, Hilary K. *Clay and bool construction near Urquhart, Morayshire.* Vernacular Building 31, 2007-08, pp 41-52. Scottish Vernacular Buildings Working Group. Edinburgh.

Murray-Wooley, Carolyn and Raitz, Karl. *Rock Fences of the Bluegrass.* University Press of Kentucky 1992.

NEARA Journal, Vol 40, Number 1, Summer 2006 pp 12-13 – *British Links*, Published by New England Antiquities Research Association, Maine, 2006.

Naismith, Robert J. *Buildings of the Scottish Countryside.* Victor Gollancz Ltd, London 1989.

Newman, Paul I. *Stone Igloos: circular pig houses in Orkney.* Vernacular Building 24, 2000, pp. 32-37. Scottish Vernacular Buildings Working Group. Edinburgh.

Nicol, Walter. *The Practical Planter or a treatise on forest planting: comprehending the culture and management of planted and natural timber, in every stage of growth: also on the culture and management of hedge fences, and the construction of stone walls etc. Second edition: corrected and improved.* London 1803.

Nicholson, John. *The Farmer's Assistant; being a digest of all that relates to agriculture, and the conducting of rural affairs; alphabetically arranged, and adapted for the United States.* Benjamin Warner, Philadelphia and Richmond, Virginia, Second Edition 1820.

Nicholson, Peter. *Practical Masonry, Bricklaying, and Plastering, both plain and ornamental: etc.* Thomas Kelly, London 1838.

Nickey, J.M. *The Stoneworker's Bible.* Tab Books, Pennsylvania, 1979, second printing 1980.

Ó Maitiú, Séamas and O'Reilly, Barry. *Ballyknockan, a Wicklow stonecutter's village.* The Woodfield Press, Dublin 1997.

O'Reilly, M and Perry, J. *Drystone retaining walls and their modifications – condition appraisal and remedial treatment.* Construction Industry Research and Information Association (CIRIA), London 2009.

Omand, D. *Fences and Dykes* – Moravian Field Club. Caithness Historical Notes 1750 – 1900. Published October 1974. Consulted online at http://www.caithness.org/caithnessfieldclub/bulletins/1974/october/historical_notes_fences_and_dykes.htm. (Accessed December 2013).

Oram, Richard and Stelfox, Dawson. *Traditional Buildings in Ireland.* Mourne Heritage Trust, Newcastle, County Down 2004.

Oxland, Evan. *A Typology of Japanese Dry Stone Masonry.* Stonechat, the branch magazine of the North Wales Branch of the Dry Stone Walling Association, No. 25, Autumn 2011, pp 3-10.

Paaswell, George. *Retaining Walls, their design and construction.* McGraw-Hill, New York 1920.

Parnell, Sir Henry Brooke. *A Treatise on Roads etc.* London 1833.

Patterson, William Hugh, MRIA. *A glossary of words in use in the counties of Antrim and Down.* The

English Dialect Society, London 1880.

Pease, Sir Alfred Edward, Bt, MA. *A dictionary of the dialect of the North Riding of Yorkshire*. Horne & Son Limited, Whitby 1928.

Perry, Roger. *The Pixie Houses of Devonshire*. Stonechat, the branch magazine of the North Wales Branch of the Dry Stone Walling Association, No. 26, Summer 2012, pp 17-18.

Pilipski, Mark. *The Fractal Lacunarity of Dry Walls and Their Matrix Stones: Why some walls look good and others do not*. Proceedings of the 8th International Drystone-Walling Congress in Visp – Valais (CH) August/September 2002.

Pleasant, Barbara. *Garden Stone: creative landscaping with plants and stone*. Storey Publishing, Massachusetts 2002.

Plunkett-Dillon, E. *The Field Boundaries of the Burren, Co. Clare*. PhD Thesis, T.C. Dublin 1985-86.

Powell, George T. *Foundations and Foundation Walls, for all classes of buildings, etc etc, with added treatise by Frederick Bauman, Architect*. Third Edition, William T Comstock, New York 1889. The reproduction copy consulted was produced by Leopold Classic Library, Las Vegas, NV, 10th July 2021.

Prevost, W A J. *The Dry Stone Dykes of Upper Annandale, with Glossary*. Transactions of the Dumfriesshire and Galloway Natural History and Antiquarian Society, Third Series, 1957, pp 84-101.

Pride, Glen L. *Dictionary of Scottish Building*. The Rutland Press, Edinburgh 1996.

Radford, Andy. *A Guide to Dry Stone Walling*. The Crowood Press, Marlborough 2001.

Rainsford-Hannay, Colonel F, CMG, DSO, of Kirkdale. *Dry Stone Dyking*. Transactions of the Highland and Agricultural Society of Scotland, Fifth Series vol. LVI, pp 23-36 William Blackwood & Sons Ltd, Edinburgh 1944.

Rainsford-Hannay, Colonel F, of Kirkdale. *Dry Stane Dyking*. The Scottish Journal of Agriculture vol XXVII, pp 80-101. HMSO, Edinburgh 1948.

Rainsford-Hannay, Colonel F, CMG, DSO. *Dry Stone Walling*. Originally published 1957. Third edition, published, without revisal by Stewartry of Kirkcudbright Drystane Dyking Committee 1976.

Raistrick, Arthur. *The Pennine Walls*. Originally published 1946. Dalesman Books, Clapham, Yorkshire. Sixth impression 1973.

Rankine, William John Macquorn. *A Manual of Civil Engineering*. Griffin, Bohn, and Company. London 1862

Reed, David. *The Art and Craft of Stonescaping*. Lark Books, New York 2000.

Reed, David. *The Art and Craft of Stonework*. First paperback edition, Lark Books, New York 2003.

Rendall, Jocelyn. *Scone house at Garso, North Ronaldsay*. Vernacular Building 20, 1997, p. 38. Scottish Vernacular Buildings Working Group. Edinburgh.

Ribble Valley Borough Council – Slaidburn Conservation Area Management Guidance public consultation December 2006. Accessed, July 2021, at https:www.ribblevalley.gov.uk/downloads/

file/3704/Slaidburn_conservation_area_management_guidance.

Riches, Anne and Stell, Geoffrey. (editors) *Materials and Traditions in Scottish Building.* Scottish Vernacular Buildings Working Group, Edinburgh 1992.

Robertson, George. *General View of the Agriculture of the County of Mid Lothian.* Edinburgh 1793.

Robertson, James. *General View of the Agriculture in the County of Perth.* Perth 1794.

Robertson, James. *General view of the Agriculture of the County of Inverness.* London 1813.

Robson, Michael. *Dykes, Ditches and Disputes. A history of boundary and field enclosures in the Borders.* Michael Robson, Isle of Lewis 2004.

Rollinson, William. *Making Drystone Walls.* Smith Settle Ltd, Otley 1998.

Romney, P (editor). *The Diary of Charles Fothergill 1805. An itinerary to York, Flamborough and the north-western Dales of Yorkshire.* Record Series, Volume CXLII (142) for the year 1982. The Yorkshire Archaeological Society, Leeds 1984.

Roy, Rob. *Stone circles: a modern builder's guide to the megalithic revival.* Chelsea Green Publishing Company, Vermont 1999.

Russel, James E. *Walks, walls and fences, plus retaining projects for steps, ramps, gates, retaining walls, privacy screens.* Creative Homeowner Press, New Jersey 1981.

Russell, Howard S. *A Long, Deep Furrow. Three centuries of farming in New England.* University Press of New England, Hanover, New Hampshire. First Edition 1976.

Ryan, Nicholas. *Stone Upon Stone. The Use of Stone in Irish Building.* The Collins Press, Cork 2005.

Ryan, Nicholas M. *Sparkling granite, the story of the granite working people of the Three Rock region of County Dublin.* Stone Publishing, Dublin 1992.

Rybczynski, Witold. *Home, a short history of an idea.* Penguin Books, New York 1987.

SPCK (Society for Promoting Christian Knowledge). *Sketches of Rural Affairs.* London 1848.

Sagui, Pat. *Landscaping with Stone.* Creative Homeowner, New Jersey 2005.

Salvadori, Mario. *Why buildings stand up – the strength of architecture.* W W Norton & Company, New York, reissued paperback 2002.

Schrank, Gilbert. *An Orkney Estate, Improvements at Graemeshall, 1827-1888.* Tuckwell Press, Edinburgh 1995.

Scully, Sean. *Walls of Aran.* Thames & Hudson, New York 2007.

Scully, Sean. *Walls of Aran* (compact edition) Thames & Hudson, New York 2019.

Seymour, John. *The forgotten arts, a practical guide to traditional skills.* Dorling Kindersley, London 1984.

Shackleton, Sir Ernest, CVO. *South. The story of Shackleton's last expedition 1914-1917.* William Heineman, London 1919.

Shadmon, Asher. (1989) *Stone, an introduction.* Intermediate Technology Publications, London, second edition 1996.

Shirreff, John. *General View of the Agriculture of the Orkney Islands.* Edinburgh 1814.

Simpson, A C and Connor, R D. *Measures, Interpreting Scots Measurement Units.* Page 104-105,

Dictionary of Scottish Building, Glen L Pride, published by The Rutland Press., Edinburgh 1996.

Sinclair, Archie. *Spittal, Caithness: home of the world's finest flagstone.* Spittal Action Centre, Caithness 1988.

Sinclair, C. *The Thatched Houses of the Old Highlands.* Oliver and Boyd Ltd, Edinburgh 1953.

Sinclair, Sir John. *The Statistical Account of Scotland drawn up from the communications of the Ministers of the different parishes, volume sixth.* Edinburgh 1793.

Sinclair, Sir John. *An Account of the Systems of Husbandry adopted in the more improved districts of Scotland etc.* Second Edition, Vol. I. Edinburgh 1813.

Sinclair, Sir John. *General Report of the Agricultural State, and Political Circumstances of Scotland.* Edinburgh 1814.

Sinclair, Sir John. *General View of the Agriculture of the County of Caithness.* London 1815.

Singer, Dr W. *General View of the Agriculture, State of Property, and Improvements, in the County of Dumfries.* Edinburgh 1812.

Slimon, Campbell. *Stells, stools, strupag – a personal reminiscence of sheep, shepherding, farming and the social activities of a Highland Parish.* Laggan Heritage, Laggan 2007.

Slocombe, Matthew. *Traditional Building materials.* Shire Publications, Oxford 2012.

Smart, S M; Allen, D; Murphy, J; Carey, P D; Emmett, B A; Reynolds, B; Simpson, I C; Evans, R A; Skates, J; Scott, W A; Maskell, L C; Norton, L R; Rossall, M J; Wood, C. *2009 Countryside Survey: Wales Results from 2007.* NERC/Centre for Ecology & Hydrology, Welsh Assembly Government, Countryside Council for Wales, 94 pp. (CEH Project Number: C03259).

Smith, James. *On agricultural fencing.* The Farmer's Magazine, No. 3, Vol VII, September 1837, pp. 163-165.

Smith, John. *General View of the Agriculture of the County of Argyll.* Edinburgh 1798.

Smith, Reverend Samuel. *General View of the Agriculture of Galloway; comprehending the two counties, viz. The Stewartry of Kirkcudbright, and Wigtonshire; etc.* London 1813.

Snow, Dan. *Listening to Stone.* Artisan, New York 2008.

Snow, Dan. *In the Company of Stone: the Art of the Stone Wall.* Artisan, New York 2001.

Sprague, Ernest H. *Stability of masonry and other structures subject to the pressure of earth and water.* Scott, Greenwood & Son, London 1915.

Stanier, Peter H. *Quarries and Quarrying,* Shire Album 134. Shire Publications, Aylesbury 1985.

Steed, Christopher. *Let the Stones Talk, glimpses of English history through the people of the moor.* AuthorHouse, Milton Keynes 2011.

Stell, Geoffrey P and Harman, Mary. *Buildings of St Kilda,* The Royal Commission on the Ancient and Historical Monuments of Scotland, Edinburgh 1988.

Stephens, Henry. *On Dry Stone-Walls as a Fence.* The Quarterly Journal of Agriculture. Volume III. February 1831 – September 1832. pp 239-255. William Blackwood, Edinburgh and London 1832.

Stephens, Henry. *The Farmer's guide to scientific and practical agriculture etc. Volume II.* Leonard Scott & Co., New York 1853.

Stephens, Henry. *The book of the farm, detailing the labours of the farmer, steward, plowman, hedger, cattle-man, shepherd, field-worker, and dairymaid. Volume II,* A. O. Moore, Agricultural Book Publisher, New York 1857.

Stephens, Henry. *The Book of the Farm.* Fourth Edition, Volume III, revised by James Macdonald, F.R.S.E., William Blackwood and Sons, Edinburgh and London 1891.

Stephens, Henry, FRSE. and Burn, Robert Scott. *The Book of Farm-Buildings, their arrangement and construction.* William Blackwood and Sons, Edinburgh and London 1861.

Stevenson, David. *The Origins of Freemasonry. Scotland's Century, 1590-1710.* Cambridge University Press, Cambridge 1990.

Stevenson, William. *General View of the Agriculture of the County of Surrey.* London 1809.

Stevenson, William. *General View of the Agriculture of the County of Dorset.* London 1812.

Stilgoe, John R. *Common Landscape of America, 1580-1845.* Yale University Press, New Haven CT 1982.

Svizzero, Serge and Tisdell, Clem. *Desert Kites: Were They Used For Hunting Or For Herding? A Review of the Recent Academic Literature.* Journal of Zoological Research. 2(4). 2018. pp 7-28.

Tabraham, Christopher. *Scottish Castles and Fortifications.* HMSO, Edinburgh 1986.

Tait, Ian. *Setting the record straight about punds.* The New Shetlander, 222, pp 23-25, Lerwick 2002.

Taylor, Dr Michael A. *Hugh Miller, stonemason, geologist, writer.* National Museums of Scotland, Edinburgh 2007.

Thomas, J J. *The illustrated annual of rural affairs and cultivator almanac, for the year 1864, containing practical suggestions for the farmer and horticulturist.* Luther Tucker & Son, Albany, NY 1864.

Thomson, John. *General View of the Agriculture of the County of Fife.* Edinburgh 1800.

Thorson, Robert M. *Stone by Stone. The magnificent history in New England's stone walls.* Walker & Company, New York 2002.

Thorson, Robert M. *Exploring Stone Walls: a field guide to stone walls.* Walker & Company, New York 2005.

Todd, S. Edwards. *A Treatise on Fencing.* Transactions of the N. Y. State Agricultural Society. Vol XVIII – 1858, pp. 460-474. Albany, 1859.

Toulson, Shirley. *Lost Trade Routes, Shire Album 108,* Shire Publications Ltd, Aylesbury 1983.

Trautwine, John C and Trautwine, John C Junior. *The Civil Engineer's Reference-book.* 21st Edition edited by John C Trautwine 3rd. C.E. Trautwine Company, Ithaca, N.Y. 1937.

Trias, Antoni Reynés and Arrom, Vicenç Sastre I. *Llibre de la pedra en sec.* Edicions de Turisme Cultural, Consell de Mallorca, Illes Balears 2002.

Trotter, James. *General View of the Agriculture of the County of West Lothian.* Edinburgh 1794.

Tucker, D Gordon. *Millstone making in Scotland.* Proceedings of the Society of Antiquaries of Scotland, Volume 114, 1984, pp 539-556.

Tucker, Luther. *Stone Walls.* The Monthly Genesee Farmer and Horticulturist. Volume 1, September 1836, No. 9, page 136. Rochester N.Y. 1836.

Tuke, John. *General View of the Agriculture of the North Riding of Yorkshire.* London 1800.

USDA Forest Service. *Trail Construction and Maintenance Notebook*. United States Department of Agriculture Forest Service, Missoula, Montana. 2007 edition.

Ure, Reverend David. *General View of the Agriculture of the County of Roxburgh*. London 1794.

Vivian, John. *Building stone walls.*, second edition 1978. Storey Books, Vermont 1976.

Vivian, John. *Building fences of wood, stone, metal and plants*. Williamson Publishing, Vermont 1987.

Walker, Bruce and McGregor, Christopher. *The Hebridean Blackhouse, a guide to materials, construction and maintenance*. Technical Advice Note 5, Historic Scotland, Edinburgh 1996.

Walker, Bruce and McGregor, Christopher. In collaboration with Rebecca Little. *Earth Structures and Construction in Scotland. A Guide to the Recognition and Conservation of Earth Technology in Scottish Buildings*. Technical Advice Note 6, Historic Scotland, Edinburgh 1996.

Walker, Bruce. In association with Christopher McGregor and Gregor Stark. *Scottish Turf Construction*. Technical Advice Note 30, Historic Scotland, Edinburgh 2006.

Walker, John. *An Economical History of the Hebrides and Highlands of Scotland*. London and Edinburgh 1812.

Warland, E G. *Modern Practical Masonry*. Originally published 1929. Volume referenced is a photocopy described on the first page as "Limited reprint by the Stone Federation, 82 New Cavendish Street, London, W1 by courtesy of Pitman books Limited." The reprint is probably dated 1953.

Watt, Bobby. *Snecked Stonemasonry*, Stonexus, the periodical publication of the Stone Foundation, number VII, winter 2006/07, pp.36-37.

Webb, Michael and Schwartzman, Arnold. *It's a Great Wall!* Watson-Guptill Publications, New York 2000.

White, Hazel and the editors of Sunset Books. *Landscaping with stone*. Sunset Books, California 2000.

Whitner, Jan Kowalczewski. *Gardening with Stone, using stone features to add mystery, magic, and meaning to your garden*. Macmillan, New York 1999.

Whitten, D G A and Brooks J R V. *The Penguin Dictionary of Geology*. Penguin Books, Middlesex 1979 reprint.

Whyte, Reverend Andrew and Macfarlan, Duncan. *General View of the Agriculture of the County of Dumbarton*. Glasgow 1811.

Wiggins, Francis S. *The American Farmer's Instructor, or Practical Agriculturalist; comprehending the cultivation of plants, the husbandry of the domestic animals, and the economy of the farm; together with a variety of information which will be found important to the farmer*. Philadelphia 1840.

Wightman, Mark and Kempe, Nick (editors). *Hostile Habitats, Scotland's mountain environment*. Scottish Mountaineering Trust, Glasgow 2006.

Wildgoose, Martin. *The Drystone Walls of Roystone Grange. The Archaeological Journal*, volume 148 (1991) pp. 205-240. The Royal Archaeological Institute, London 1991.

Williams, David B. *Cairns: messengers in stone*. The Mountaineers Books, Seattle 2012.

Williamson, Tom. *Living Landscapes – Hedges and Walls*. The National Trust, London 2002.

Wilson, Reverend John M. *The Rural Cyclopedia, or a general dictionary of agriculture etc. Volume IV*.

A Fullerton and Co., Edinburgh and London 1849.

Wilson, Peter (editor). *Building with Scottish Stone.* Arcamedia, Edinburgh 2005.

Winchester, Angus J L. *Dry Stone Walls, history and heritage.* Amberley Publishing, Stroud, Gloucestershire 2016.

Wood, Pamela (editor). *The Salt Book, lobstering, sea moss pudding, stone walls, rum running, maple syrup, snowshoes, and other Yankee doings.* Anchor Books, Anchor Press/Doubleday, New York 1977.

A Working Man. *Reminiscences of a Stonemason.* John Murray (publisher), London 1908.

Wormer, Andrew. *Stonescaping idea book.* Taunton Press, Connecticut 2006.

Wright, Frank Lloyd. *The Natural House.* Plume. New York 1970.

ACKNOWLEDGEMENTS

I extend thanks and appreciation to:

Christine Robinson, former Director of Scottish Language Dictionaries, Edinburgh, for initial encouragement.

Ian Spring of Rymour Books and Paul Philippou of Tippermuir Books for designing and publishing this book

Staff of Historic Scotland (now Historic Environment Scotland) for their assistance in preparing early drafts.

Tracy Benfield, Viv Halcrow, Tessa Jones and Judith Bullivant for practical help with word processing. Drew McClelland for assistance with mapping, illustrations and the geography of southwest Scotland. Dominique McKay for help with later drafts.

Alison Shaw at the Dry Stone Walling Association HQ for access to an original list, compiled from their membership, which confirmed I was on the right track with some of the more 'provincial' terms.

Dykers, wallers, masons, naturalists, cartographers and members of drystone walling associations in Great Britain, Ireland, Canada, Australia, USA and Europe who reviewed drafts, shared knowledge and confirmed there are fundamental similarities in all stone activity including:

Bobby Watt, Norman Haddow, Jane Wooley, Robin Callander, Colin Purvis, Sunny Wieler, Karl Opanowicz, Jacqui Simkins, Iain Macmillan, Richard Tufnell, Ellen Hughson, Ian Dewar, Jonathan Swale, Pete Moore (SNH), Pete Moore (RSPB),Tracey Blackwell, Ken Curran, Mich Wegener, Jim Ramsay, Alan Ash, Brian Post, Meg Telfer, Helmut Schneider, Bruce Curtis, Dave Goulder, Eric Landman, Lisa Norton, Anne-Berit Ø Borchgrevink, Irwin Campbell, Bob and Helen Powell, Flora Hunter, Bruce Munday, Geoff Duggan, Peggy Perazzo, Richard Love, Elin and Dan Snow, Isabel Duncan, Neil Rippingale, Pat McAfee, Sean Adcock, Kathryn Laws, Sebastian Rico.

George Duff and The Highland Council who granted leave to pursue a Winston Churchill Travelling Fellowship in 2001. Staff at Baile nan Gàidheal (Cape Breton), Ros Strong, Sue Carlson, Donald Parker and members of New England Antiquities Research Association for showing me drystone structures in the woods of northeast North America.Prestige Copy & Print, Northgate, Seattle, for printing early drafts. Staff of Fed-Ex, 45th Street, Seattle, for help with diagrams and images.

A skilled librarian is a valuable guide. I thank staff of Reference Services at the National Library of Scotland, Edinburgh; The University of Oslo Library; The National Library of Wales, Aberystwyth; The Scottish Poetry Library, Edinburgh; staff at The Highland Folk Museum, Kingussie (now the Highland Folk Park, Newtonmore); Dr Ian Tait of Shetland Museum and Archives; staff of Mary Evans Picture Library.

The staff of Dumfries and Galloway Libraries provided valuable assistance translating the poem 'Davie the Dyker', the original inspiration for this volume. I also thank Drew McClelland and Mrs Elizabeth McClelland of Inverness for their help in the translation, and her stirring recording of the poem in southwest Scotland dialect.

Roy Bartlett Henderson kindly allowed me to include 'The Mole Song', and Stuart Campbell and Tom Hubbard made some additional suggestions.

Nicolette Stessin, who has, since 2001, graciously accepted roadside delays wherever there are lunkies or spectacular copes.

Finally, Jim Aitken, my father. Together, we built a dyke out of impossible stone at the Balloch in Glen Isla and were sent out to the back of the hirsel to repair collapsed sections.

APPENDIX ONE: DAVIE THE DYKER

'Davie the dyker' is a well-known poem, often referred to in drystone dyking circles, especially the line — 'a rickl'd rood ne'er left his han" — an aspiration for all dykers.

The original eighteen verse poem was included in *The Scottish Gallovidian Encyclopedia* by John Mactaggart (1824) as part of the definition for 'dyke'. He gives no specific title for the poem.

What follows is a 'translation' for the modern reader, with the assistance of the online version of *The Dictionary of the Scots Language* (https://dsl.ac.uk/), *Jamieson's Scottish Dictionary* (a nineteenth-century forerunner of DSL) the works of F. Rainsford-Hannay, and those familiar with the dialect of southwest Scotland.

F. Rainsford-Hannay quoted verses 5-8 in an article entitled 'Dry Stone Dyking' in volume 56 (1944) of the *Transactions of the Highland and Agricultural Society of Scotland*. He included a variation of verse 8 in a 1948 article entitled 'Dry Stane Dyking' in *Scottish Agriculture*, volume 27, published by HMSO in 1948. Verses 5-11 and 14-15 were included in his *Dry Stone Walling,* first published in 1957 (reprinted 1976) with the title 'Elegy to Davie the Dyker by John McTaggart of Borgue.'

Readers will note the different spellings for 'Mactaggart', spelt as found in each of the sources. This variation is typical for such a surname and the lack of standardised spelling until relatively recently.

The following pages are a transcript of the original poem, in original dialect, idiom and spelling, with a modern 'interpretation' opposite.

Mactaggart starts with a definition for 'dyke':

> Dyke – A fence. Anciently dykes were built for confining mankind to certain portions of the earth; as those the Romans built in Scotland, to keep at bay the daring Scots; and that one by the Chinese to hold out Tartars, considered one of the wonders of the world; now they are only used to inclose brute cattle; some people make a trade of building dykes, one of these must have been *Davie the Dyker,* as the following Poem doth testify.

The final lines are interpreted as confirming Heaven would never reward wrongdoing or include sinners. The inscription on the gravestone could therefore be:

Davie the dyker, flesh and bone, is stretched beneath this stone.
His soul has surely gone to Heaven.
He earned his place.

What horrid news are thae we hear,
Owre true they be, we fear, we fear,
Our een begin to brew the tear,
  Wi' sabbing speed;
Puir Davie wham we lo'ed dear,
Is smother'd dead.

thae – those – referring to the rumours of an accident.
Ower true – too true, overwhelmingly tragic.
een – eyes.
brew – to form.
sabbing – sobbing. 'Sab' is also the term for a heavy downpour of rain or a thunder shower. This emphasises the nature of the tears.
Puir – poor, in the sense of unfortunate, rather than penniless.
lo'ed – loved.
smother'd – smothered, suffocated, overwhelmed by the rockfall.

The sclateban o' the quarry shott,
Afore our worthy out o't got,
And crush'd him lifeless on the spot,
Without remorse;
Sae he was carried to his cot,
  Ane mangl'd corse.

sclateban – a band of slate, a bed of rock in a quarry.
shott – DSL, under 'S(c)hute', gives us an old meaning – 'Of earth, sand, a bank etc: to fall, or slide, suddenly downwards; to collapse.' An alternative reading, less likely, is that the gunpowder charge, or shot, used to dislodge rock, was fired before Davie reached safety.
Afore – before.
worthy – referring to Davie, a well-known, and well-respected person.
out o't got – got out of it.
Without remorse – with no regrets.
Sae – so, therefore, as a result.
cot – bed.
corse – corpse, body.

O! dreadfu' shocking news thae be,
The body we'll awa and see,
Upon his back, and eke his knee,
  We aft war borne;
Us boys about may lang for he,
  Lament and mourn.

thae be – those are, it is, referring to the news.
awa – away.
eke – also, as well as.
We aft war borne – we were often carried.
Us boys about may lang for he, Lament and mourn – we boys will mourn Davie for a long time. In old Scots, according to DSL, the word 'about' is – 'In verse placed after the

noun.' This gives the line rhythm and emphasises the noun.

He never fley'd us frae nor fun
The bluchton and the billet gun –
Ay glakes that birl'd in the wun,
  Did Davie make;
By him too dragons ne'er the sun,
  Their tails did shake.

fley'd – frightened, scared or chased.
frae – from.
fley'd us frae nor fun – 'nor' may be a misprint in the original for 'oor', meaning 'our'. With that assumption the phrase makes more sense – he never chased us from our fun – never interrupted the children's play.
bluchton – a peashooter made from the hollow dried stems of the mugwort plant (Artemisia vulgaris). Hawthorn seeds (Crataegus monogyna) were used as ammunition.
billet gun – a pop-gun fashioned from hollowed-out stems of the 'boor tree' or 'bourtree' – the elder (Sambucus nigra). For its definition DSL's source states 'alder' (Alnus glutinosa) instead of 'elder'. Alder is unsuitable, it has solid wood stems. Elder stems are full of pith.
Ay – yes, used here as emphatic confirmation – yes!
glakes – a child's shiny toy, in this case a windmill which spins in the breeze.
birl'd – birled, spun.
wun – wind.
dragons ne'er the sun – literally 'dragons near the sun' – flying a kite.

And was there ever ere his like,
At bigging o' a strang Stanedyke,
He was na fractious, dip na fyke,
  For meikle doon;
He sought for, through-ban's that wad rike,
  And capes wad croon.

was there ever ere his like – was there ever anyone as good as him?
bigging – the act of building.
a strang Stanedyke – a strong stone dyke.
na – not.
fractious – peevish, cranky, inclined to make trouble.
dip – to question or investigate. Although Rainsford-Hannay (1957:95) tells us 'dip' does not occur in any Scottish dictionary, DSL does include 'dip', with definitions of '…to question' and '…to make investigation (in a mat-

ter)…'

fyke – fussy, overly concerned with petty cares. 'dip na fyke' can be translated as 'unconcerned with trifles'.

For meikle doon – Rainsford-Hannay (1957:95) has a footnote: 'A very obscure line. 'doon' is a variation of 'dool', sorrow. So Davie's home troubles, 'meikle doon', did not affect his work.'

through-ban's – throughbands or throughstones. These long stones lie across the width of the dyke, approx. half way up the face, and often protrude a few inches. They strengthen and stabilise the structure.

rike – to reach or extend.

capes – copes, copestones.

croon – to crown, to top off.

His dykes had ne'er the sleek'd skin,
Ne'er fair without and fause within,
He didna batter, line and pin,
  To please the e'e;
There ne'er was heard a clanking din,
  Whar bigged he.

His dykes had ne'er – his dykes never had.

the sleek'd skin – this could describe the dyke as having a slick, smooth, outer face. Here it is used in the Scots sense of 'sleekit' – untrustworthy, sly, cunning – where the nice exterior disguises a poorly crafted interior.

Ne'er fair without and fause within – literally 'never fair on the outside and false on the inside'.

He didna batter, line and pin – these are elements, and tools, of the building process, to ensure the structure is built strong, and good looking.

To please the e'e – to please (or deceive) the observer. Here it means to build with the facade in mind, rather than the overall strength of the dyke.

Clanking din – the loud noise of falling stone.

Whar bigged he – where, or what, he had built.

A rickl'd rood ne'er left his han',

rickl'd – poor work, loosely thrown togeth-

His dykes for centries will stan'
A slap wi clutter's never fa'en
  In ane o'em pet;
May they the name o' he puir man,
  For ages get.

er, little more than a hurriedly piled heap or 'rickle'.

rood – an old term for a measurement representing a reasonable day's work for a dyker, usually 6 or 7 yards, depending on the stone and custom of the area.

ne'er – never.

centries – centuries, hundreds of years, a long time.

slap – DSL defines this well – 'A gap or opening in a wall, fence, hedge or the like, whether intentionally or accidentally caused, a breach…'

clutter – a Scots term for anything in a precarious or dilapidated state, for example a badly built stone dyke. DSL has another meaning – 'the noise of much confused speaking'. This leads us to consider the word more akin to 'clatter', and the sound it brings to mind.

fa'en – fallen.

A slap wi' clutter's never fa'en – literally 'a section of dyke has never noisily fallen.' Definitely more eloquent in the original!

In ane o' em – in any one of them, in any of his dykes.

pet – can mean to sulk or take offence, or be a local variation of the word 'put'. In this case it seems more likely a term of endearment, used to get the rhyme.

May they the name o' he puir man, For ages get – may he be long remembered as their builder.

Ower moor and dale for mony a year,
May Davie's famous dykes appear,
Ne'er bilged out wi' wather-wear,
  But just the same;
As whan puir cheel he left them there –
  To bear his name.

Ower – over.

mony – many.

Ne'er – never.

bilged – bulged, perhaps caused by bad foundations or poor stone disintegrating in frost and rain.

wather-wear – weather wear, weathering.

whan – when.

puir cheel – poor fellow – a term of fondness and endearment.

Nae wadder fleet can ower them jump,
If e're they try't back on their rump,
They will recoil wi' whulting bump,
  E'en Rigling Rallions;
Wi' rattling doddles arset stump,
  Our down gae brallions.

wadder – a wether, a castrated male lamb. A rough looking sheep, often with misshapen horns and shaggy fleece.
fleet – nimble.
ower – over.
If e're they try't – if ever they try it (to jump over the dyke).
whulting – large, emphatic. Imagine the dull thud of a heavy beast hitting the ground.
E'en – even.
Rigling –a rig or rigg, a ram with one testicle. An inferior specimen.
Rallions – the *Scottish Gallovidian Encyclopedia* provides an unflattering definition – 'raullion or rullion – a rough ill-made animal.'
rattling doddles – the dung-encrusted wool at the sheep's rear. These hard pellets rattle together when the sheep runs.
arset – knocked backwards.
stump – DSL has 'stump' as 'a short stocky person or animal'.
Our down gae – literally 'over down go' - knocked down or knocked back.
brallion – DSL's definition is excellent – 'A careless unthinking person devoid of taste and clumsy as a bull in a china shop … an unwieldy creature.' The word is linked to the Irish *breallan,* a foolish person. This verse emphasises the stubborn determination of an animal with little brain power. Rainsford-Hannay (1957-96) omitted part of this verse with the note – 'The last two lines of this verse are diverting but un-becoming to this highly respectable work.'

Had he been wi' the chaps lang syne,
Wha wad the ancient Scots confine,
(The Romans war they if we min')

Had he been wi' the chaps lang syne – if Davie had been with the folks long ago.
Wha wad the ancient Scots confine – lit-

Wi' meikle dyke;
Brave Grahm wad been waur to haud in,
The norlan tyke.

erally 'who would' – who tried – to restrict the Scots.
war – were.
min' – mind, remember.
The Romans war they if we min' – the Romans, we recall, were those who tried to confine the Scots.
meikle – big, large, extensive.
Brave Grahm - The Antonine Wall was built by the Romans around 140AD-150AD along a line between the rivers Clyde and Forth in Central Scotland. Legend says the wall was breached by the founder of Clan Graham and the Roman legions were driven out of Scotland. Because of this the Antonine wall was often called 'Graeme's Dyke'.
haud – hold.
wad been waur to haud in – 'waur' usually means 'worse' or comparatively bad. DSL gives us another, more fitting, definition – 'To get the better of, worst, beat, overcome, outdo…' This phrase therefore translates to – 'Grahm would have been got the better of, and held back by the dyke' (if Davie had built it).
norlan tyke – northern rogue or rascal.

Auld Agricola had na ane,
In a' his core cud bed a stane,
Let learn'd historians write, and grane
Out what they like;
Wi' our puir Davie they had nane,
Cud big a dyke.

Auld Agricola – a reference to the Roman general Gnaeus Julius Agricola (40 - 93AD) whose campaigns, including forays into Scotland, were a major part of the conquest of Britain.
had na ane – had not one, had no one.
core – corp, army.
cud – could.
bed a stane – to lay a stone, as when building a wall.
learn'd – learned
grane – to complain or grumble. We can imagine academics 'grinding out' their personal opinions.

nane – none, no one.
Wi' our puir Davie – compared to our poor Davie.
big – build.

The thing is queer to think o't yet,
For a' the fowk the Romans beat,
Our countrymen they ne'er cud get
  Squeez'd 'neath their pride;
Our Claymores gaed them a' the pet,
  Sair tools to bide.

to think o't yet – to think of it nowadays.
fowk – folk, people, nations.
cud – could.
Squeez'd 'neath their pride – squeezed beneath their pride, conquered.
Claymore – a heavy two-edged sword used by the Scots.
gaed them a' the pet – gave them all an upset, upset them all.
Sair tools – serious/dangerous weapons.
bide – DSL includes 'bide' as a noun meaning 'pain' or as a verb meaning 'to endure'. – emphasising the success of the claymore against the Romans.

But Davie, what's a' this to thee,
It wunna change the stern decree,
Whilk sinners us we sadly see,
  For ever mair;
In thy cauld den thou soon will't be,
  Our hearts are sair.

wunna – will not.
the stern decree – the unavoidable reality.
Whilk – which.
For ever mair – for evermore. Confirming the certainty of death for everyone.
cauld den – a cold den – a grave.
soon will't be – soon will be.
sair – sore.

Wi' you nane cud the Gellock wield,
The yellest craigs for you boud yeal'd,
What hoolochs down ye clantering reel'd,
  At ae gude prize;
And junrells till the echoes peal'd,
  O' munstrous size.

Wi' you nane cud – 'with you no one could' – no one was as good as you.
Gellock – DSL defines this as an iron crowbar or pinch bar. Rainsford-Hannay (1957-97) transcribes this word as 'gelloch' and wrongly defines it as 'a 14 lb. hammer'.
wield – use or employ as a tool.
yellest craigs – the roughest most unyielding rock faces. JSD, under the word 'yeld', has 'yell' defined thus – 'Applied as an epithet to hard rocks. A rock is said to be yell when it will not quarry but with gunpowder.'

boud yeal'd – to burst open, to yield or succumb to pressure.
hoolochs – Mactaggart defines 'hooloch' as 'A deep, roaring, rumbling noise, eg a fall of heavy stones, as on a hillside, an avalanche…'
clantering reel'd – crashed down with great din. DSL defines 'clanter' as – 'A jarring noise, such as proceeds from clogs walked about with in a house'.
ae gude – one good.
prize – a variation of 'prise', applying leverage.
junrells – a large irregular mass of stone.
echoes peal'd – the sound echoed round the quarry, like a peal of bells, or a roll of thunder.
munstrous – monstrous, huge.

Wi' jumper too, ye whiles wad bore,
And make the rocks wi' powther roar,
Whilk scar'd the pellocks frae the shore
  Wi' smacking fin;
What Maukins too wad scud afore
  The dunnering din.

jumper – a long pointed chisel, used with a heavy hammer for drilling into rock. The holes were filled with gunpowder and fired, to break up the rock face.
whiles – sometimes.
wad – would.
powther – powder, gunpowder.
Whilk – which.
scar'd – scared.
pellocks – porpoises. Rainsford-Hannay (1957:97) transcribes this word as 'pellochs'.
What – an exclamation to emphasise what follows.
Maukins – DSL defines this as a hare, most likely the brown hare (Lepus europaeus). Rainsford-Hannay (1957:97) transcribes this word as 'mankins', and as 'mawkins', oyster = catchers' in the footnotes. Oystercatchers (Haematopus ostralegus) are a black and white wading bird with a long orange beak. DSL also has a less likely definition – the domestic cat. Interpreting the word as 'hare' creates a nice contrast with the porpoise.
scud – to move briskly, to flee in leaps and bounds from the explosion.

dunnering – deafening, thunderous.

O' war we rhymesters sae profite,
As epitaphs on thee to write,
To thee the muses should indite
(Our honest Cronnie);
Ane that nae saul alive cud wyte,
'Twad be sae bonny.

O' war we – if we were.
rhymesters – poets.
profite – proficient, skilful.
indite – compose, write down.
Cronnie – friend, cronie.
wyte – to find fault with.
Ane that nae saul alive cud wyte – one that no soul alive could dispute.
Twad – it would.

But hoh-ance we'er a' unfit,
The attempt wad only be a skit,
For want o' pith then, we'll submit –
'Na faith' (cries out,
Ane honest muse) come hae at it,
'We'll do ne'er doubt.'

hoh-ance – an expression of acceptance like 'oh well'.
we'er a' – we are all.
wad – would.
skit – a practical joke, a foolish action.
want o' – for want of, lacking.
pith – strength, vigour, resolution.
Na faith – an exclamation urging self-belief.
Ane honest muse – an earnest would-be poet.
come hae at it – let's attempt it!
We'll do ne'er doubt – we'll surely manage.

Our dyker Davie, flesh and bane,
Is streeked un'erneath this stane,
'But sure to heaven his saul is gane,
'Gif ony gangs;
The rulers there, o' never ane,
Deserving't wrangs.

Bane – bone
Streeked – stretched.
Un'erneath – underneath.
gane – gone
Gif ony gangs – if anyone goes.
Deserving't – deserving or meriting.
wrang – a wrong, an injury.

APPENDIX TWO: THE MOLE SONG

'The Mole Song' was written by Roy Bartlett Henderson, a Perth-based singer and songwriter. The tune is 'Big Rock Candy Mountain'.

As I strode through the countryside
I came upon a dyker
A man with a terrible tale to tell
To an unsuspecting hiker.

He laid his hand upon my arm
I was trapped by his glitt'ring ee
My thermos dropped from my nerveless hand
As he told this tale to me…

Chorus:
When moles get into a drystane dyke
It's an unkind variorum
For they wreak their havoc right and left
With an absence of decorum.

Now a dyker's duty is specialised
And he knows just what's involved
Carefully defined in ev'ry possible respect
And to do with building walls……

…Or with fixing them when they fall down
Or patching up their holes
Or (now we come to the central theme)
Wrestling with predatory moles.

Chorus: For when moles, etc

There's some who think that a drystane dyke
Is a random heap of stone
That's not a theory I'd support
Nor one I could condone.

Imagine two free-standing walls
Expertly filled with rubble
Then imagine moles with beavering paws
Intent on causing trouble…

They scoop the earth up from below
With no regard for cost
Then along comes another really terrible thing –
Twelve degrees of frost…

…Which mixes with the upthrown earth
To create a kind of porridge
Whose effect at minus centigrade
Is best described a horrid….

…The mixture swells and swells and swells
As the temp'rature gets colder
'Til a thundering, tumbling avalanche
Leaves not one standing boulder

Chorus: For when moles, etc

Earth lay hard as iron
Water like a stone
Four months building a bloody wall
And now it's simply gone.

No albatross nor sea-snake
Can save a dyker's soul
When all he wants is a well-made dyke
Without a mole to thole.

Chorus: For when moles, etc

The proper study of mankind
Was in bygone days a goal
But a twenty-first century dyking man
Just wants to beat the mole.

With his story told, the spell was snapped
And I stumbled away in the night
Had he really been embarrassed by a plethora of moles
Or was he just talking …rubbish!

Chorus: For when moles, etc